AF473730

STEFAN BOGNER • JAN KARL BAEDEKER

TRAUMSTRASSEN IM SCHNEE • SNOW-CAPPED DREAMS

ESCAPES
WINTER

DELIUS KLASING VERLAG

VORWORT / FOREWORD

Stefan Bogner

Auch wenn die Sommermonate traditionell die Zeit der Alpenpässe sind – visuell hat der Winter in den Bergen noch mehr zu bieten. Denn während die einen Passstraßen in den Winterschlaf fallen und sich ihre Serpentinen gerade noch als Silhouetten unter der Schneedecke abzeichnen, ziehen sich die Kurven und Kehren der geräumten und offen gehaltenen Alpenübergänge wie schwarze Tuschelinien über den weißen Grund. Bei meinen winterlichen Reisen in die Berge weiß ich bis zuletzt oft nicht, was meine Kamera und ich geboten bekommen werden: den monochromen White-Out oder den ultimativen kalligrafischen Kontrast. Es ist diese saisonale Reduktion auf das Wesentliche – die Kurve als Teil der alpinen Topografie, das schwarze Asphaltband in seiner reinsten Form –, die mir die künstlerische Leistung der Architekte und Ingenieure stets aufs Neue vor Augen führt.

The summer months in the Alps are traditionally the busiest on the passes – but visually, wintertime in the mountains holds a far greater appeal. While the pass roads go into hibernation and their snaking contours only appear as silhouettes beneath a blanket of snow, the corners and bends of the cleared and open Alpine crossings stretch like black ink marks over the white landscape. During my winter trip in the mountains I often don't know what's in store for me and my camera around the next corner: the monochrome whiteout or the ultimate calligraphic contrast. It is this seasonal paring-down to the very essence – the curves as a part of the Alpine topography, the black strip of asphalt in its purest form – that constantly reminds me of the artistic achievement of the architects and engineers.

PROLOG

Jan Baedeker

Und dann beginnt es zu schneien. Erst fallen die Flocken noch spärlich auf ausgedörrte Wiesen und leuchtend gelbe Lärchenwipfel, schiefergraue Felsen und stille, dunkle Seen herab. Doch das Schneetreiben wird rasch dichter, bis man kaum mehr die nächste Kurve erkennt und die Serpentinen, die den Berg noch vor wenigen Augenblicken so markant gezeichnet haben, unter einer kalten weißen Decke verschwinden. Mit dem ersten Schnee kehrt in den Alpen die Ruhe ein. Die Urlauber in ihren Wohnmobilen sind längst in ihre Täler zurückgekehrt – und mit ihnen die dröhnenden Rennmaschinen und Sportwagen der Kurvenjäger. Noch klingen irgendwo aus dem Nebel ein paar Kuhglocken, doch auch sie werden bald nicht mehr zu hören sein.

Es ist diese Zeit zwischen dem ersten Schnee und dem Beginn der Wintersportsaison, in der die Bergwelt ihren ursprünglichen Charakter offenbart: unwirtlich und rau, einsam und still, sich in ihrer monochromen Monotonie aus weißen Gipfeln, weißen Tälern, weißen Nebelschwaden, dunklen Nadelwäldern und tiefschwarzem Gestein allen Sinnen widersetzend. Interessanterweise zeigen die meisten Bilder, die uns in Zeitschriften, Büchern und Werbekampagnen begegnen, die Alpenwelt entweder im Hochsommer mit blühenden Bergwiesen, oder als Winterwunderlandschaft mit geschlossener Schneedecke – lediglich der Himmel ist zu jeder Jahreszeit tiefblau. Die Wochen und Monate der Transformation hingegen scheinen sich mit ihren Wolken- und Schneefetzen dem medialen Ideal der Alpen zu widersetzen und sind deshalb visuell nur wenig dokumentiert. Gerhard Richters Davoser Gemälde Monstein von 1981 gehört zu den wenigen, dafür umso eindrücklicheren Porträts dieser Zwischenzeit, wenn die Sonne höchstens noch als diffuses Leuchten durch die Wolken dringt. „Es gibt für mich keinen Unterschied zwischen einer Landschaft und einem abstrakten Bild", hat Richter einmal gesagt. Und tatsächlich: Hier oben, jenseits der Baumgrenze, wenn der Schnee langsam die Kontraste auslöscht, verschwimmen für den Betrachter die Grenzen zwischen Gegenständlichkeit und Abstraktion.

Noch kann man unter dem Schnee hier und da eine Serpentinenkurve erahnen – doch die meisten Alpenpässe fallen nun in ihren Winterschlaf, aus dem sie erst im späten Frühjahr wieder erwachen. Wenn Sonne und Schneefräsen die schwarzen Asphaltbänder im Mai oder Juni zurück ans Tageslicht bringen, hat man die schönsten Serpentinen nochmals für kurze Zeit für sich allein, bevor die Blechkarawane sich wieder den Berg hinaufschiebt. Doch für viele der großen Passstraßen, vom Gotthard bis zum Stilfser Joch, sind die Sommermonate eigentlich die Ausnahme, die Abweichung von ihrem tiefgekühlten Aggregatzustand. Seit die meisten Berge der Alpen von Tunneln durchlöchert werden, durch die man bequem fast jedes Tal und Skigebiet erreicht, gibt es nur noch eine Handvoll Pässe, die auch in den Wintermonaten schneefrei gehalten werden. Umso erstaunlicher erscheint einem angesichts dieses physikalischen Härtetests die ingenieurtechnische Leistung des Straßenbaus. Es ist paradox, dass wir heute vor allem die modularen Fertighäuser und vergänglichen Bungalows der Architekturmoderne bewundern, während die beständigsten architektonischen Wunderwerke dieser Zeit – die Hochstraßen, aber auch die zahllosen Tunnels und Brücken – von uns weiterhin wie selbstverständlich als commodities betrachtet und befahren werden. Auf Stefan Bogner, der für sein Mammutprojekt Curves wie kein anderer Fotograf die Straßen der Alpen dokumentiert hat, üben die „schlafenden Schönheiten" gerade in der Nebensaison eine unwiderstehliche Anziehungskraft aus. Wenn der Nebel klamm in den Wäldern hängt, der erste Schnee sich über Felsen und Almen legt oder die alpine Landschaft sich in der Frühjahrssonne in einen weißbraunen Flickenteppich verwandelt, findet er seine authentischsten Motive. Mit seinen antizyklischen Ausflügen in diese kaum dokumentierte Saison der Bergwelt schließt sich vorerst auch sein Bilderzyklus der schönsten Alpenpässe.

Für einen Serpentinensammler und Berufungsreisenden wie Stefan Bogner, der von spektakulärer Straßenarchitektur und wilden, menschenverlassenen Landschaften geradezu besessen ist, darf damit allerdings noch nicht Schluss sein. Und so hält auch dieses Buch als Kür noch einen Ausblick bereit: eine Winterreise ans Ende der Welt. An welchem anderen Ort könnte es im Dezember und Januar schließlich so finster, archaisch und kalt sein wie auf Island, jener sagenumwobene Vulkaninsel am Rande des Polarkreises? Dass hier ein Fahrweg durch Eis, Schnee und die ewige Nacht führt, dessen Spur man einmal um die Insel herum folgen kann, gehört zu den Leistungen der Straßenbaukunst, vor deren Schönheit und Sinnlichkeit wir uns mit diesem Band nochmals verneigen möchten.

PROLOGUE

Jan Baedeker

And then it begins to snow. At first the flakes fall gently on arid pastures, icing the tips of the dazzlingly yellow larches, the granite-grey rocks and the serene, dark lakes. But the flurries quickly become thick and heavy until you can barely see the next corner, and the switchbacks that were snaking their way up the mountains so distinctly just a moment ago have disappeared under a crisp, white blanket. With the first snowfall tranquillity settles over the Alps once again. The tourists in their campervans have long returned to the valleys – and so have the blaring racers and sports cars of cornering zealots. And the muted jingling of cow bells floating out of the fog will soon fall silent, too.

It is this time, between the first snow and the start of the winter sport season, that the mountainscape shows its true character: inhospitable and harsh, lonely and still, in its monochrome monotony of white peaks, white valleys, white wisps of fog, dark conifer forests and inky black rocks defying the senses. Interestingly, most of the images in magazines, books and advertising campaigns portray an Alpine world in midsummer, with flowers carpeting the mountain meadows, or

a winter wonderland quilted in snow – and the skies are a deep blue, no matter what season. The weeks and months of the metamorphosis, with its cloudscapes and snowscapes, obviously go against the media ideal of the Alps and thus are not as well documented visually. Gerhard Richter's Davos 1981 painting Monstein is one of the few but all the more dramatic portraits of these bridging months, when the sun filters through the clouds as a diffused light.

"For me, there is no difference between a landscape and an abstract image," Richter once said. And indeed, up here above the tree line when the snow slowly erases the contrasts, the observer's boundaries become blurred between reality and abstraction. Here and there under the snow, hints of the serpentine corners can be seen – yet most of the Alpine passes now go into a hibernation from which they will only awaken in late spring. Before snowploughs and sun finally bring the black ribbons of tarmac to light again in May or June, we have but a short time to enjoy the beautiful switchbacks alone before the sheet-metal cavalcade pushes its way back up the mountain. But for many of the major mountain passes, from Gotthard to Stelvio, the summer months are in fact the exception, the divergence from their frozen physical state. Since most of the mountains in the Alps are riddled with tunnels through which drivers can easily reach every valley and ski area, there are only a handful of passes that are kept free of snow during the winter months. Bearing in mind the physical test of endurance required to build them, this makes the engineering feats of road construction seem all the more astonishing. It's a paradox that we pay particular attention to the modular prefabricated houses and impermanent bungalows of modern architecture, while the most enduring architectural marvels of that time – the high mountain roads and the countless tunnels and bridges – are unquestionably regarded as commodities.

For Stefan Bogner, who has documented the Alpine roads like no other photographer in his mammoth project Curves, these "sleeping beauties" have an irresistible pull in the off-peak season. When fog wraps its fingers through the forests, when the first snow swathes the rocks and meadows, or when the spring sun turns the Alpine landscape into a white and brown patchwork carpet, this is where he finds his authentic motifs. With his anti-cyclical excursions during this rarely-documented season of the mountain world, his pictorial documentation of the most beautiful Alpine passes comes full circle. For a zigzag devotee and vocational traveller such as Stefan Bogner, who is completely captivated by the spectacular road architecture and wild, uninhabited landscapes, it can never be enough. Thus, this book concludes with an extra icy flurry: a winter journey to the end of the world. After all, where else can it be as dark, archaic and cold in December and January as in Iceland, the mythical volcanic island at the edge of the Arctic Circle? A road circumnavigating the island through ice, snow and eternal light represents an achievement in the art of road construction, to which sinuous beauty we would like to pay homage once again with this book.

ALPEN / ALPS

1 — Großglockner-Hochalpenstraße ... 14
2 — Drei-Zinnen-Straße ... 30
3 — Passo di Giau ... 40
4 — Passo di Falzarego / Passo di Valzarola ... 52
5 — Passo Fedaia ... 62
6 — Passo Pordoi ... 72
7 — Passo Sella ... 86
8 — Passo Gardena ... 98
9 — Passo dello Stelvio ... 110
10 — Timmelsjoch-Hochalpenstraße ... 122
11 — Kaunertaler Gletscherstraße ... 134
12 — Silvretta-Hochalpenstraße ... 144
13 — Flexenpass ... 154
14 — Berninapass ... 164
15 — Flüelapass ... 176
16 — Furkapass ... 188
17 — Sustenpass ... 200
18 — Grimselpass ... 216
19 — Nufenenpass ... 230
20 — St. Gotthardpass ... 242

ISLAND / ICELAND

21 — Hringvegur ... 258

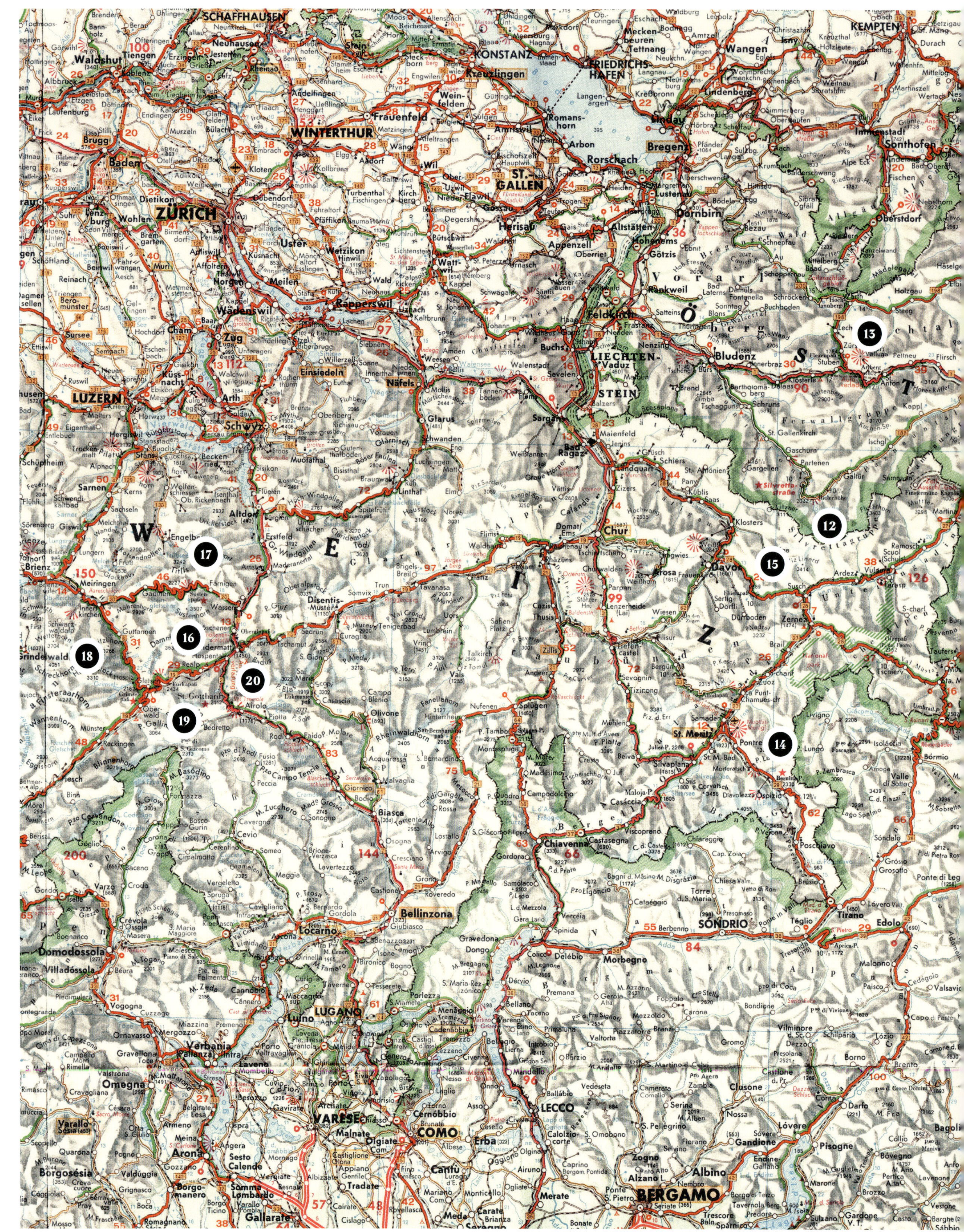
SCHAFFHAUSEN
KONSTANZ
FRIEDRICHS-HAFEN
KEMPTEN
Waldshut
Kreuzlingen
Frauenfeld
WINTERTHUR
Romanshorn
Lindau
Wangen
Bregenz
Immenstadt
Sonthofen
Brugg
Baden
ST. GALLEN
Rorschach
ZÜRICH
Dornbirn
Oberstdorf
Appenzell
Altstätten
Hohenems
Uster
Rapperswil
Feldkirch
Bludenz
Zug
LIECHTENSTEIN
Vaduz
Einsiedeln
Näfels
Glarus
LUZERN
Schwyz
Sargans
Bad Ragaz
Landquart
Sarnen
Altdorf
Chur
Klosters
Davos
Arosa
Meiringen
Brienz
Andermatt
Disentis
Thusis
Zernez
Airolo
Splügen
St. Moritz
Pontresina
Olivone
Biasca
Chiavenna
Bormio
Bellinzona
Locarno
Lugano
SONDRIO
Tirano
Domodossola
Verbania
Como
LECCO
VARESE
BERGAMO
Gallarate
Arona
Borgomanero
Varallo
Omegna
Edolo
W
E
I
Z
Ö
S
T
12
13
14
15
16
17
18
19
20

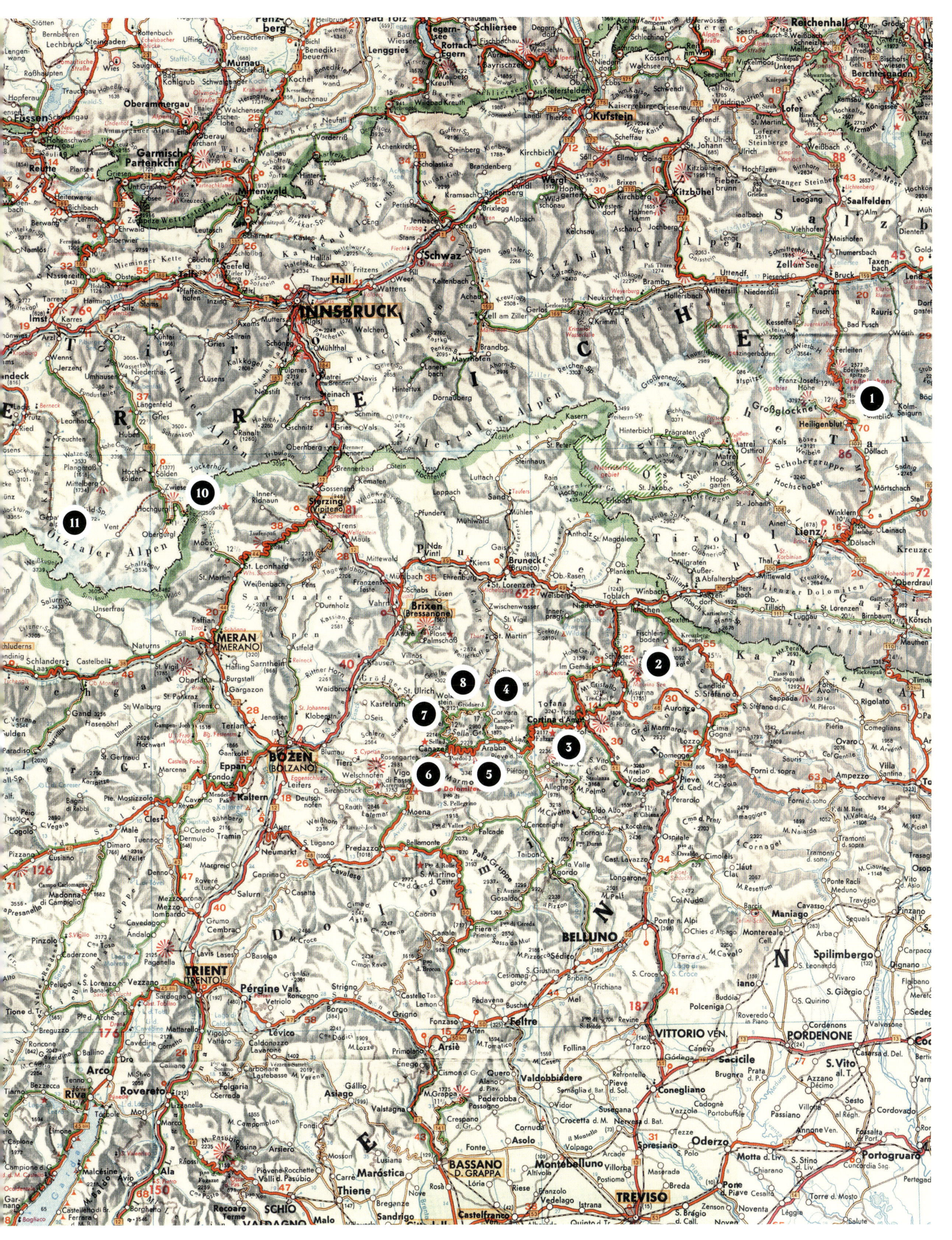
1
2
3
4
5
6
7
8
9
10
11
INNSBRUCK
Reichenhall
Berchtesgaden
Schliersee
Kufstein
Kitzbühel
Saalfelden
Zell am See
Füssen
Reutte
Garmisch Partenkirchen
Oberammergau
Mittenwald
Schwaz
Hall
Imst
Lienz
Heiligenblut
Großglockner
Sterzing (Vipiteno)
Brixen (Bressanone)
Bruneck (Brunico)
Toblach
Cortina d'Ampezzo
MERAN (MERANO)
BOZEN (BOLZANO)
Kaltern
TRIENT (TRENTO)
Pergine Vals.
Levico
Rovereto
Arco
Riva
BELLUNO
Feltre
VITTORIO VEN.
PORDENONE
Spilimbergo
Maniago
Conegliano
BASSANO D. GRAPPA
TREVISO
Oderzo
Montebelluno
Castelfranco
Thiene
Schio
Asiago
Arsiè
Ötztaler Alpen
Zillertaler Alpen
Tirol

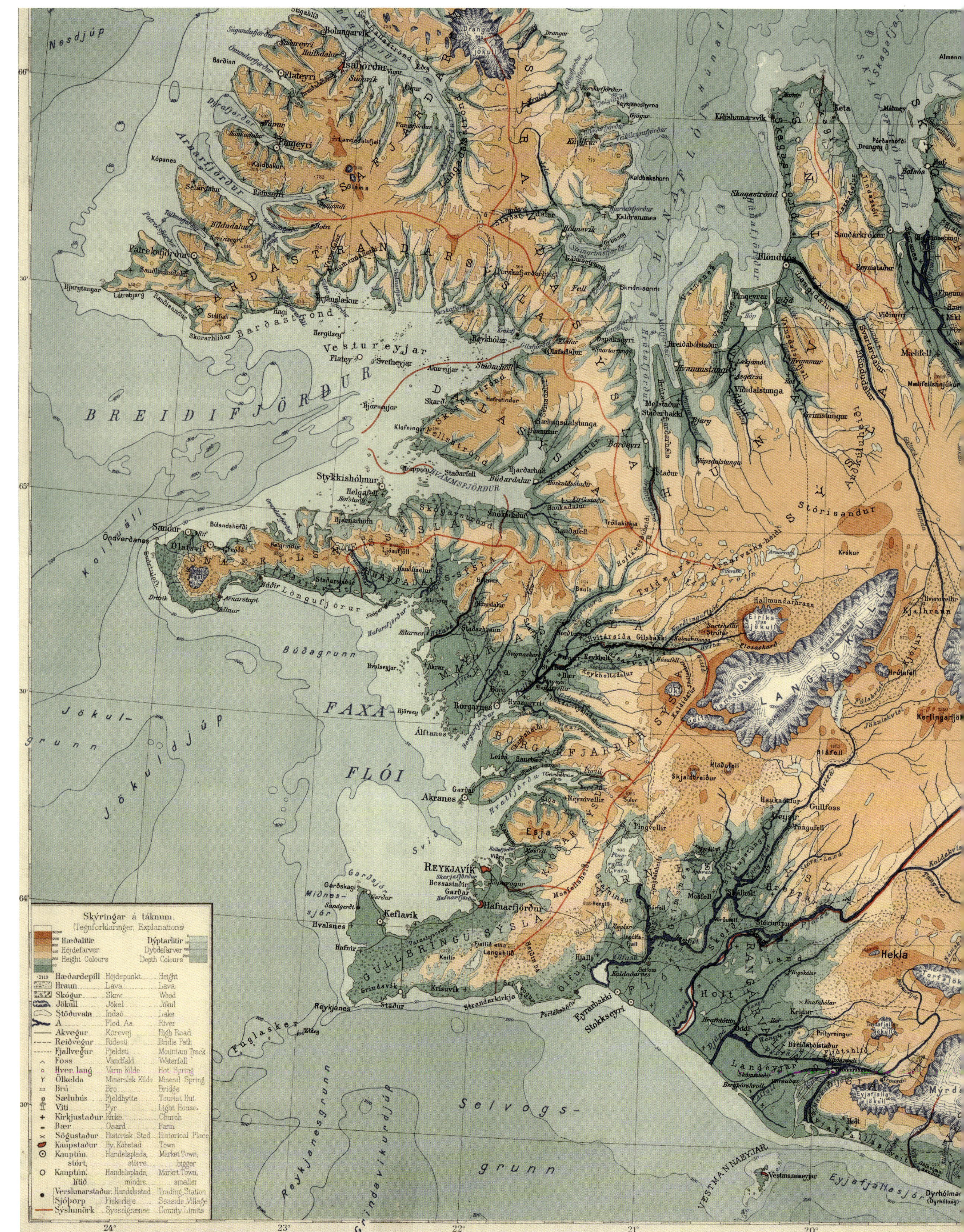
Skýringar á táknum.
(Tegnforklaringer, Explanations)
Hæðalitir Højdefarver Height Colours
Dýptarlitir Dybdefarver Depth Colours
Hæðardepill Højdepunkt Height
Hraun Lava Lava
Skógur Skov Wood
Jökull Jökel Jökul
Stöðuvatn Indsö Lake
Á Flod, Aa River
Akvegur Körevej High Road
Reiðvegur Ridesti Bridle Path
Fjallvegur Fjeldsti Mountain Track
Foss Vandfald Waterfall
Hver, laug Varm Kilde Hot Spring
Ölkelda Mineralsk Kilde Mineral Spring
Brú Bro Bridge
Sæluhús Fjeldhytte Tourist Hut
Viti Fyr Light House
Kirkjustaður Kirke Church
Bær Gaard Farm
Söguståður Historisk Sted Historical Place
Kaupstaður By, Köbstad Town
Kauptún, stórt, Handelsplads, större, Market Town, bigger
Kauptún, lítið Handelsplads, mindre Market Town, smaller
Verslunarstaður Handelssted Trading Station
Sjóþorp Fiskerleje Seaside Village
Sýslumörk Sysselgrænse County Limits
Nesdjúp
BREIÐIFJÖRÐUR
Vestur eyjar
FAXA FLÓI
Jökul-grunn
Jökuldjúp
Búðagrunn
Selvogs-grunn
Reykjanesgrunn
Grindavíkurdjúp
Fuglasker
Eyjafjallasjór
VESTMANNAEYJAR
Vestmannaeyjar
Dyrhólmar
LANGJÖKULL
Eiríks jökull
Hekla
Stórisandur
REYKJAVÍK
Hafnarfjörður
Keflavík
Akranes
Borgarnes
Stykkishólmur
Ísafjörður
Patreksfjörður
Blönduós
Sauðárkrókur
Esja
Gullfoss
Geysir
Þingvellir
Eyrarbakki
Stokkseyri
Reykjanes
Grindavík
Flatey
Flateyri
Þingeyri
Bolungarvík
Bíldudalur
Ólafsvík
Sandur
Hvammstangi
Skagaströnd
Drangajökull
Skjaldbreiður
Hvítársíða
Látrabjarg
24°
23°
22°
21°
20°
66°
65°
64°

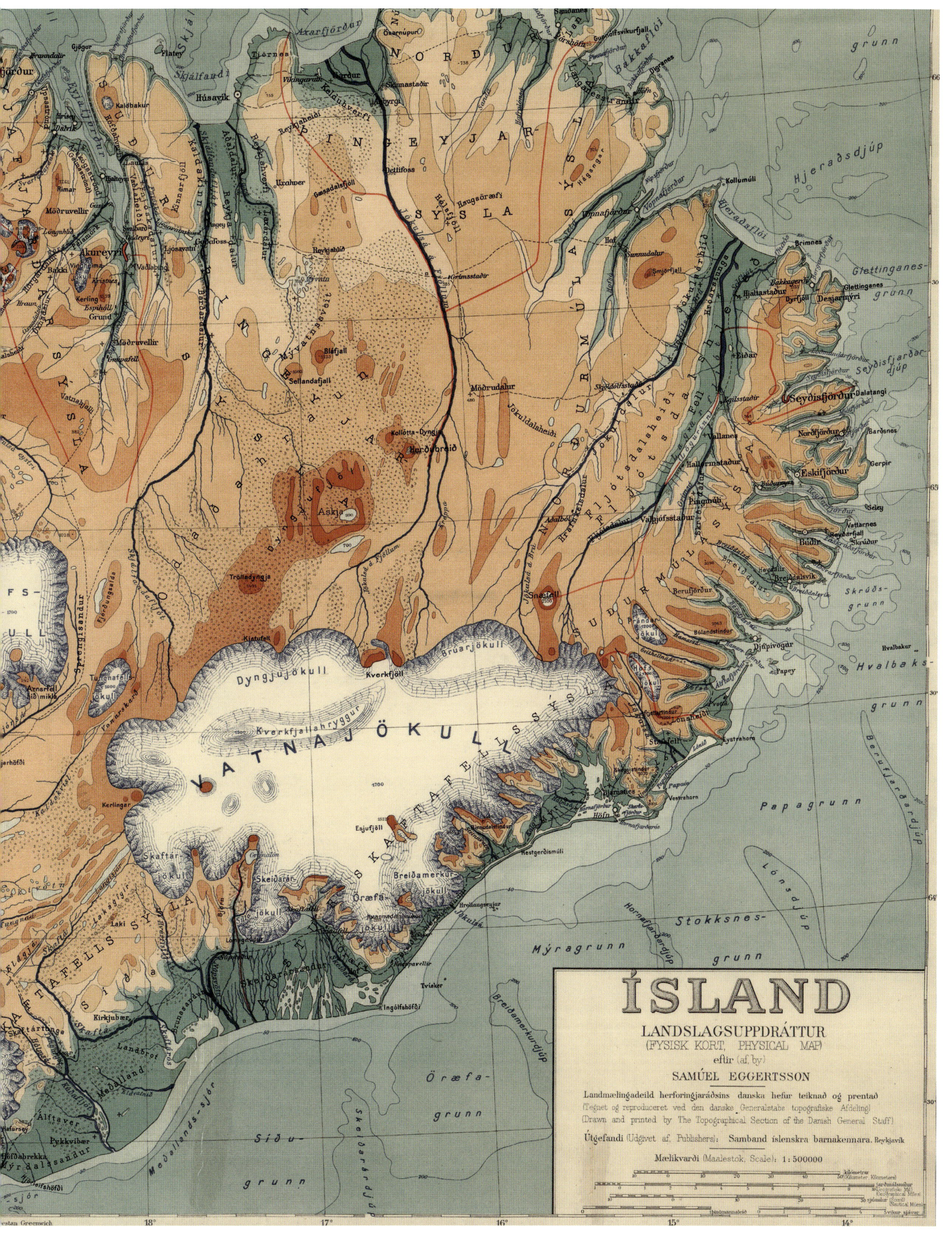
ÍSLAND
LANDSLAGSUPPDRÁTTUR
(FYSISK KORT, PHYSICAL MAP)
eftir (af, by)
SAMÚEL EGGERTSSON
Landmælingadeild herforingjaráðsins danska hefur teiknað og prentað
(Tegnet og reproduceret ved den danske Generalstabs topografiske Afdeling)
(Drawn and printed by The Topographical Section of the Danish General Staff)
Útgefandi (Udgivet af, Publishers): Samband íslenskra barnakennara, Reykjavík
Mælikvarði (Maalestok, Scale): 1:500000
kílómetrar (Kilometer, Kilometers)
jarðmálsmílur (Geografiske Mil, Geographical Miles)
sjómílur (Sømil, Nautical Miles)
þingmannaleið
vikur sjávar
VATNAJÖKULL
Dyngjujökull
Brúarjökull
Kverkfjöll
Kverkfjallahryggur
Skeiðarárjökull
Breiðamerkurjökull
Öræfajökull
Askja
Akureyri
Húsavík
Seyðisfjörður
Eskifjörður
Djúpivogur
Höfn
Skjálfandi
Axarfjörður
Vopnafjörður
Héraðsflói
Héraðsdjúp
Bakkaflói
Papey
Ingólfshöfði
Stokksnes-grunn
Mýragrunn
Pepagrunn
Lónsdjúp
Hornafjarðardjúp
Breiðamerkurdjúp
Öræfagrunn
Síðugrunn
Skeiðarárdjúp
Meðallandssjór
Hvalbaksgrunn
Berufjarðardjúp
Skrúðsgrunn
Glettinganesgrunn
Seyðisfjarðardjúp
Ódáðahraun
Sprengisandur
Mývatnssveit
Jökuldalsheiði
Fljótsdalsheiði
Laki
Snæfell
Trölladyngja
Herðubreið
Dettifoss
Goðafoss
18°
17°
16°
15°
14°

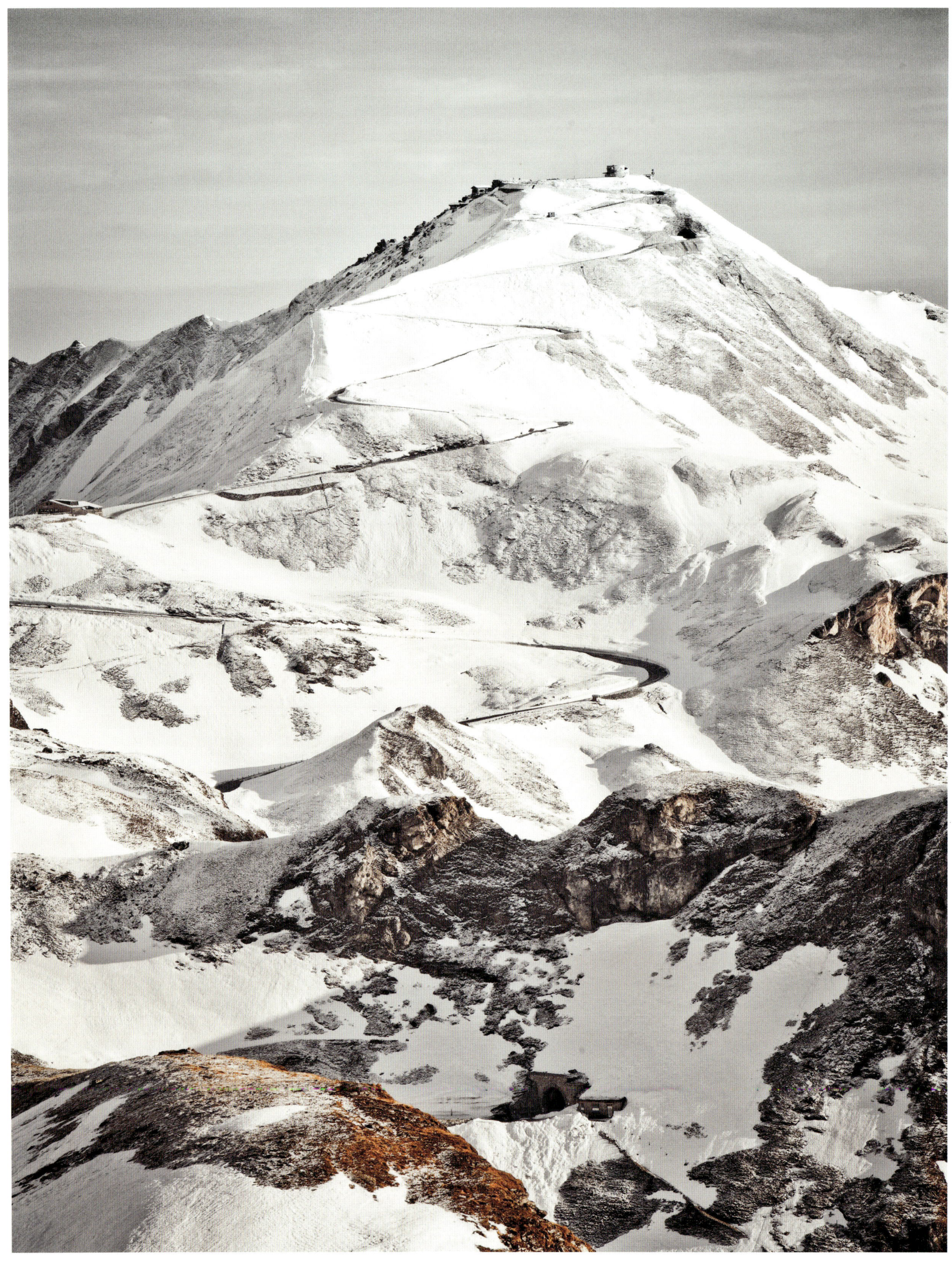

Großglockner-Hochalpenstraße

— 2504 Meter
Wintersperre meist von Oktober bis Mai

Als höchstgelegene befestigte Passstraße Österreichs und Zufahrt zum Wahrzeichen des Landes, dem 3798 Meter hohen Großglockner mit seinem eindrucksvollen Pasterzengletscher, gehört die 47,8 Kilometer lange, unter kaiserlich-habsburgerischer Regie erbaute Großglockner-Hochalpenstraße zum Pflichtprogramm jedes ambitionierten Alpenreisenden. Im Sommer stauen sich auf der Panoramastrecke meist die Touristen, im Winter türmt sich derweil der Schnee – im Jahr 1953 wurde gar eine Rekordschneehöhe von 21 Meter gemessen.

— 2,504 metres
Usually closed in winter from October to May

As the highest paved pass road in Austria and the gateway to the country's landmark – the 3,798-metre Großglockner with its stunning Pasterze Glacier – the 47.8-kilometre Grossglockner High Alpine Road, commissioned by the Imperial Habsburgs, has become a must for every zealous mountain motorist. In summer the panoramic route is clogged mostly by tourists, while in winter snowdrifts are piled high – in 1953, a record snow depth of 21 metres was measured.

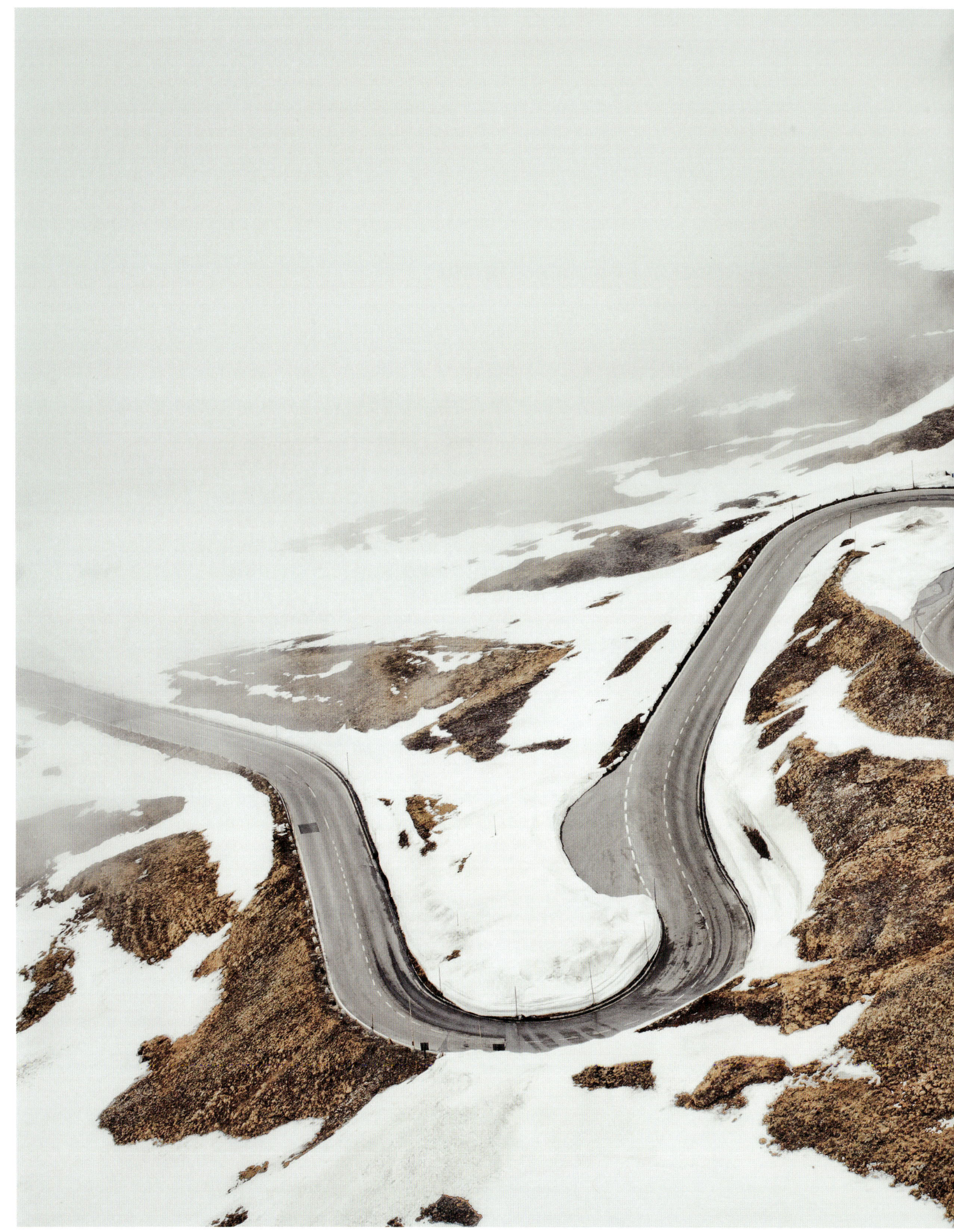

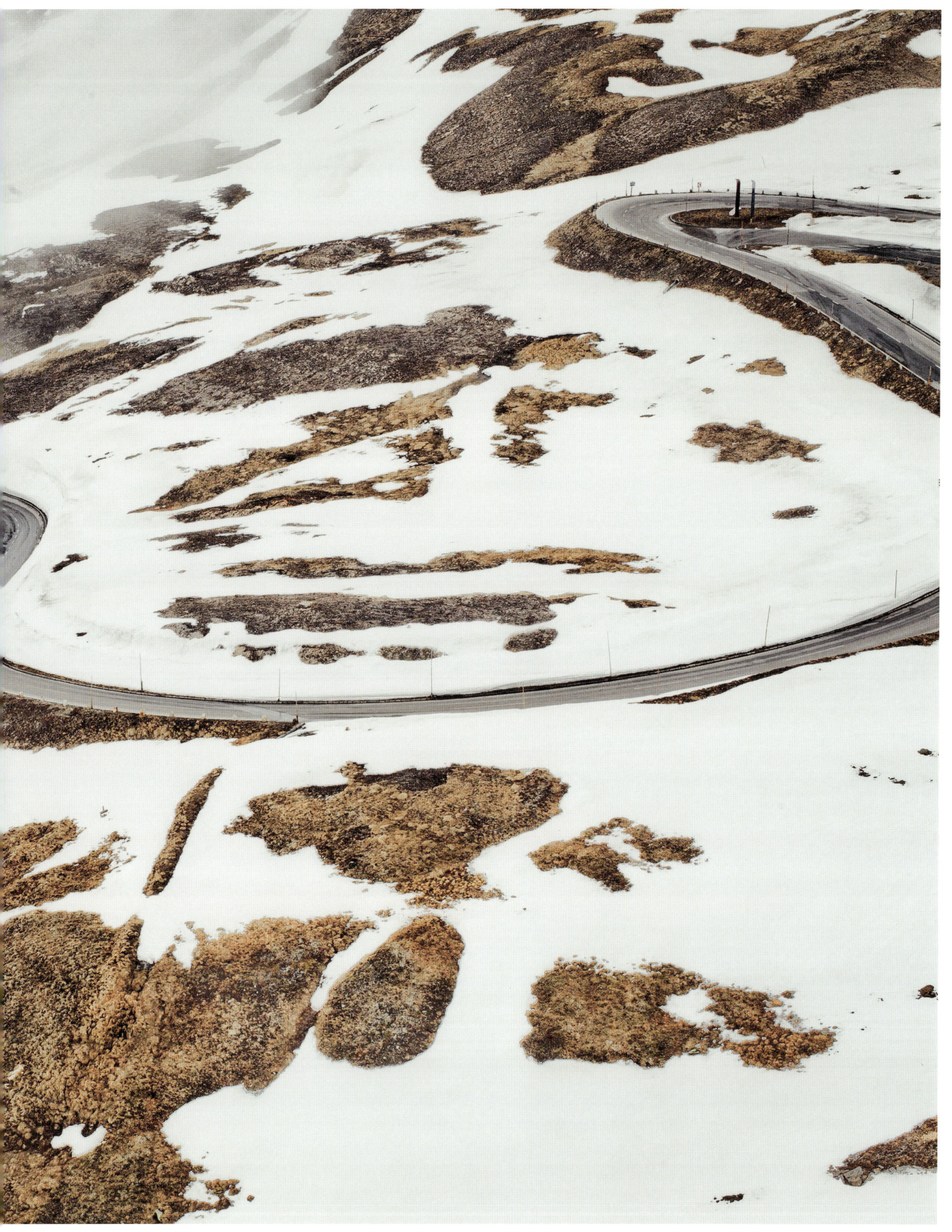

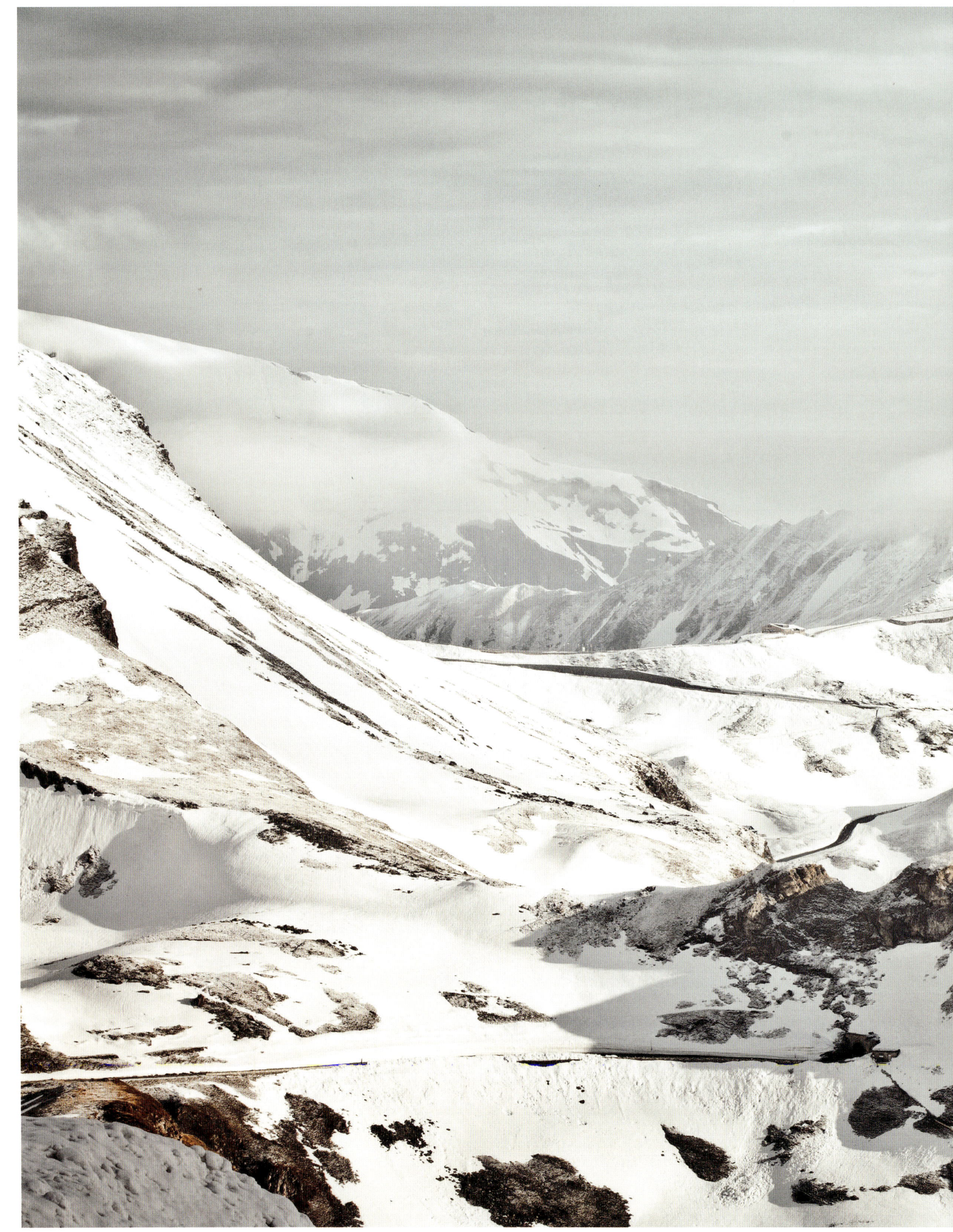

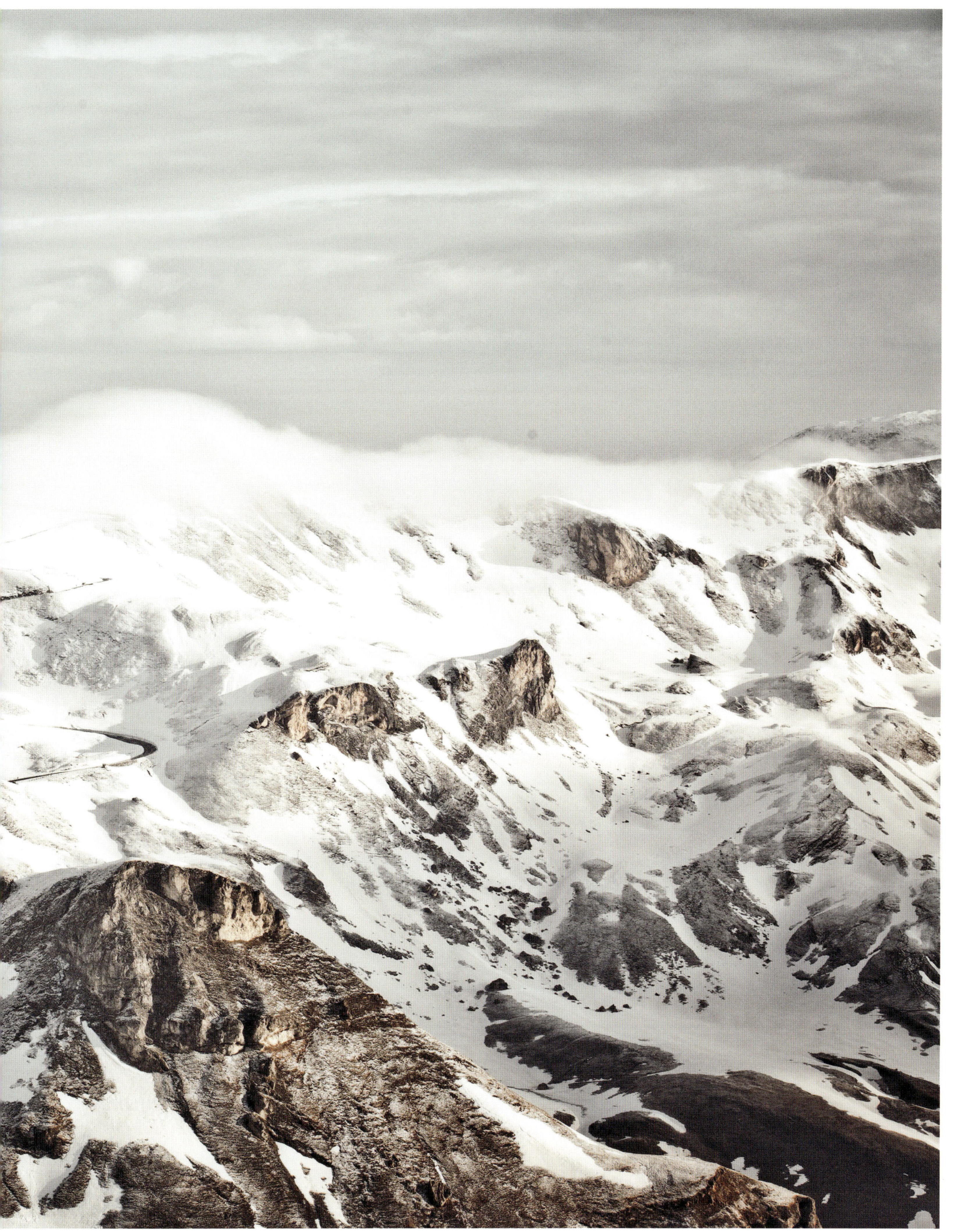

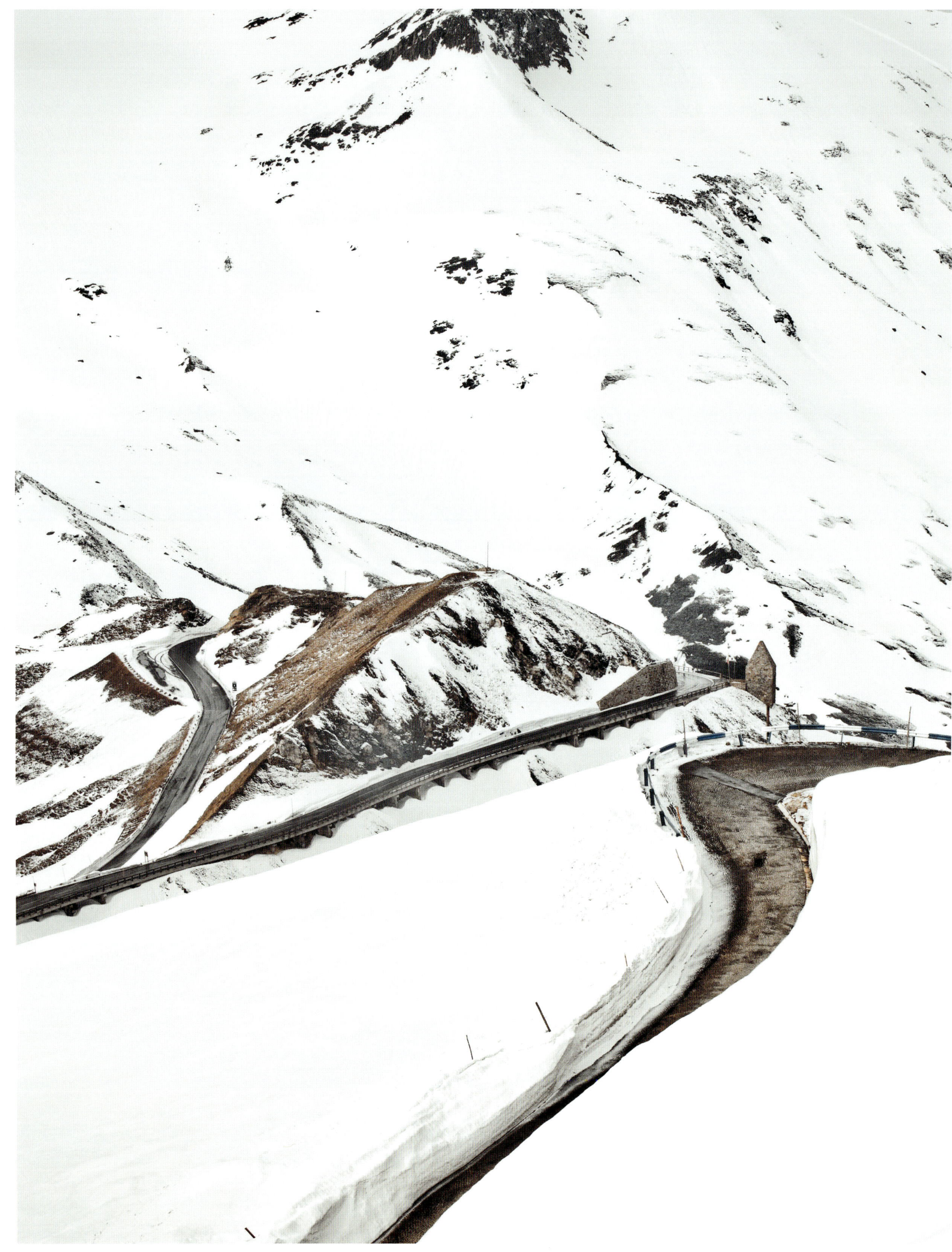

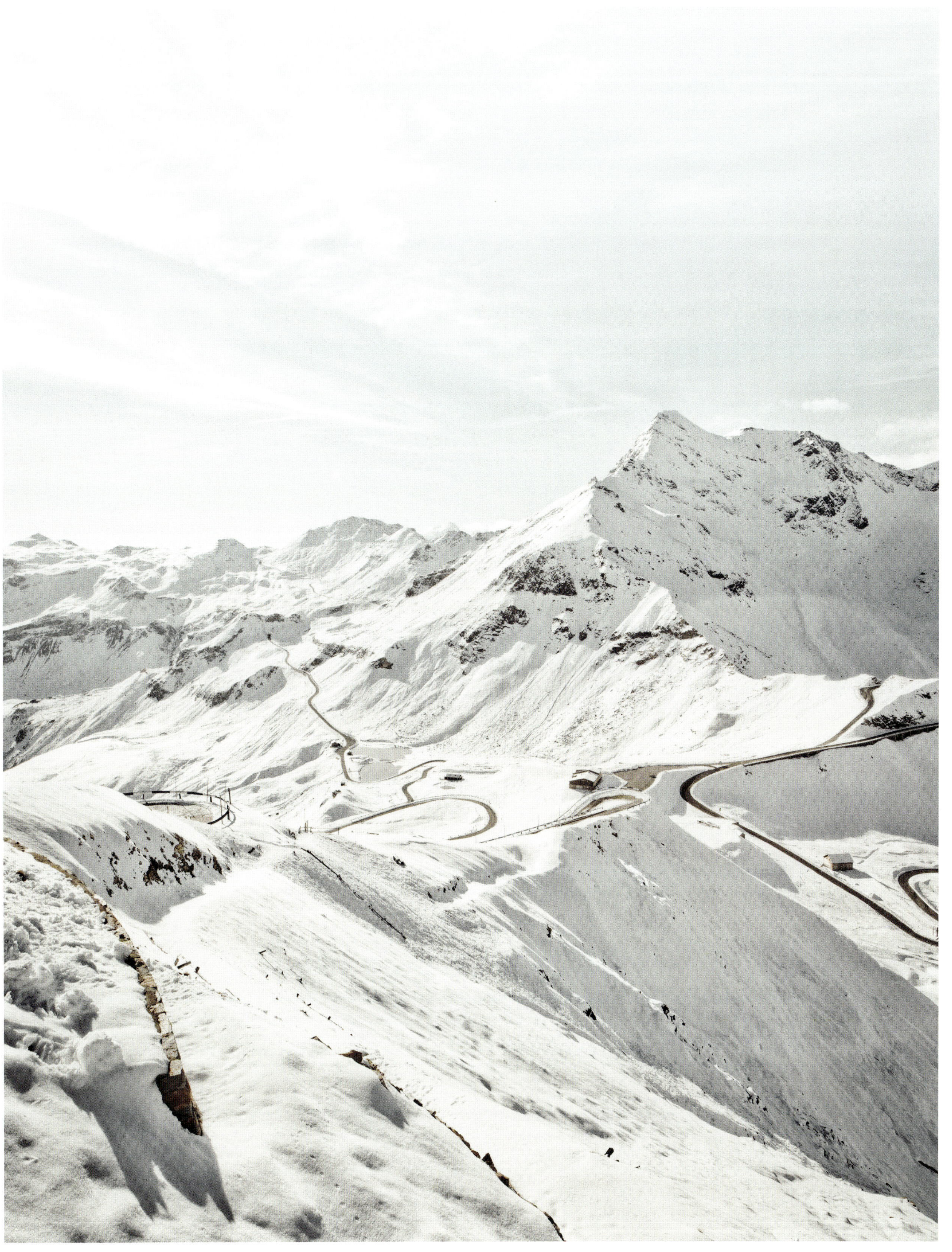

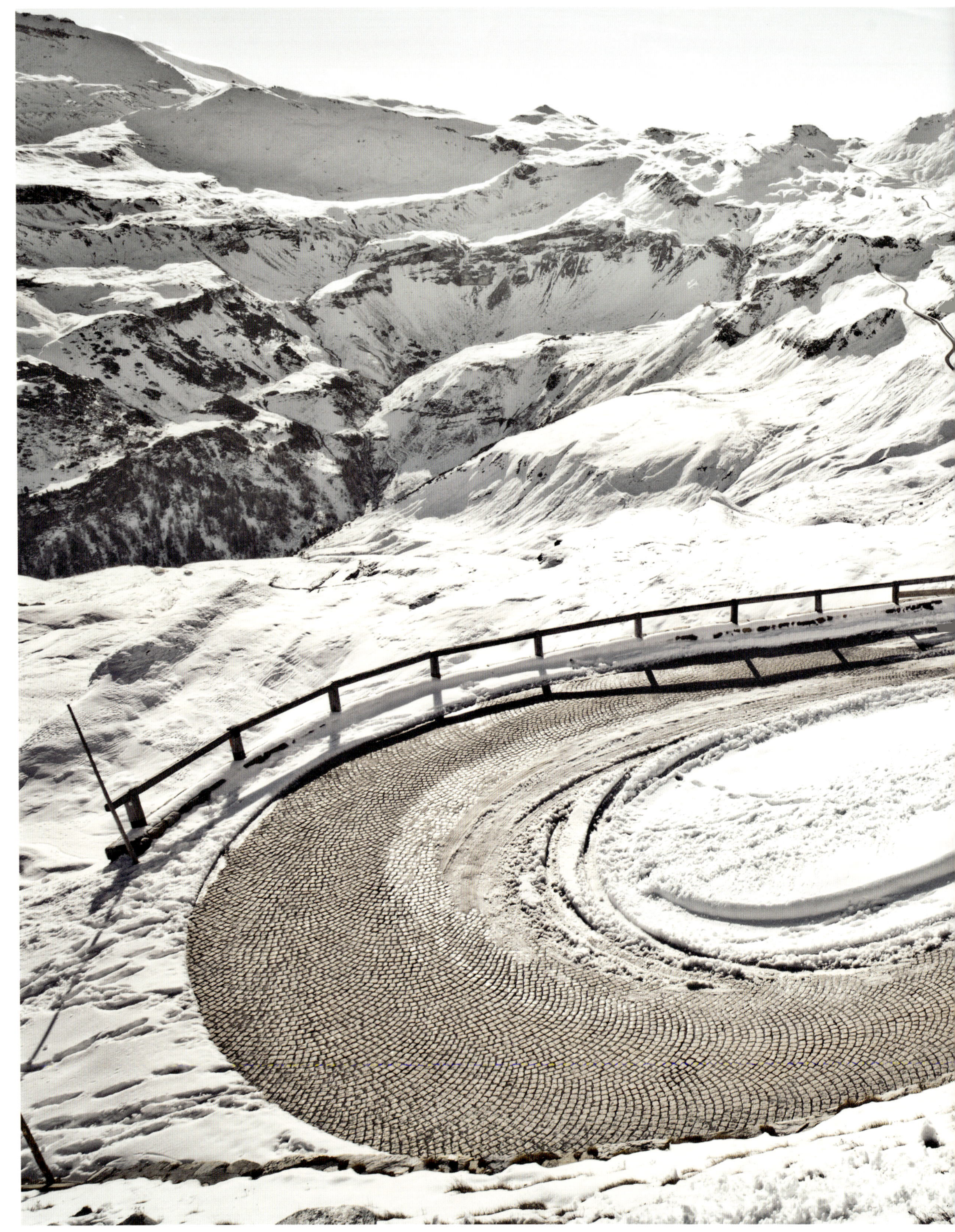

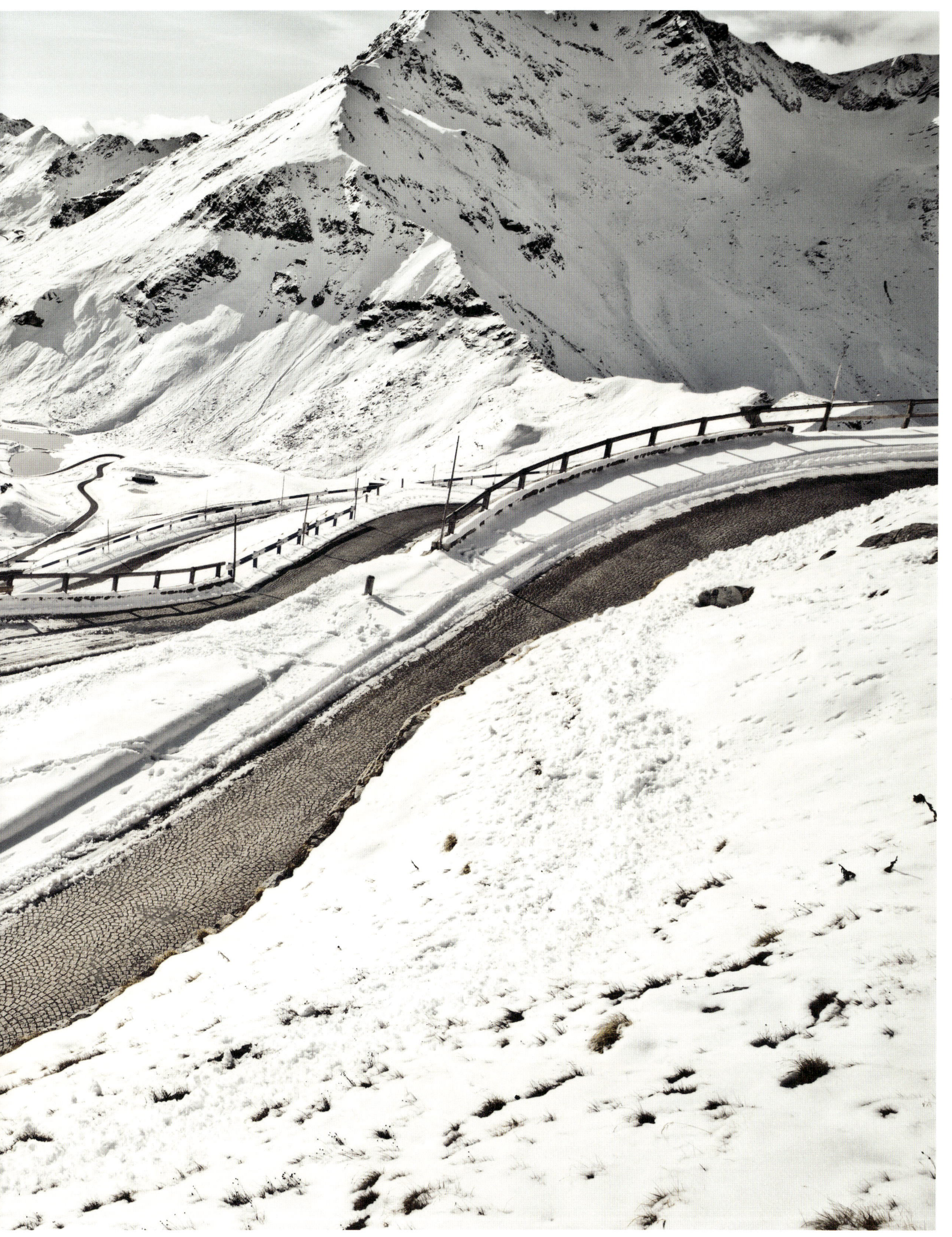

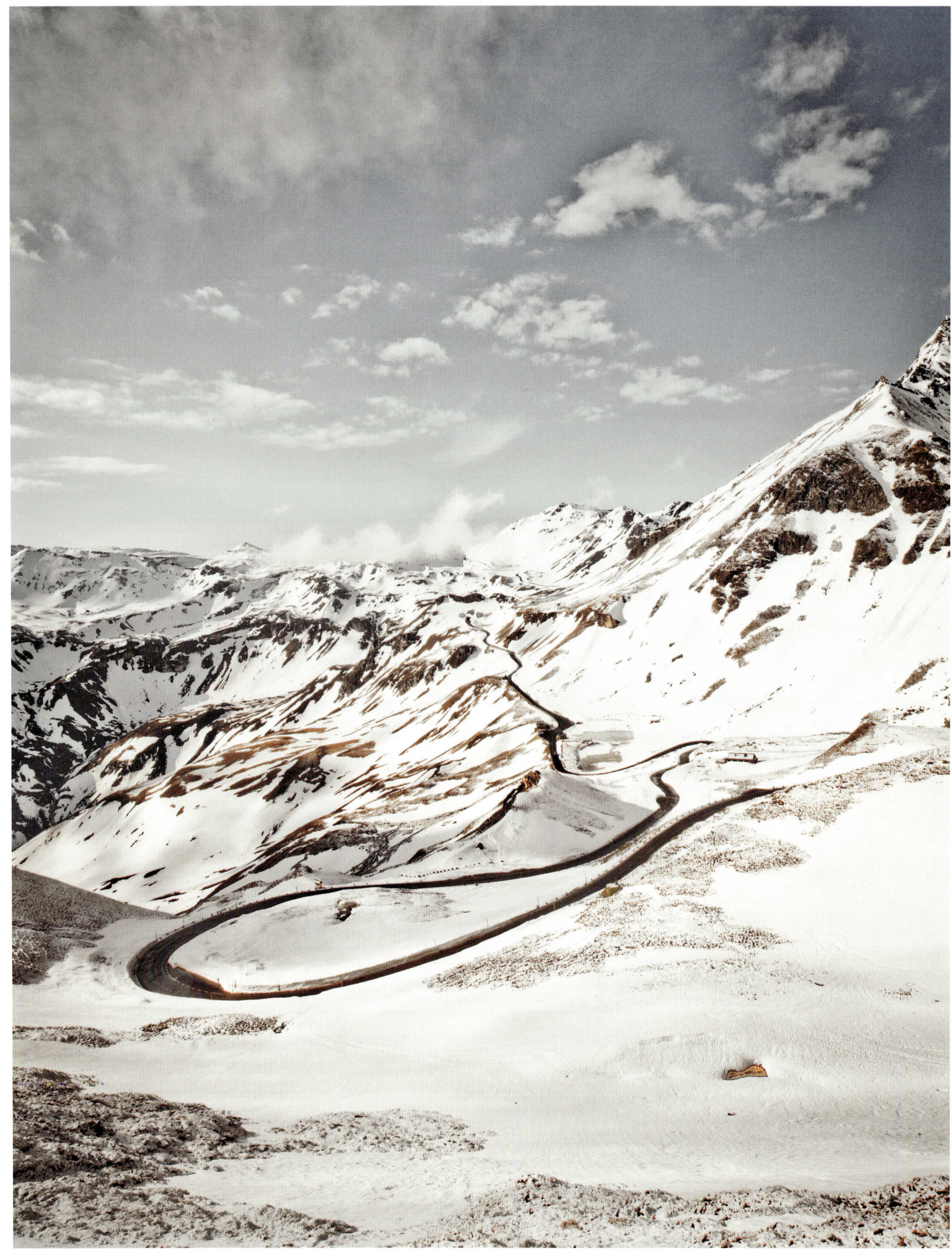

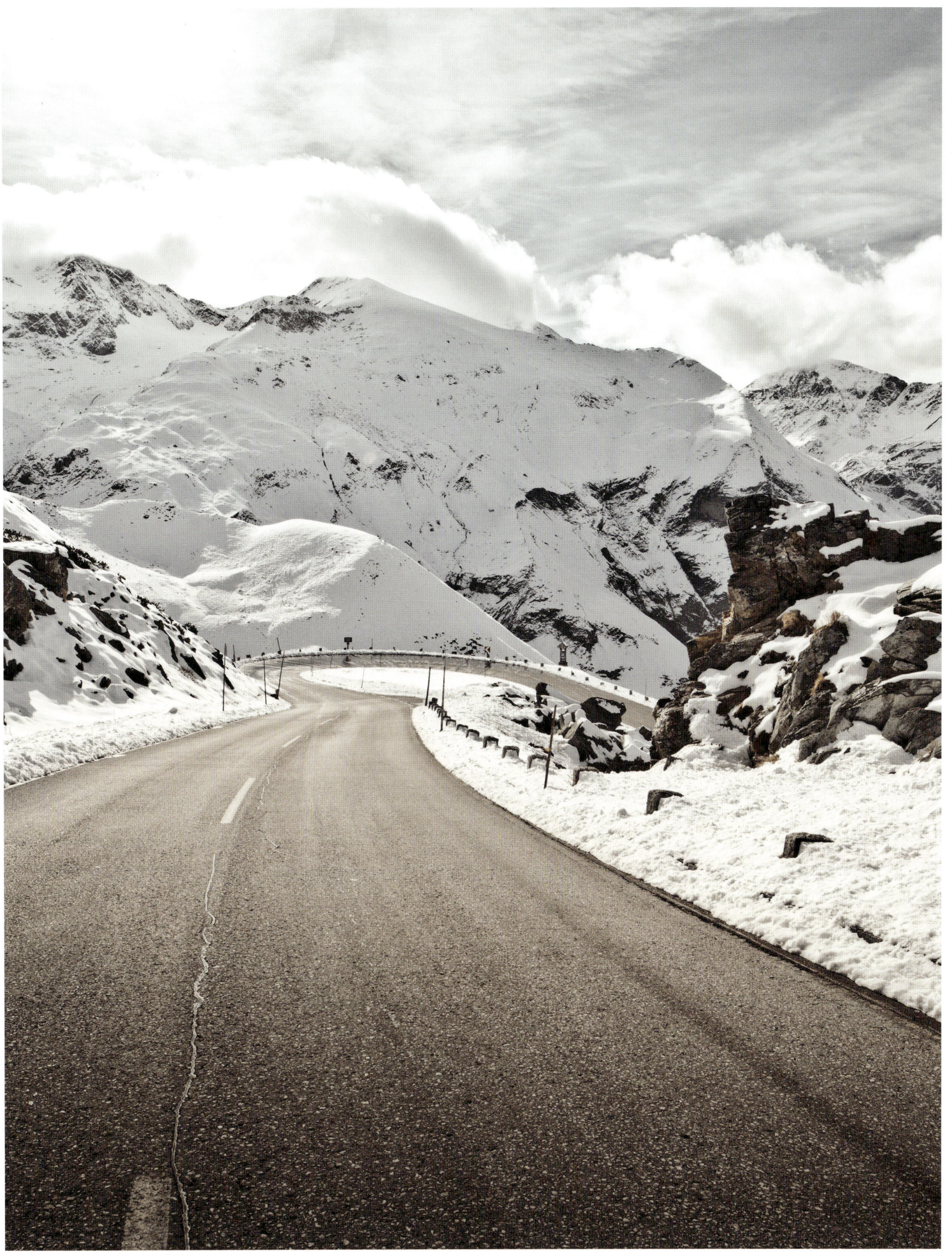

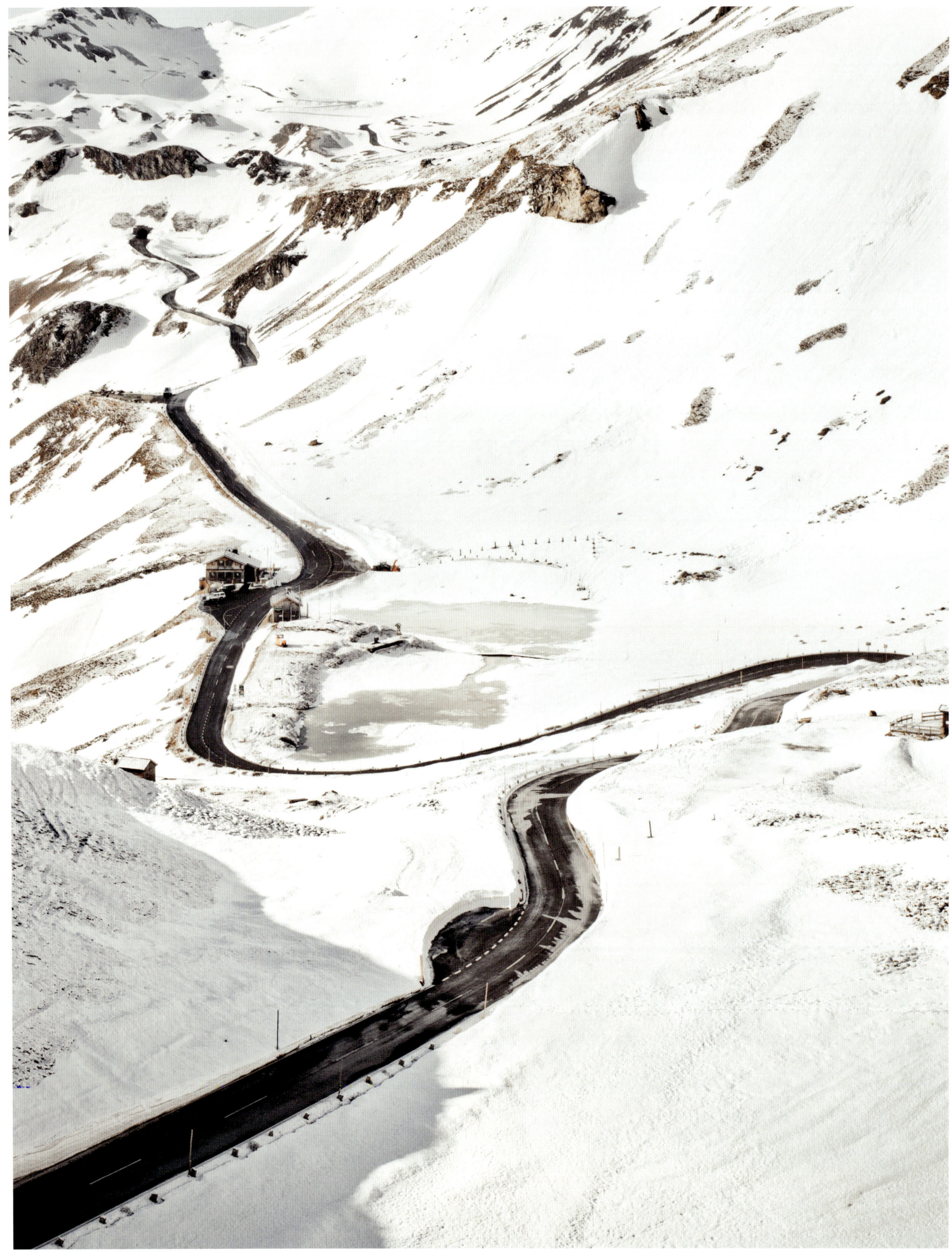

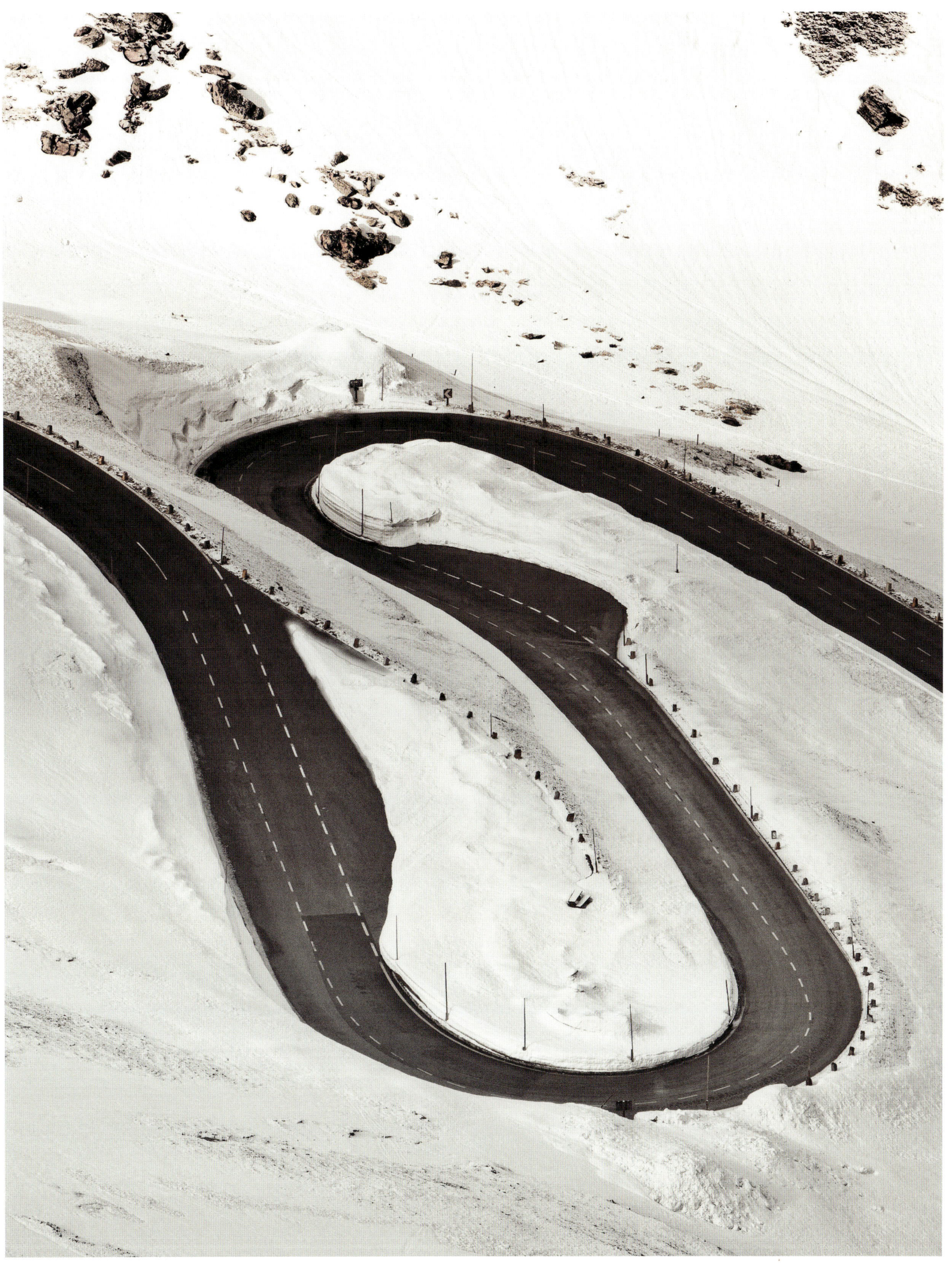

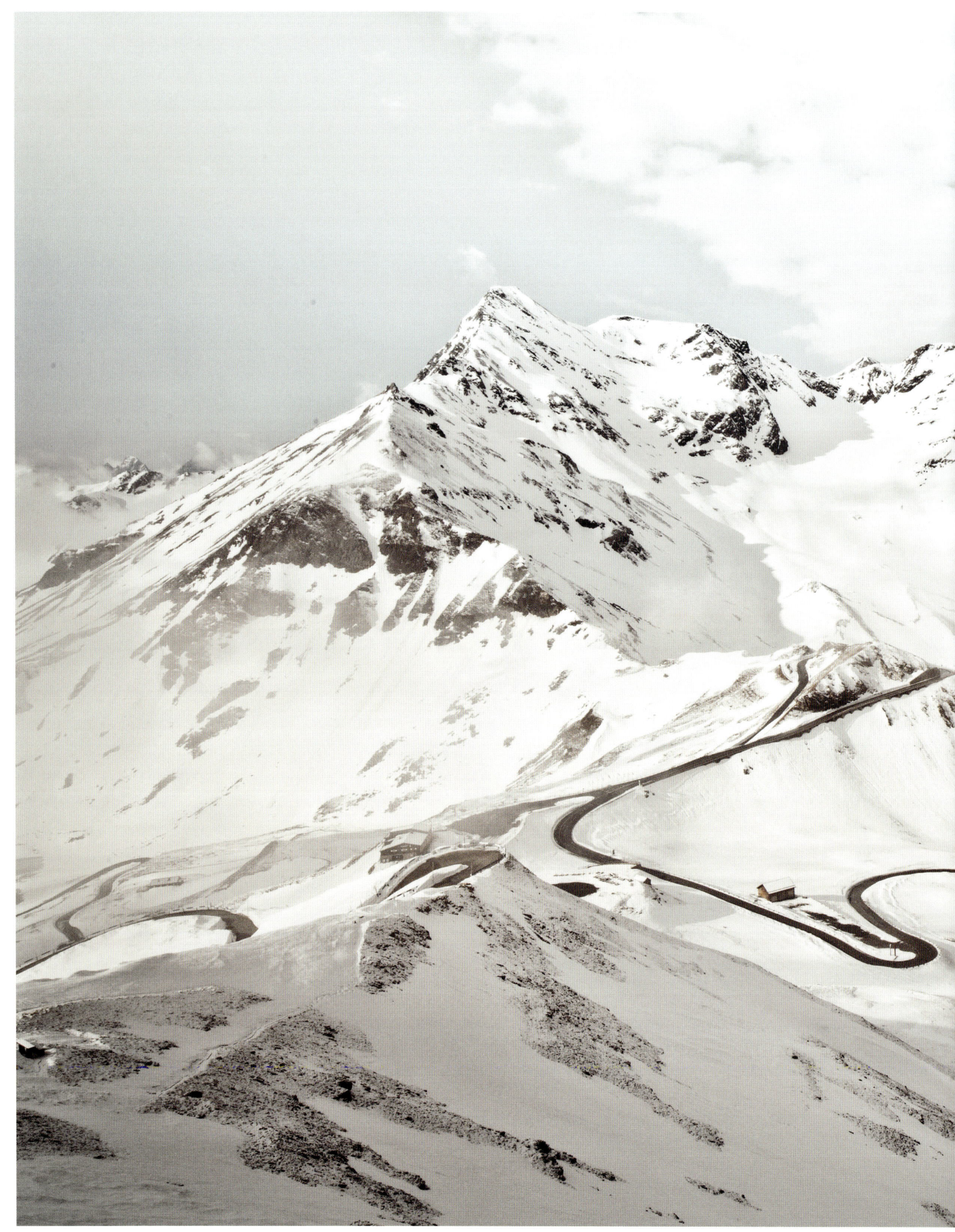

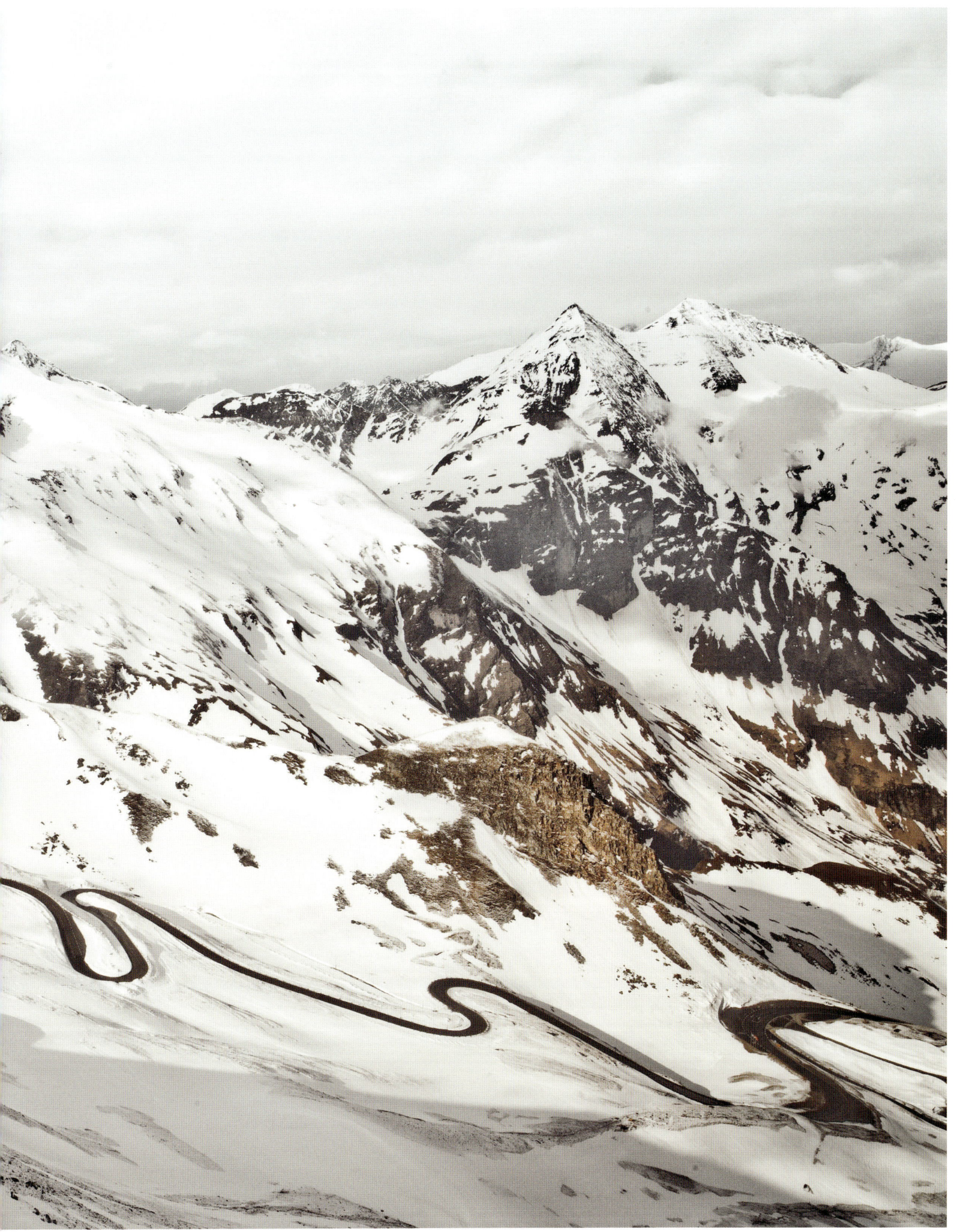

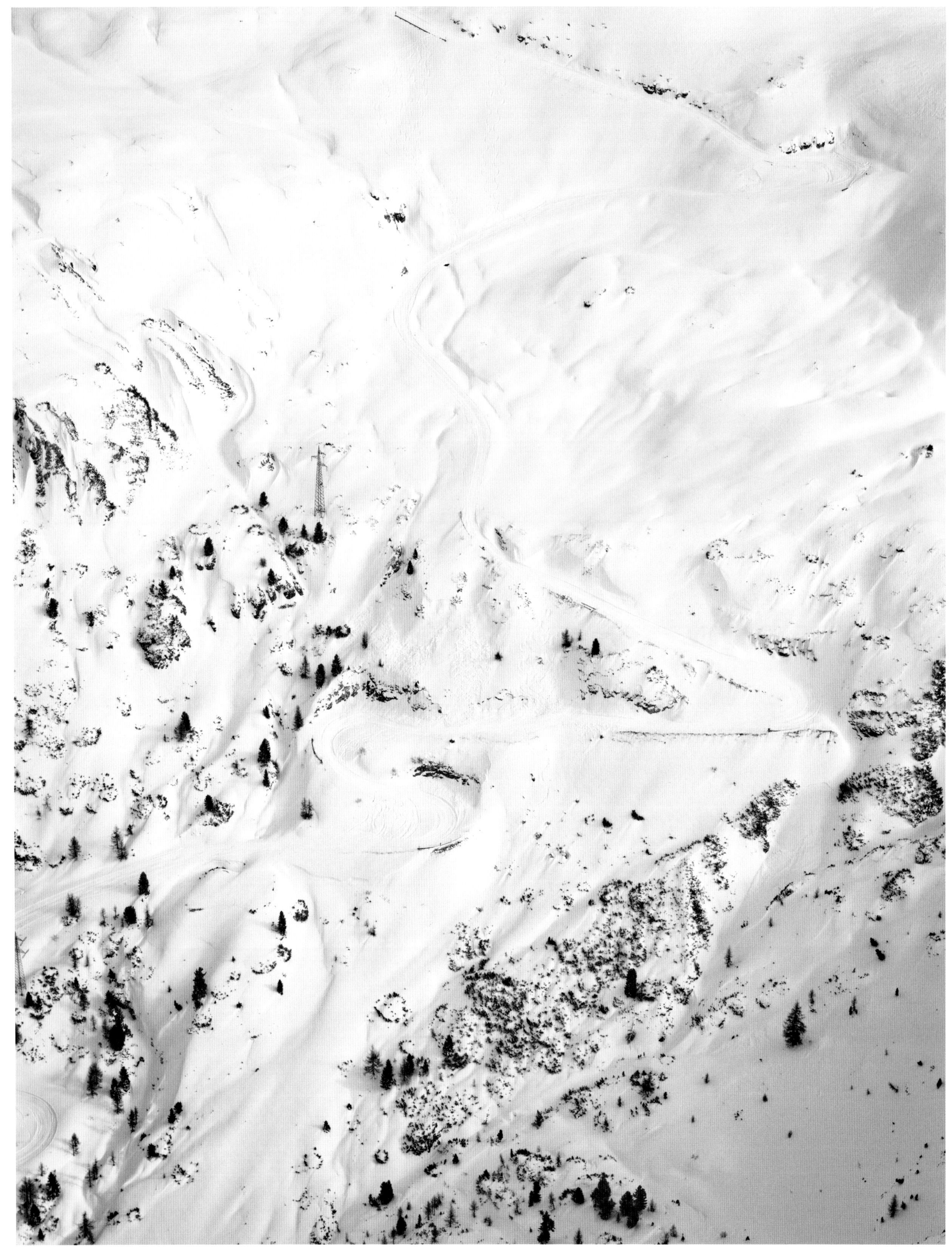

Drei-Zinnen-Straße

— 2320 Meter
Wintersperre meist von Oktober bis Mai

Die Drei Zinnen mit ihren senkrechten Felswänden und schroffen Schuttkegeln gehören zu den eindrucksvollsten, aber auch unzugänglichsten Gipfeln der Alpen. Schon im Sommer führt einen die Stichstraße vom Misuriasee südlich des Massivs nur bis zum Rifugio Auronzo auf 2320 Meter Höhe – im Winter bleibt das Wahrzeichen der Dolomiten den mächtigen Steinadlern und einigen ausdauernden Schneeschuhwanderern vorbehalten.

— 2,320 metres
Usually closed in winter from October to May

The Drei Zinnen (Three Battlements), with their vertical cliffs and precarious scree slopes, are among the most impressive but also the most inaccessible peaks of the Alps. In summertime, one of the dead-end roads from Lago de Misuri south of the massif reaches only as far as the Auronzo hut at 2,320 metres – in winter the landmark of the Dolomites is reserved for the majestic golden eagles and the hardiest of snowshoe hikers.

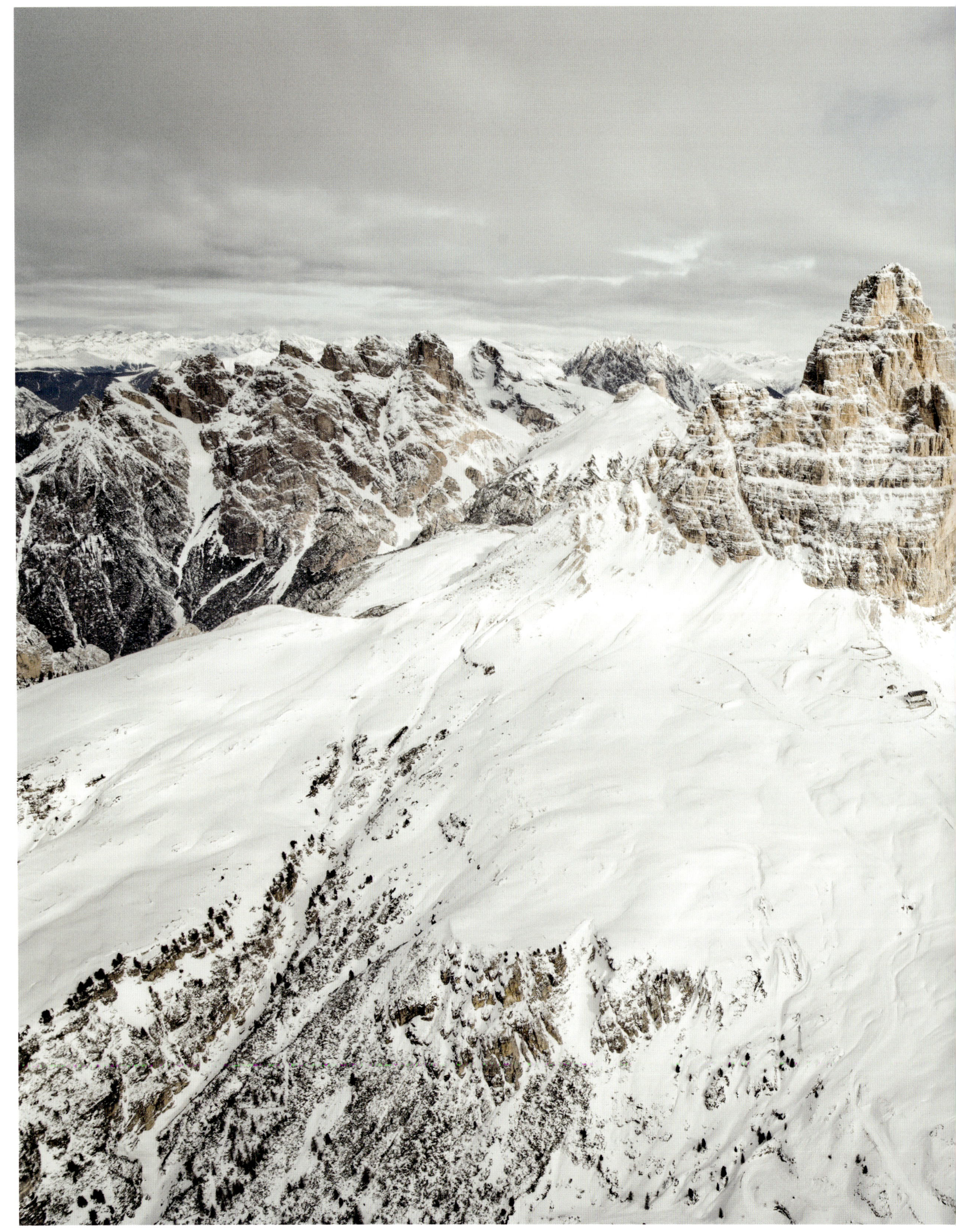

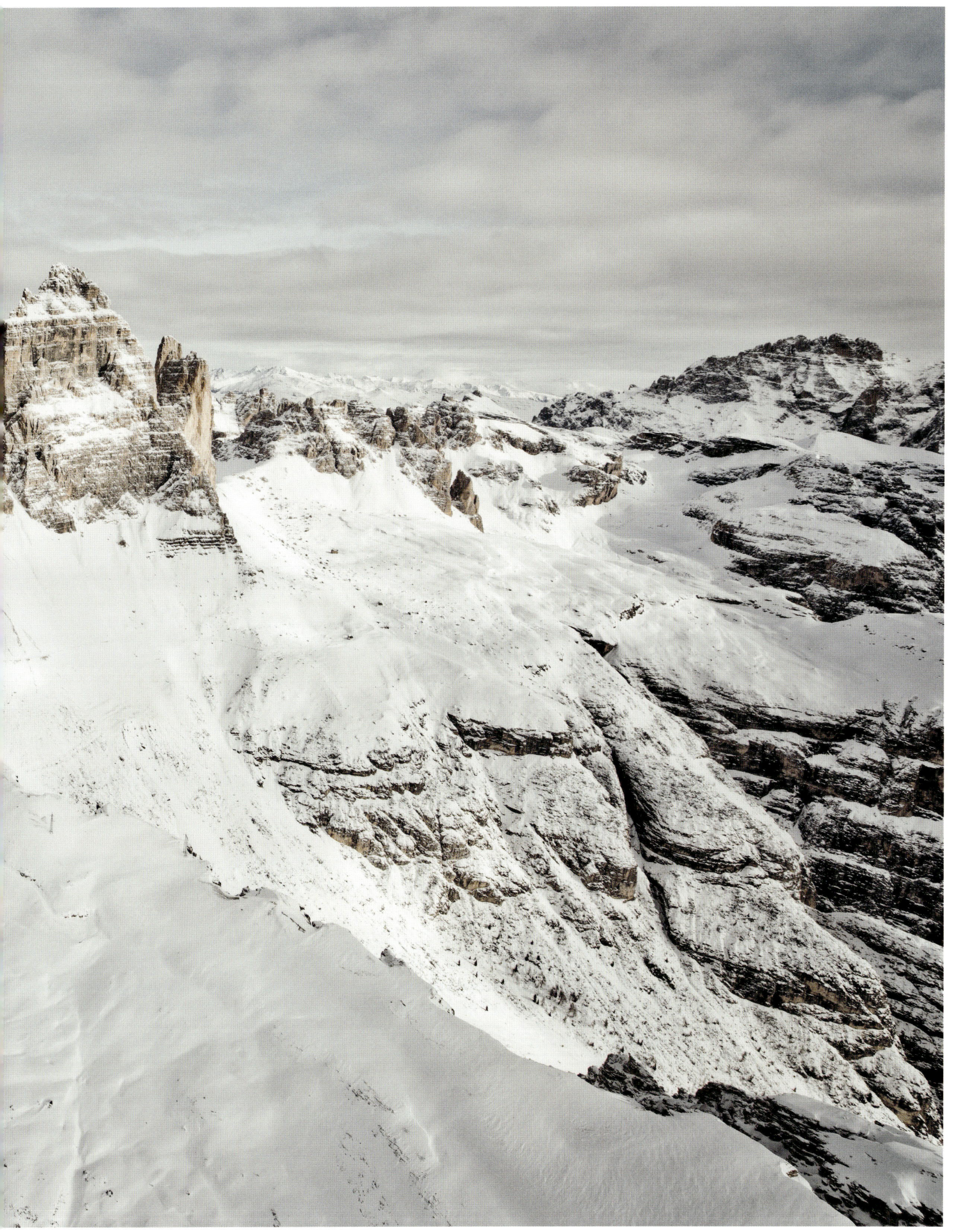

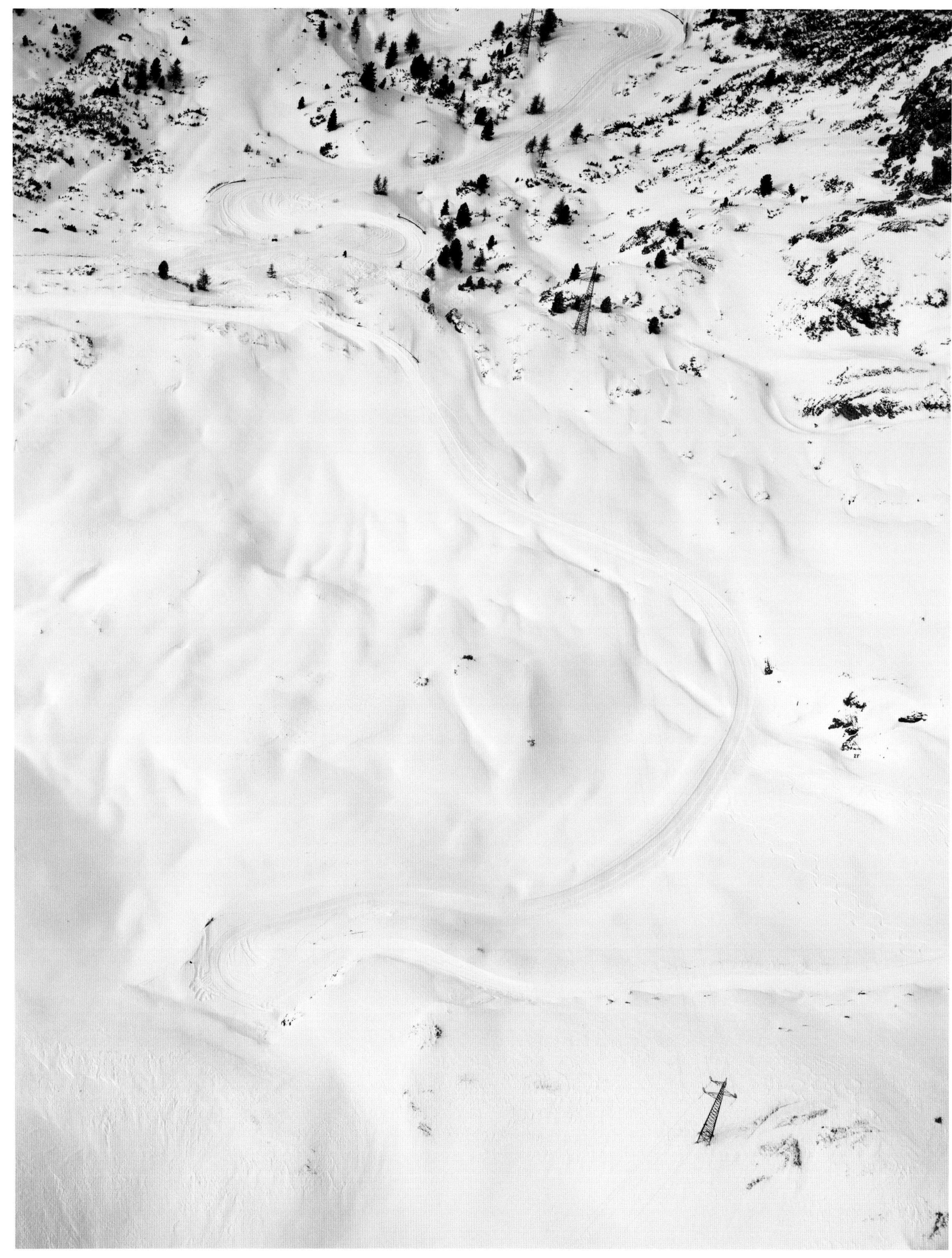

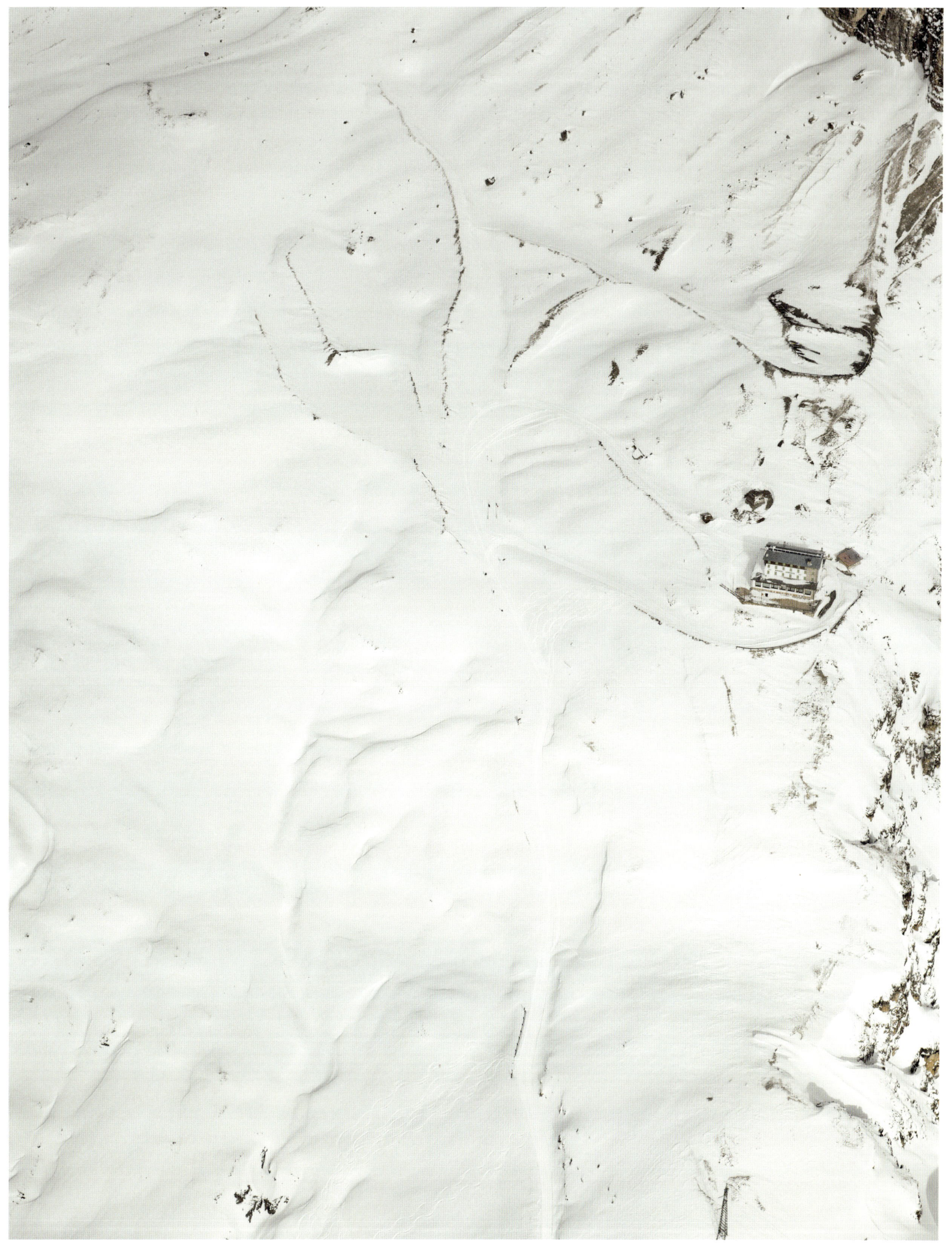

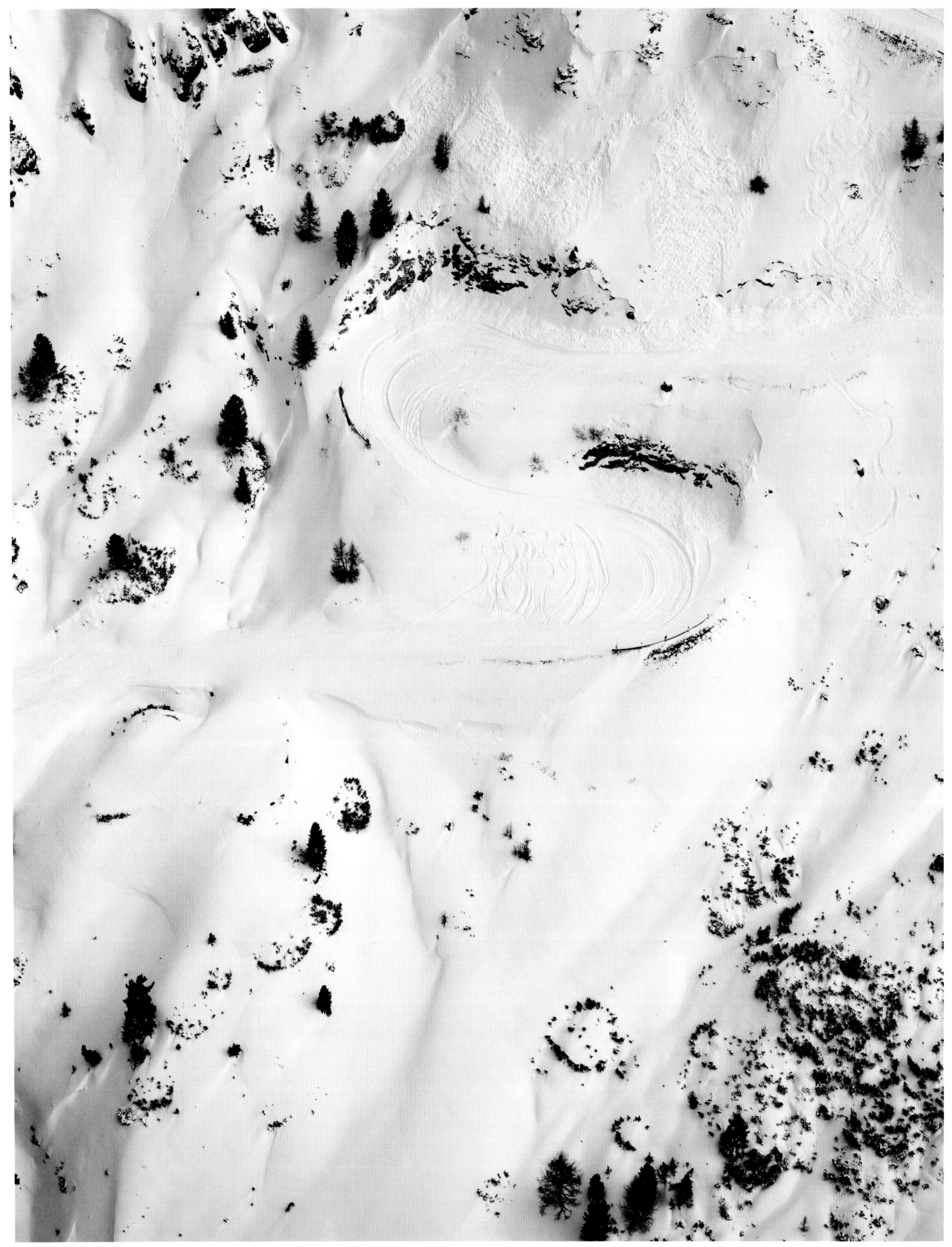

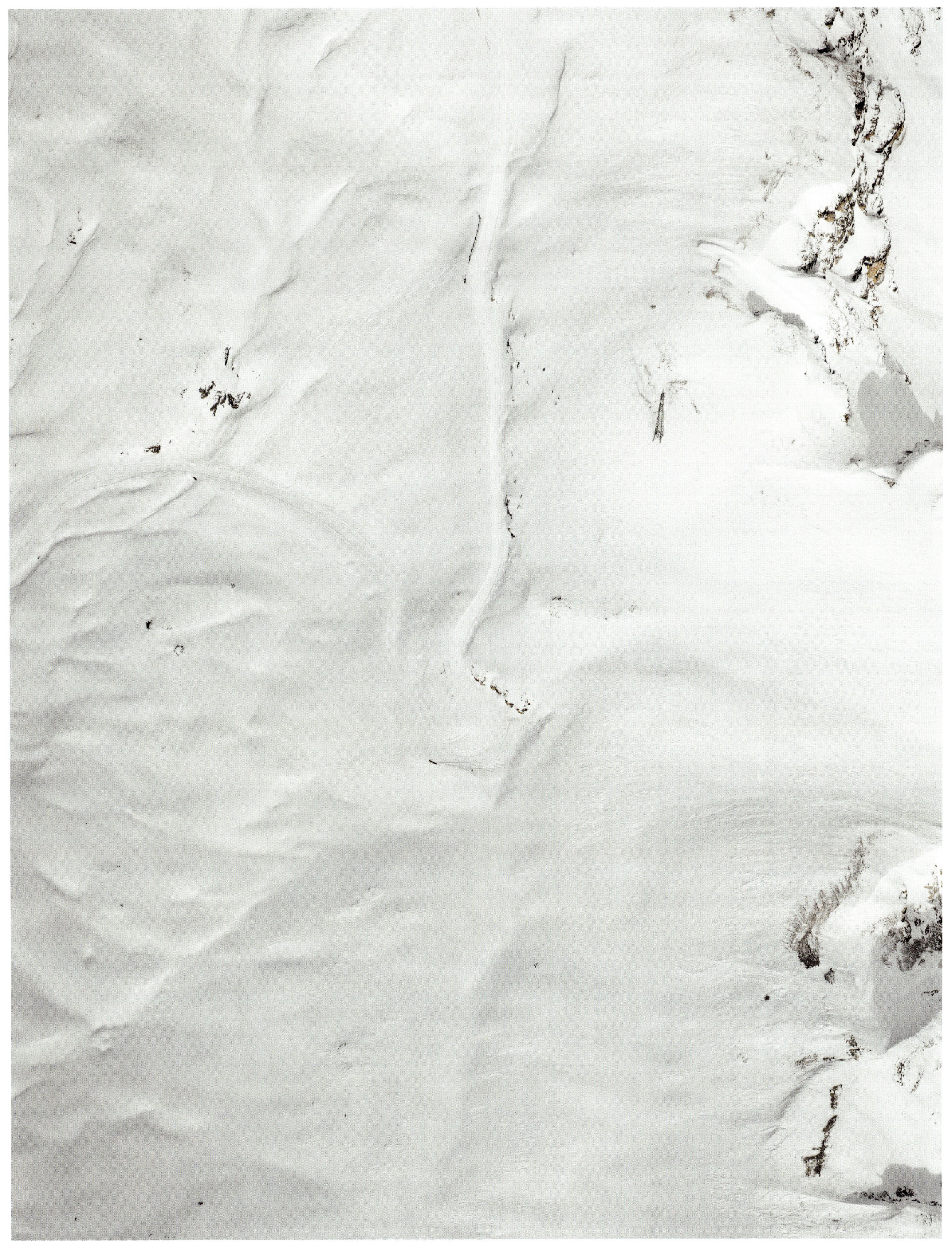

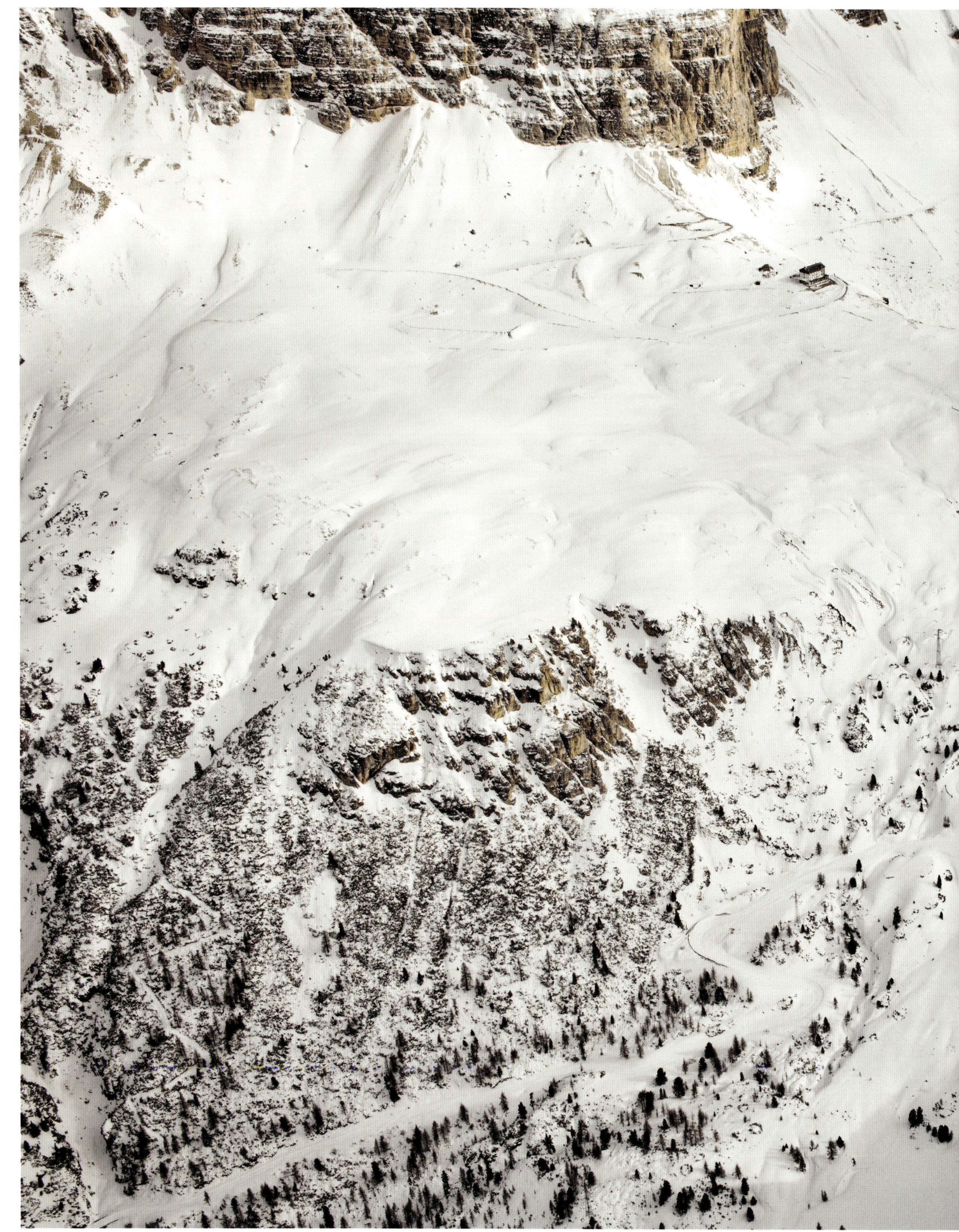

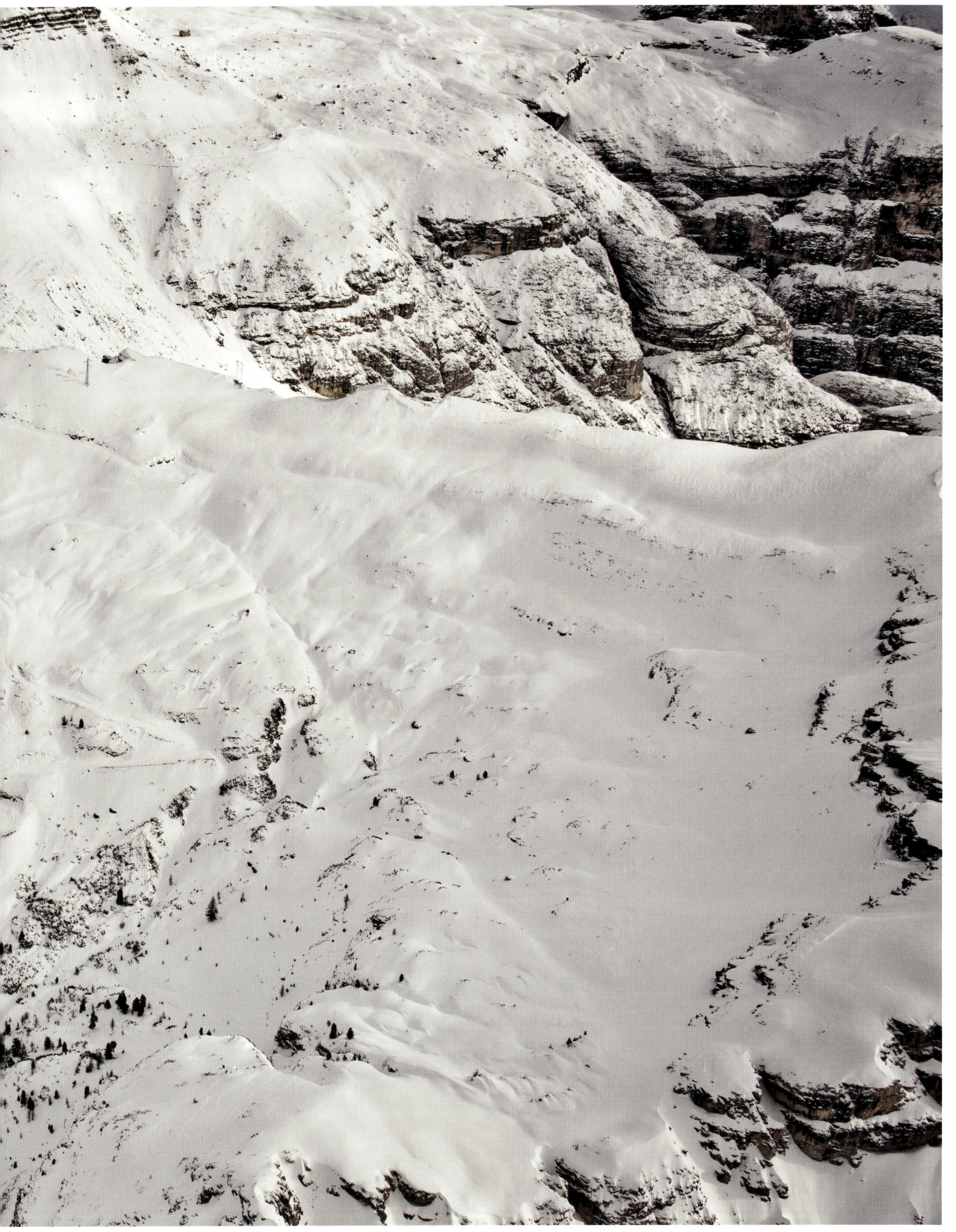

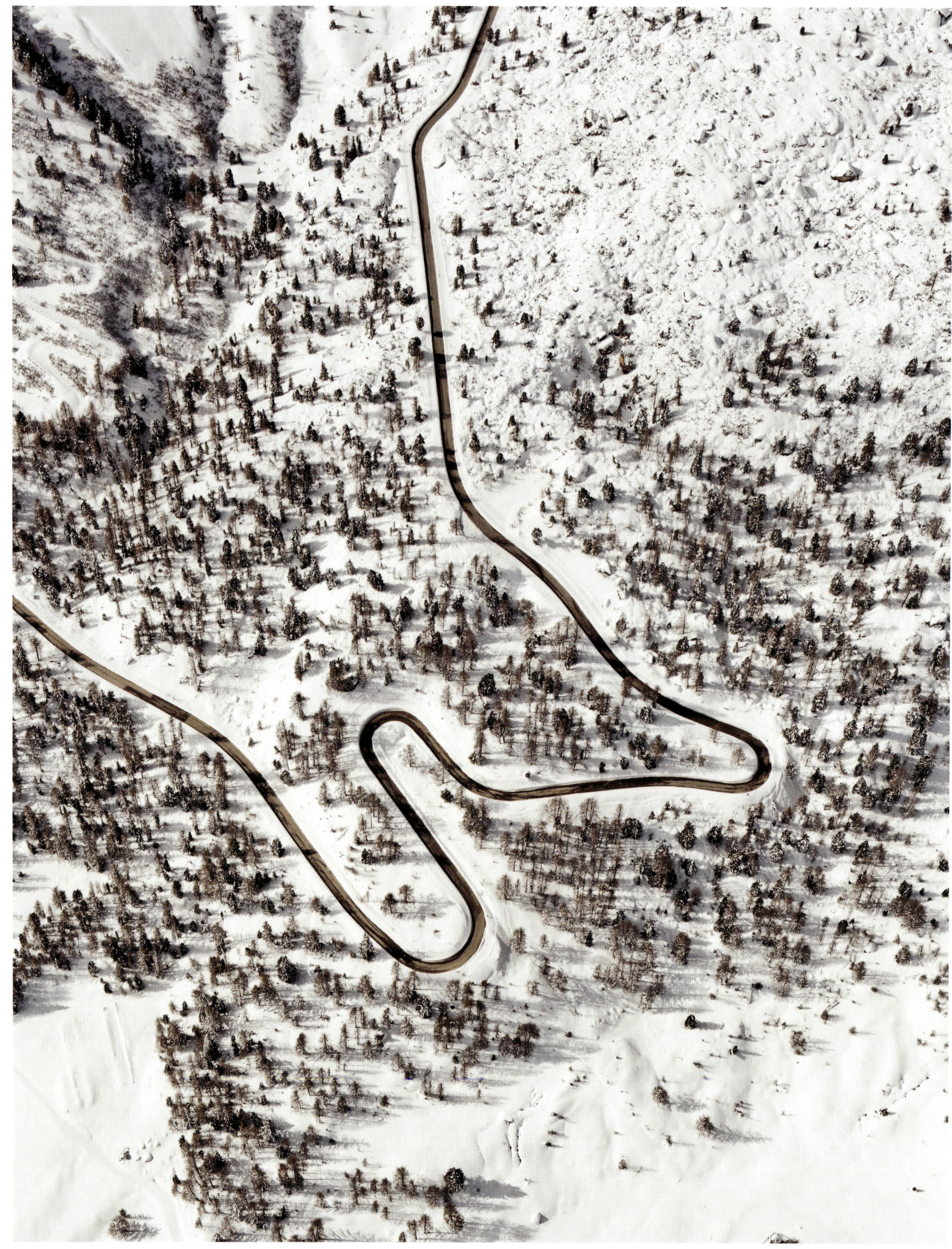

Passo di Giau

— 2236 Meter
Meist ganzjährig geöffnet

Einige der schönsten Ausblicke und spannendsten Kurven der Dolomiten bietet der Passo di Giau. In 55 schwungvollen Kehren führt er vom mondänen Wintersport-Basislager Cortina d'Ampezzo nach Selva di Cadore. Schon im Sommer ist die Fahrt durch die urtümliche Landschaft ein eindrückliches Erlebnis. Im Winter, wenn man um sich herum nur weiß verschneite Gipfel erblickt, kann man sich auf der Passstraße gar fühlen wie der letzte Mensch auf Erden.

— 2,236 metres
Usually open all year round

Some of the most stunning views and exciting corners of the Dolomites can be found on the Passo di Giau. In 55 sweeping bends, it leads from the chic winter sport base camp Cortina d'Ampezzo to Selva di Cadore. Even in summer, the journey through the primeval landscape is an impressive experience. On the mountain pass in winter, surrounded by white-capped peaks, it can feel like you are the last person on Earth.

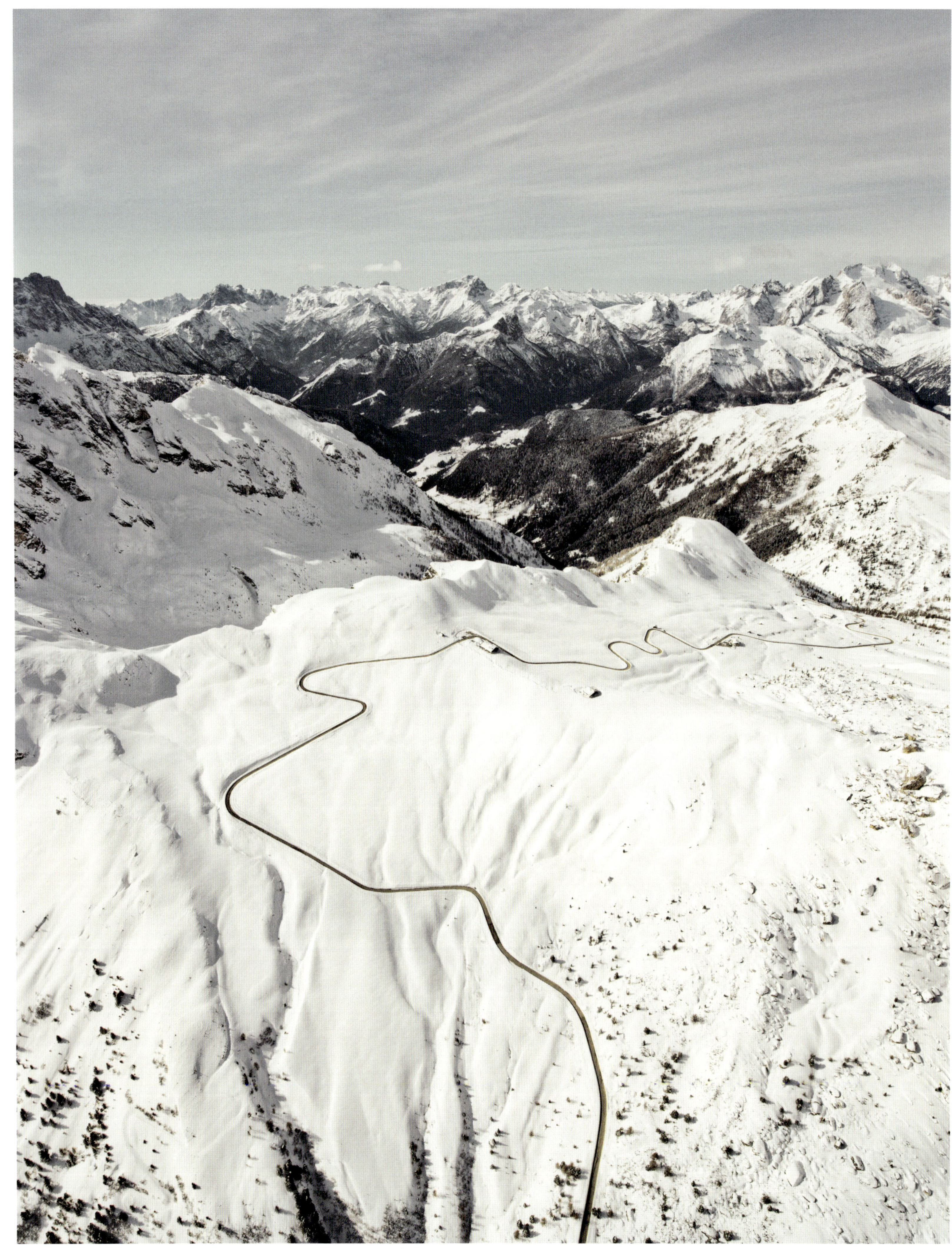

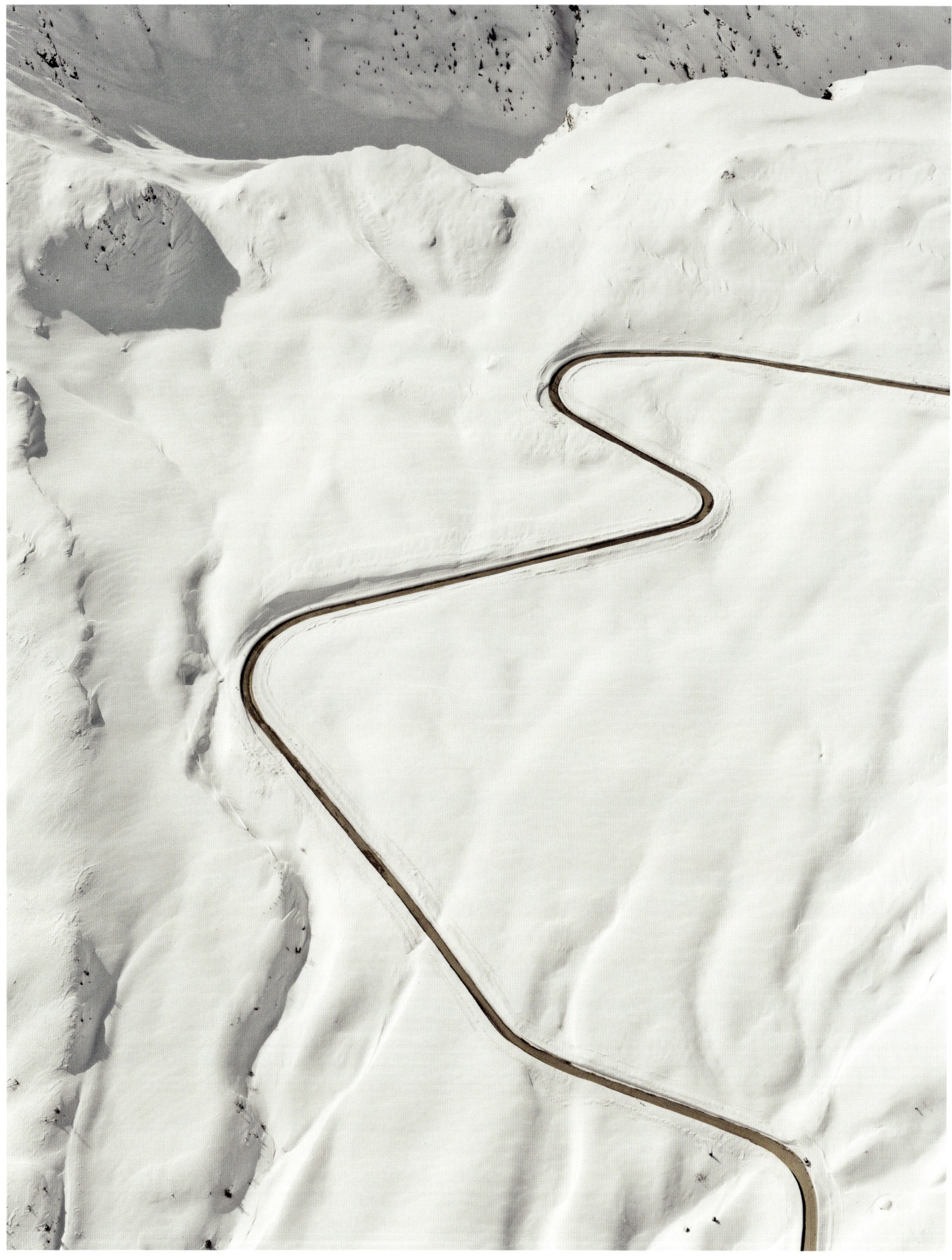

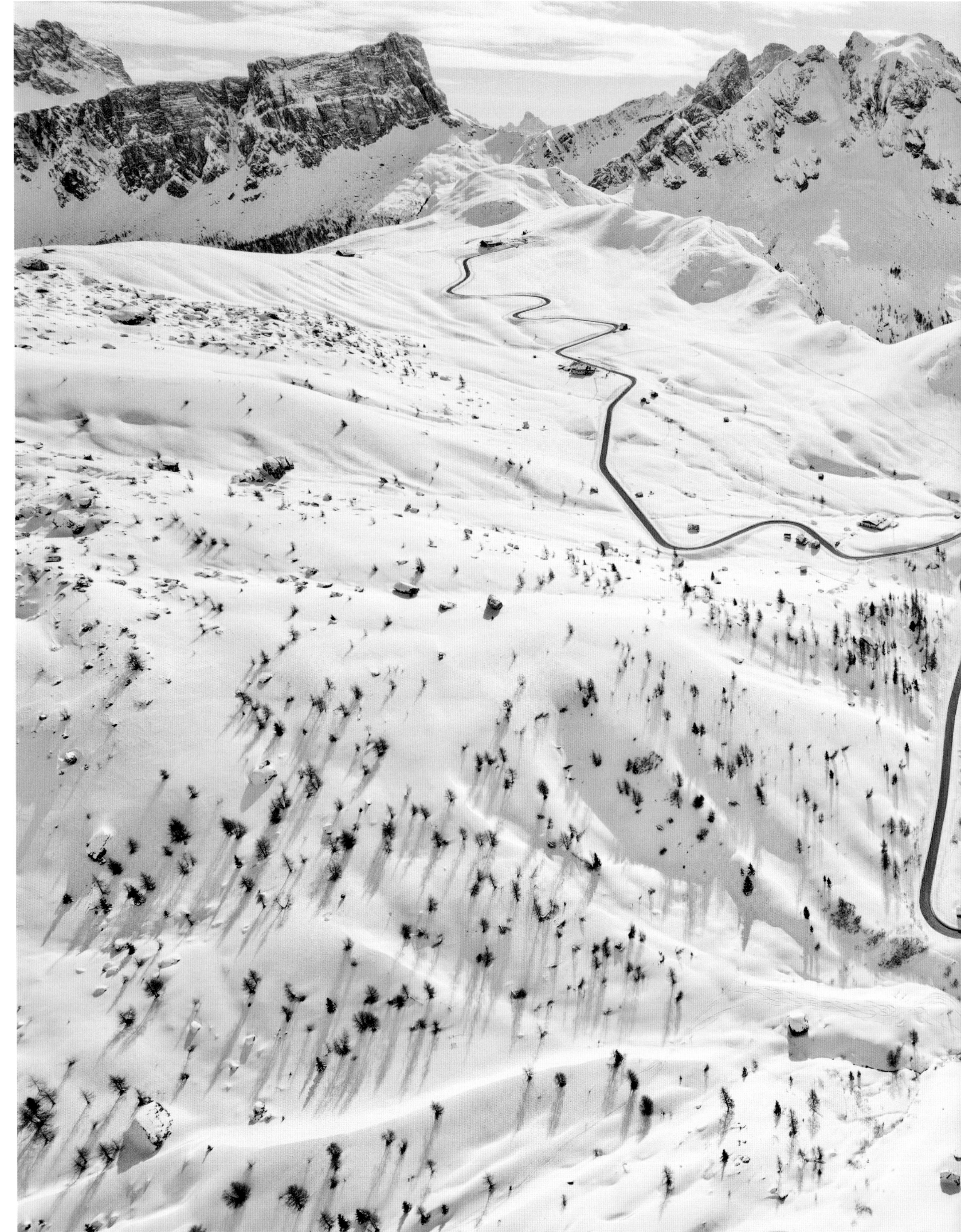

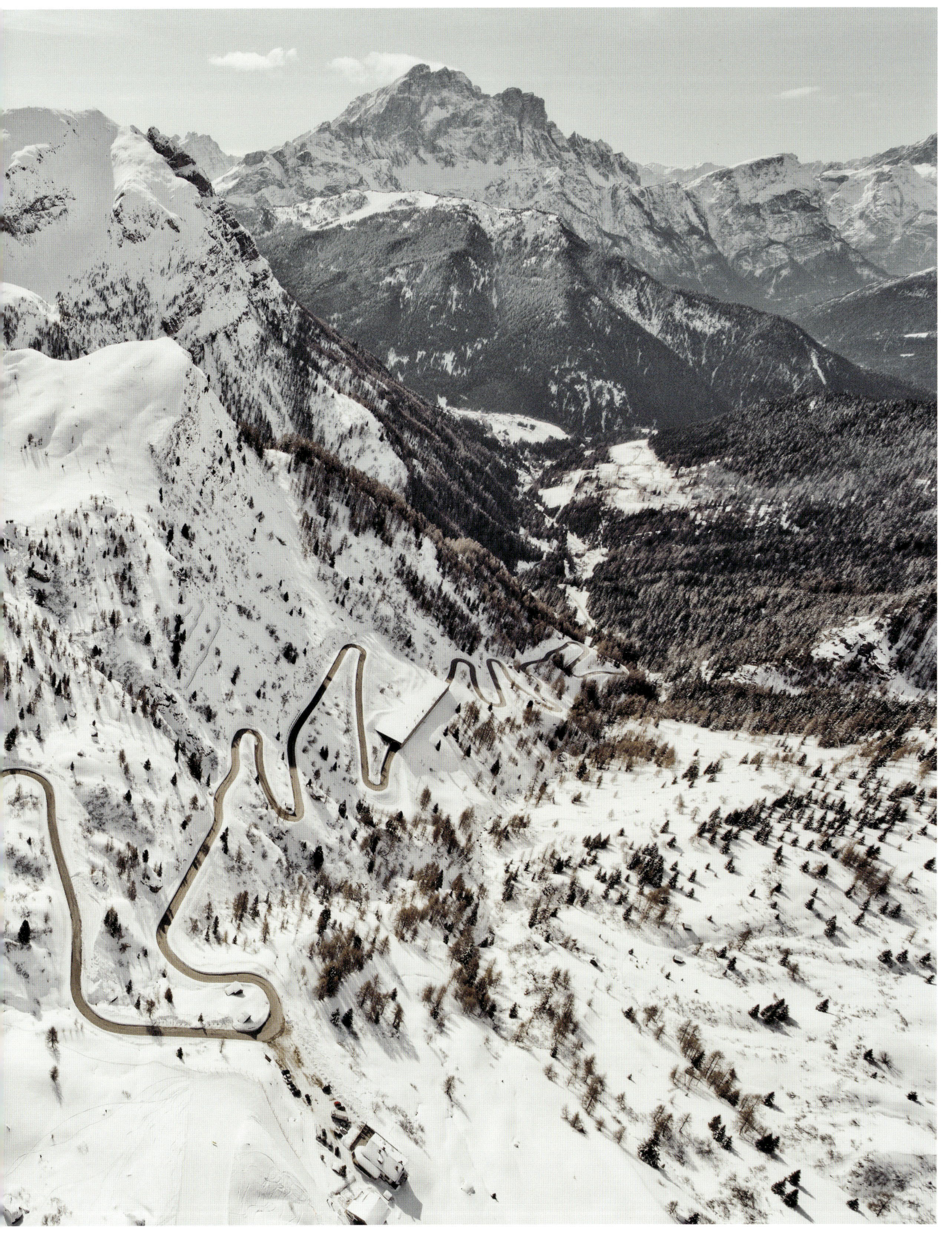

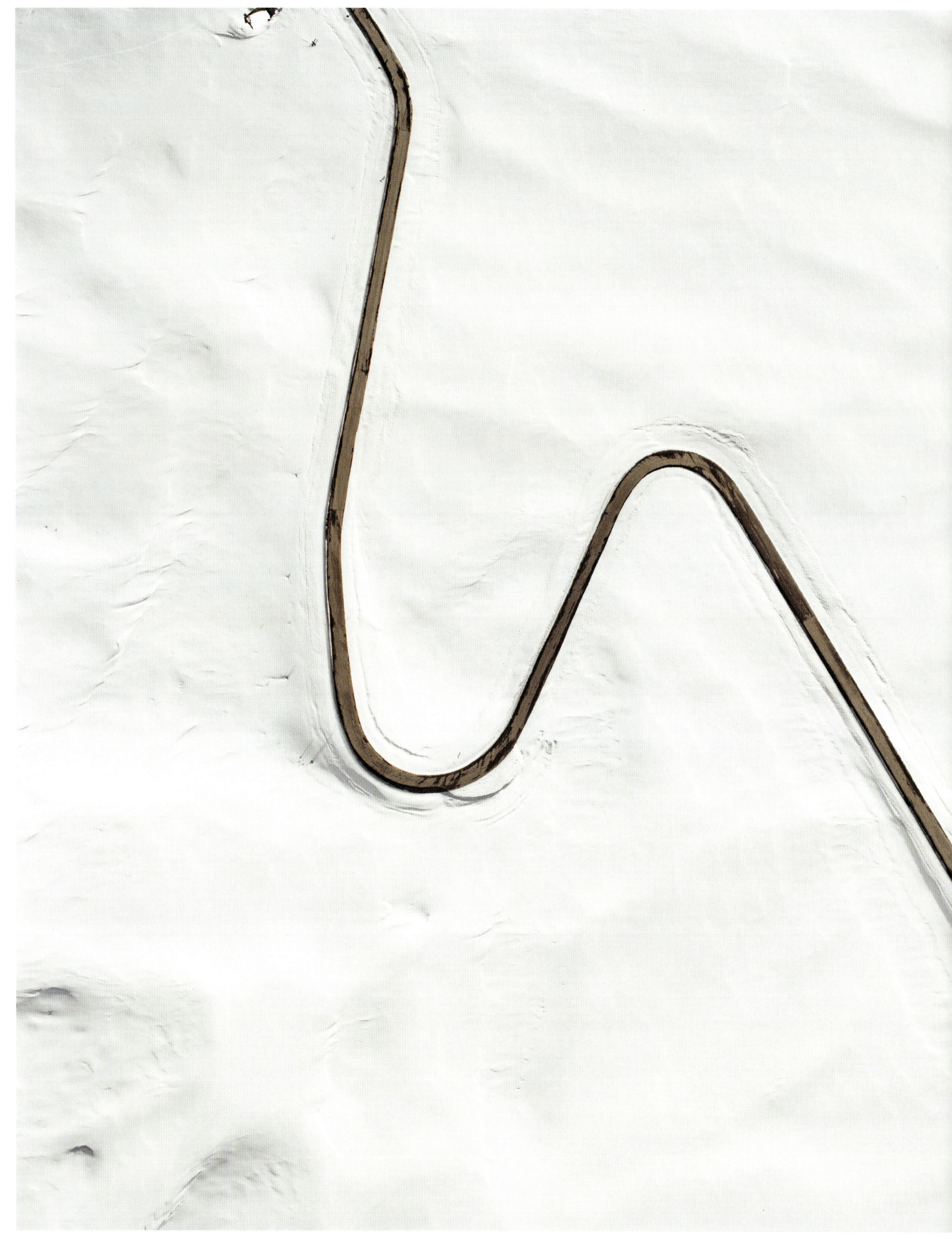

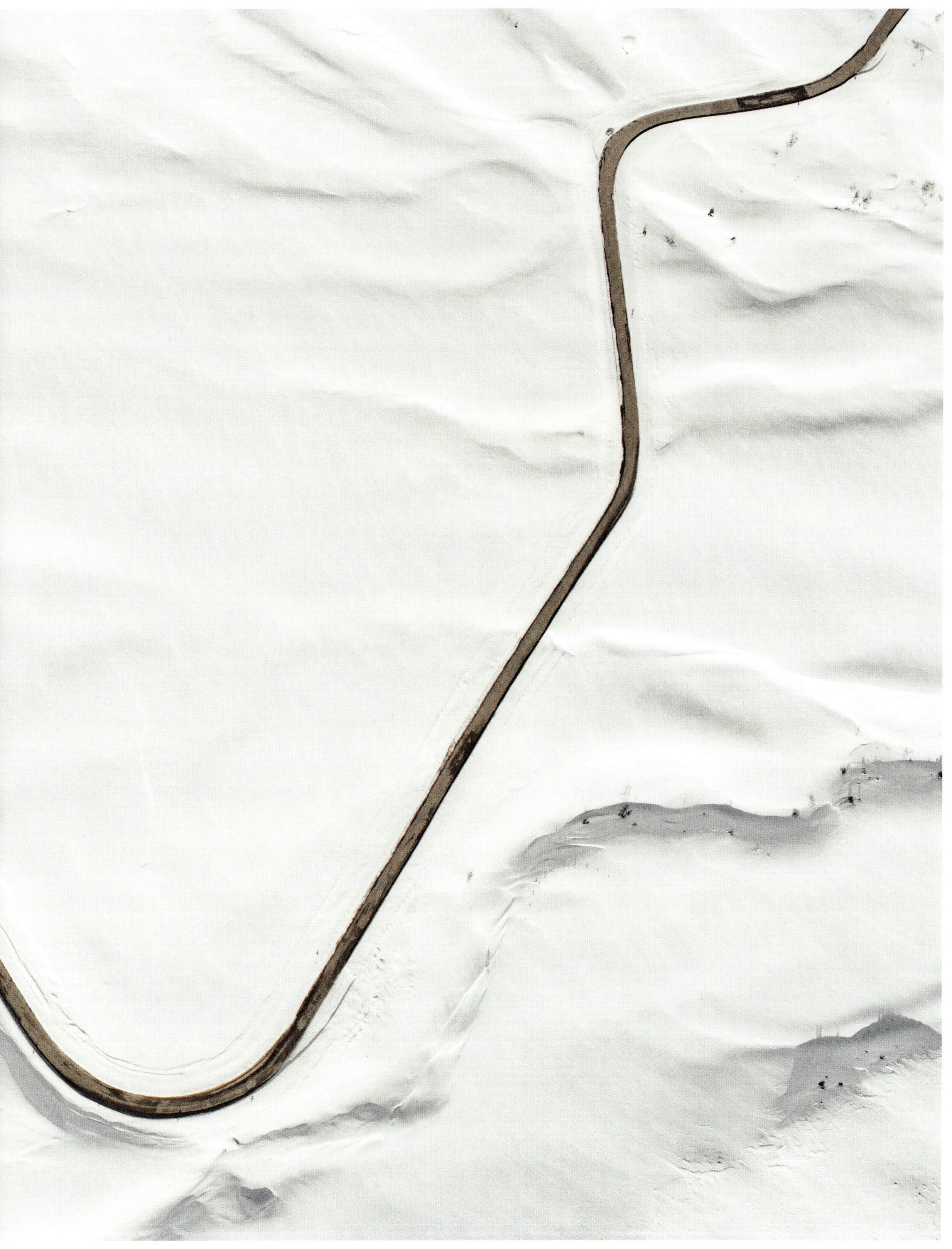

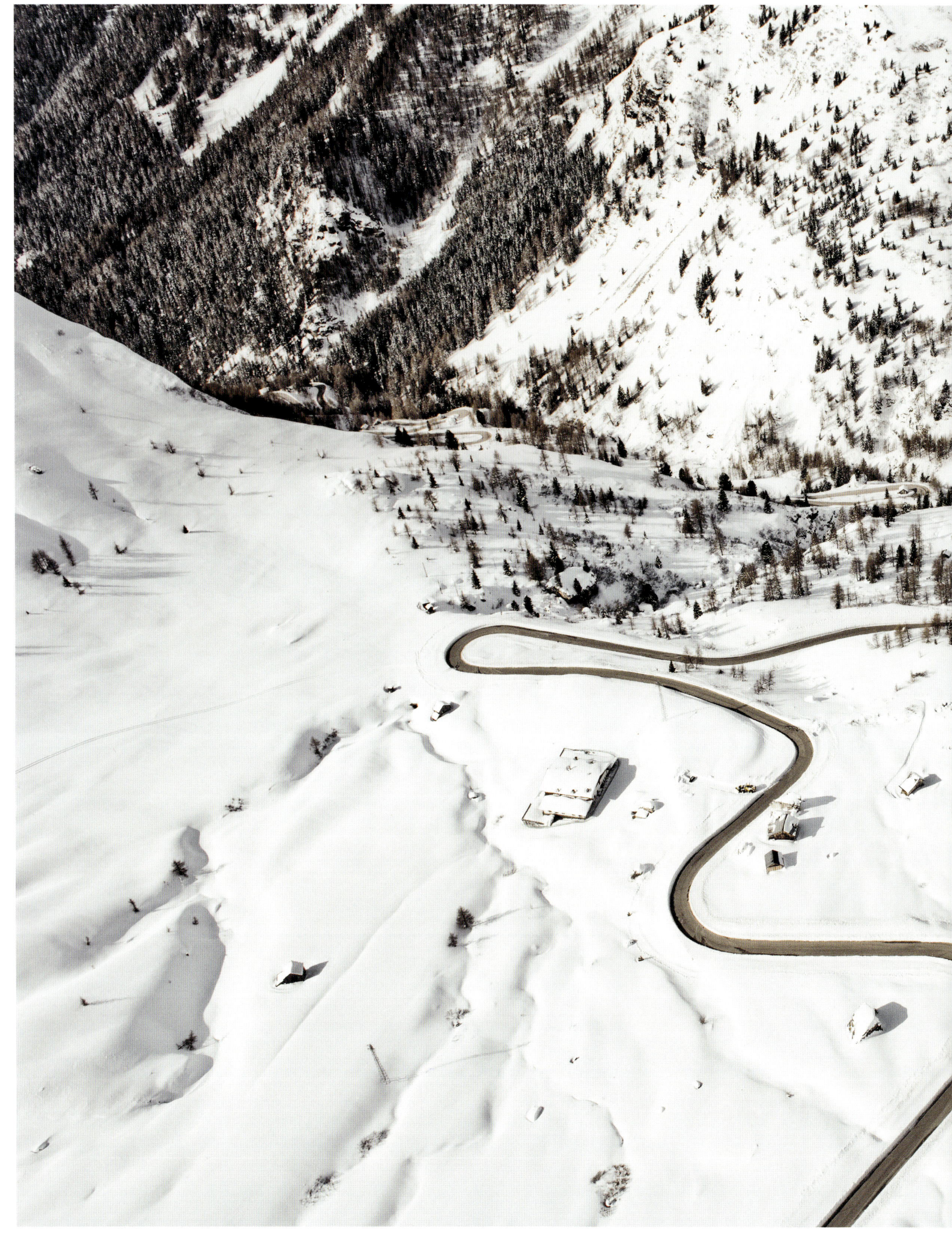

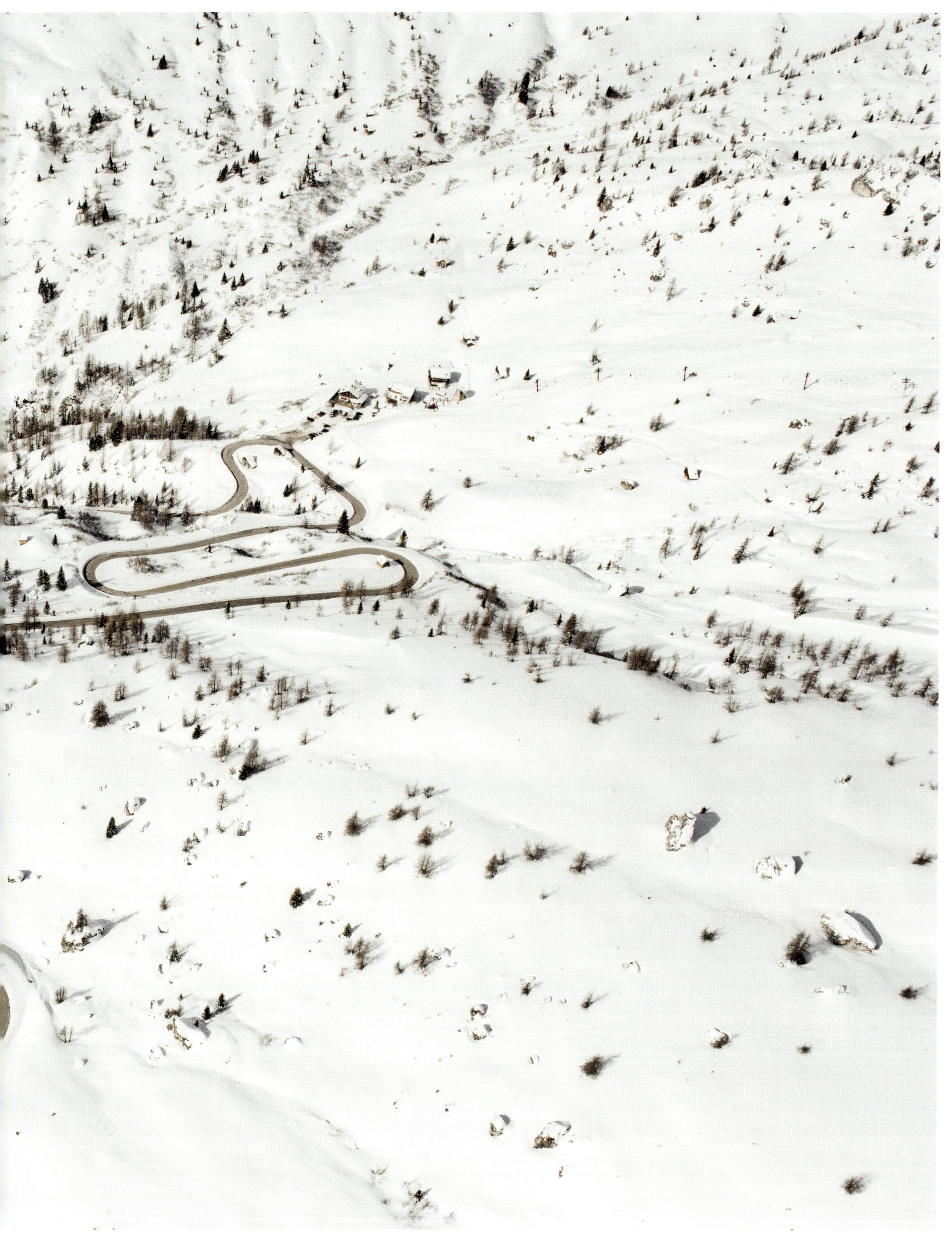

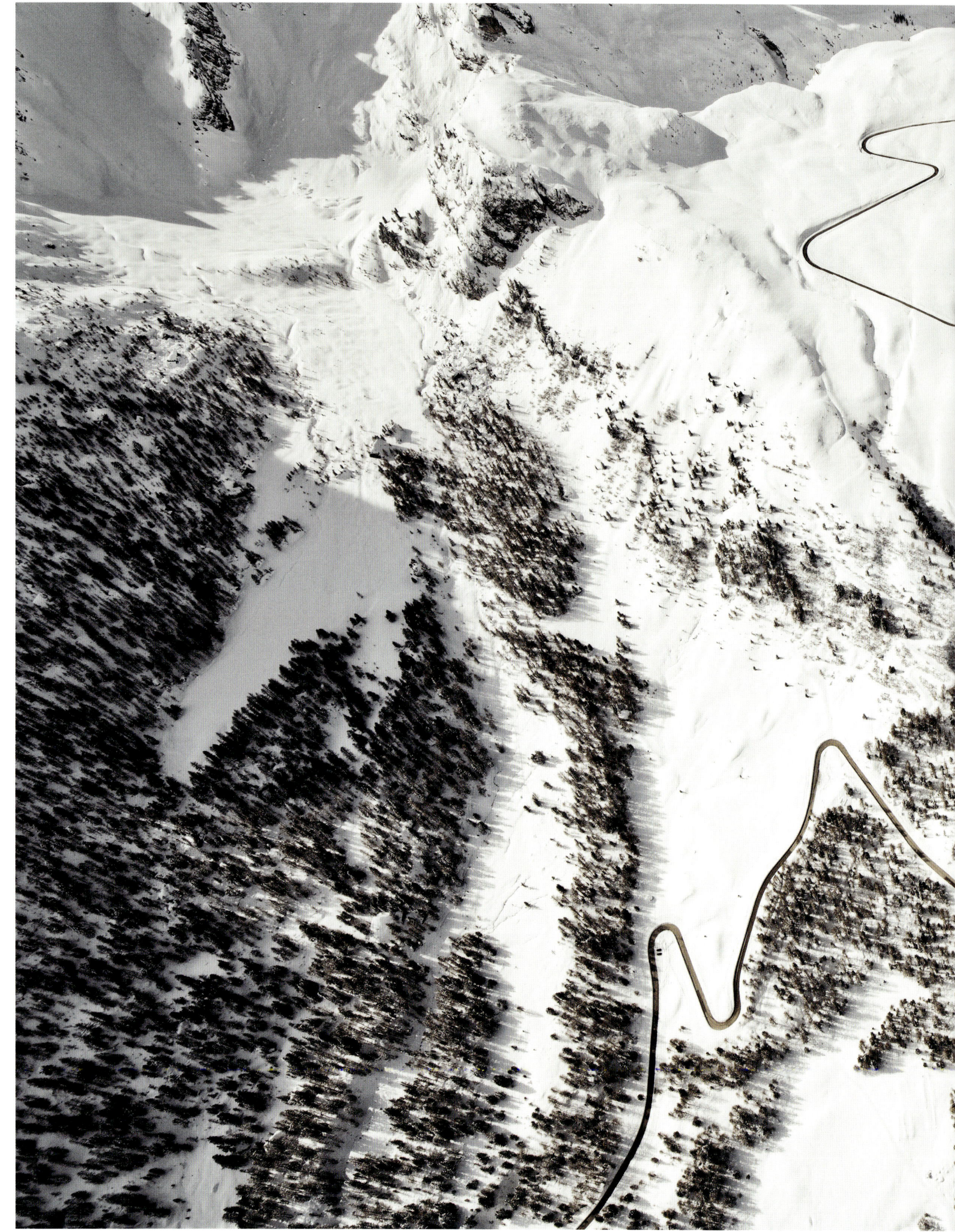

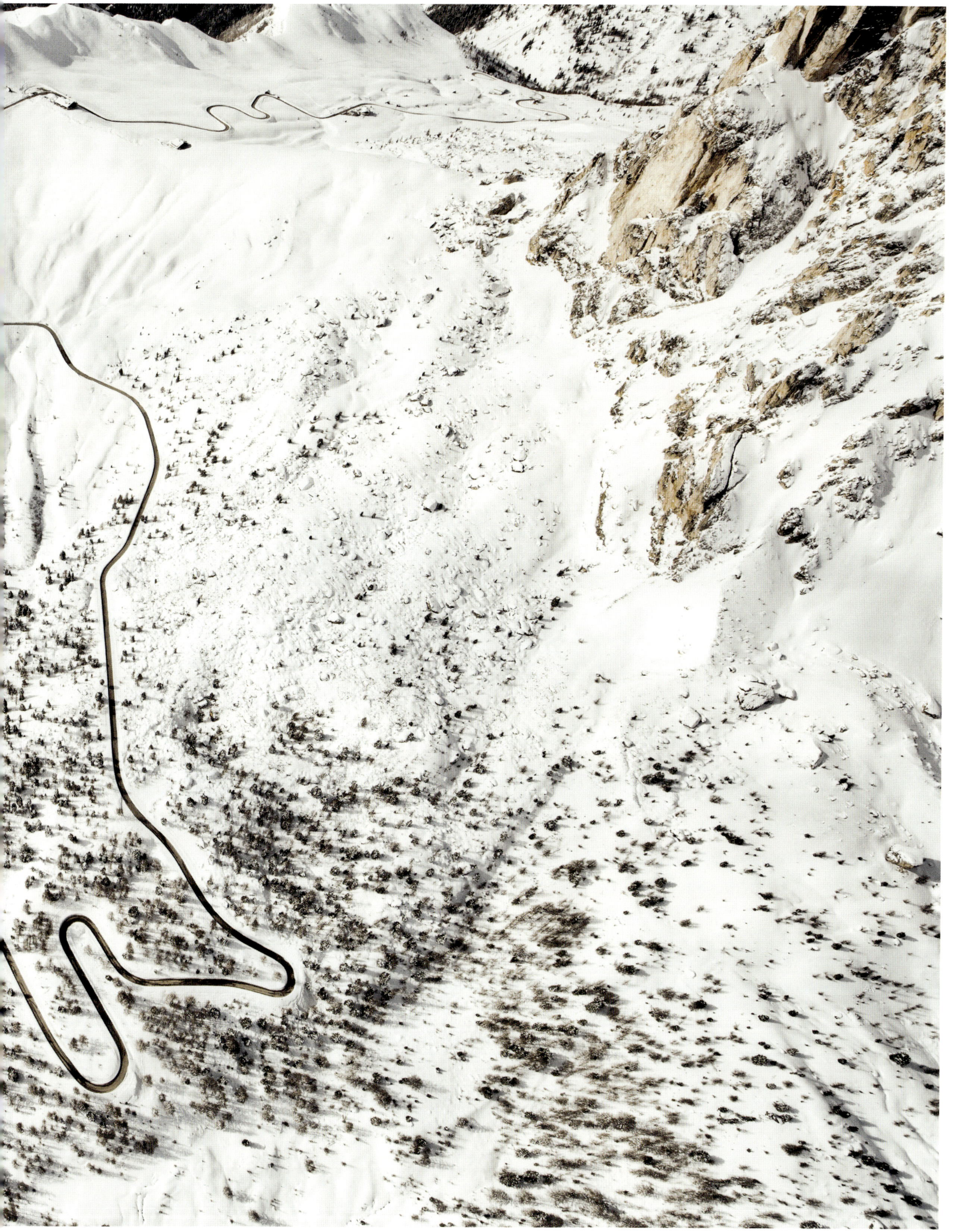

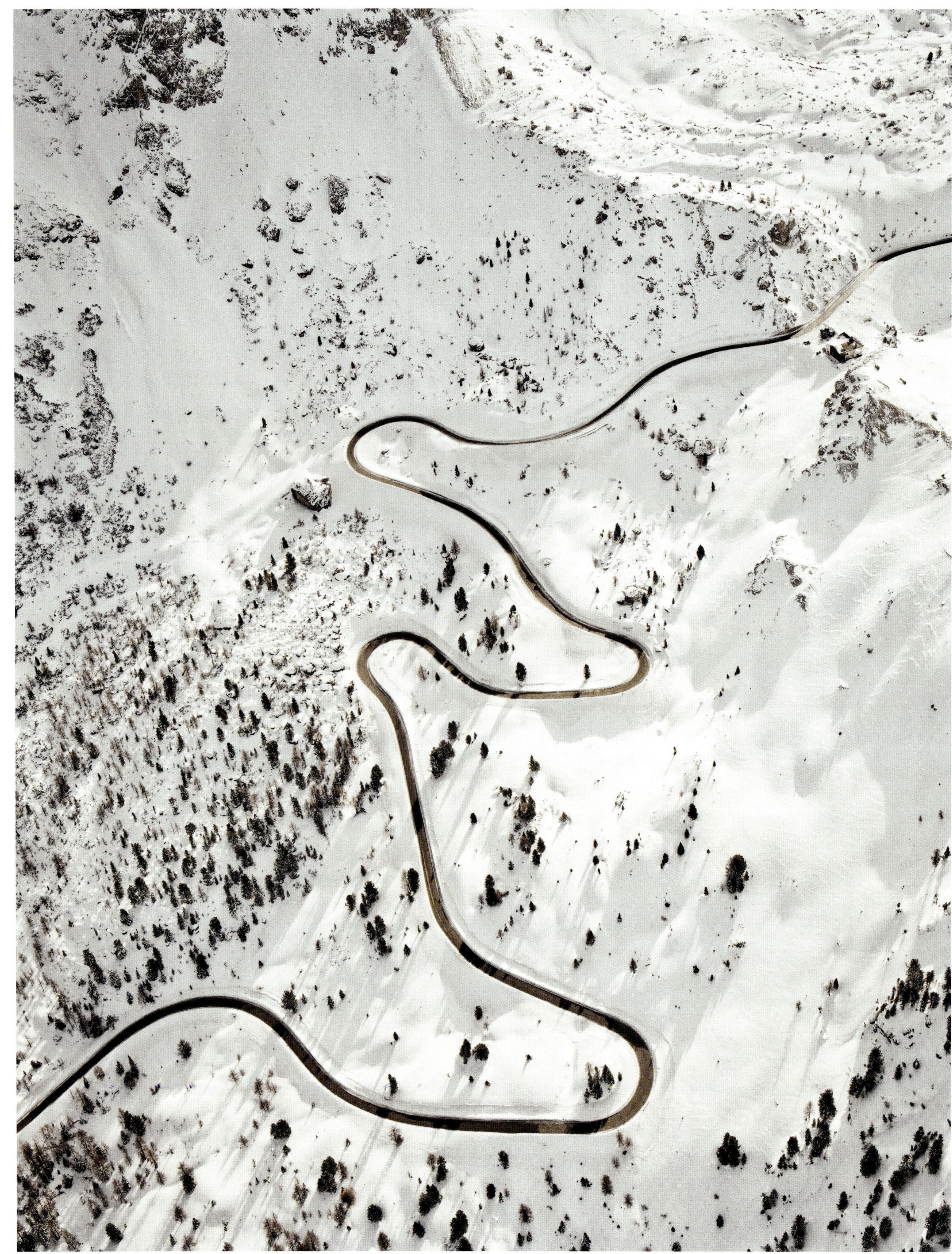

Passo di Falzarego / Passo di Valzarola

— 2105 Meter
Wintersperre meist von Dezember bis April

In 38 traumhaften Kehren verbindet der Passo di Falzarego (Falzaregopass), der 1909 als letztes Teilstück der Großen Dolomitenstraße eröffnet wurde, die Orte Andráz im Cordevole-Tal und Cortina d'Ampezzo im Valle del Boite. Einst dröhnten hier legendäre Rennfahrer wie Tazio Nuvolari bei der Coppa d'Oro delle Dolomiti durch die Serpentinen, heute verdirbt einem der Verkehr meist die Sekundenjagd. Nur im Frühjahr, wenn der Schnee schon vom Asphalt getaut ist, die Wohnmobile jedoch noch in den Carports schlummern, hat man den Passo di Falzarego noch für sich allein. Und was könnte es Schöneres geben für sportliche Fahrer als ein privates Tête-à-tête mit diesen Kurven?

— 2,105 metres
Usually closed in winter from December to April

In 38 magnificent corners, the Passo di Falzarego (Falzarego Pass), which opened in 1909 as the last section of the Great Dolomite Road, links the towns of Andraz in the Val Cordevole with Cortina d'Ampezzo in the Valle del Boite. Legendary racing drivers such as Tazio Nuvolari once contested the Coppa d'Oro delle Dolomiti here. Today, traffic hinders the hunt for fast times. Only in springtime, when the snow thaws on the asphalt and the campervans are safely tucked away in garages, can one still enjoy solitude on the Falzarego Pass. And what could be better for sporty drivers than to tackle these corners Tête-à-tête?

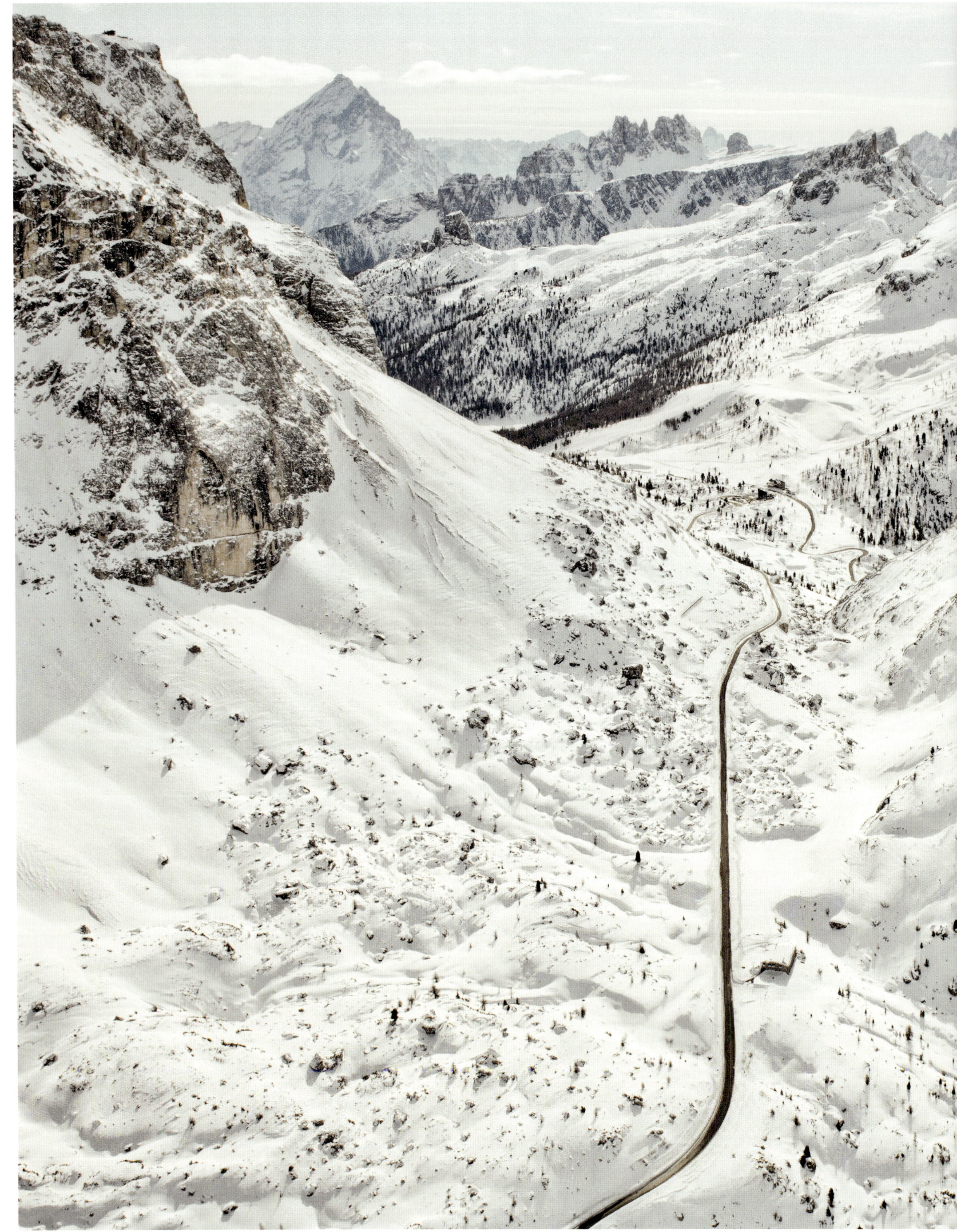

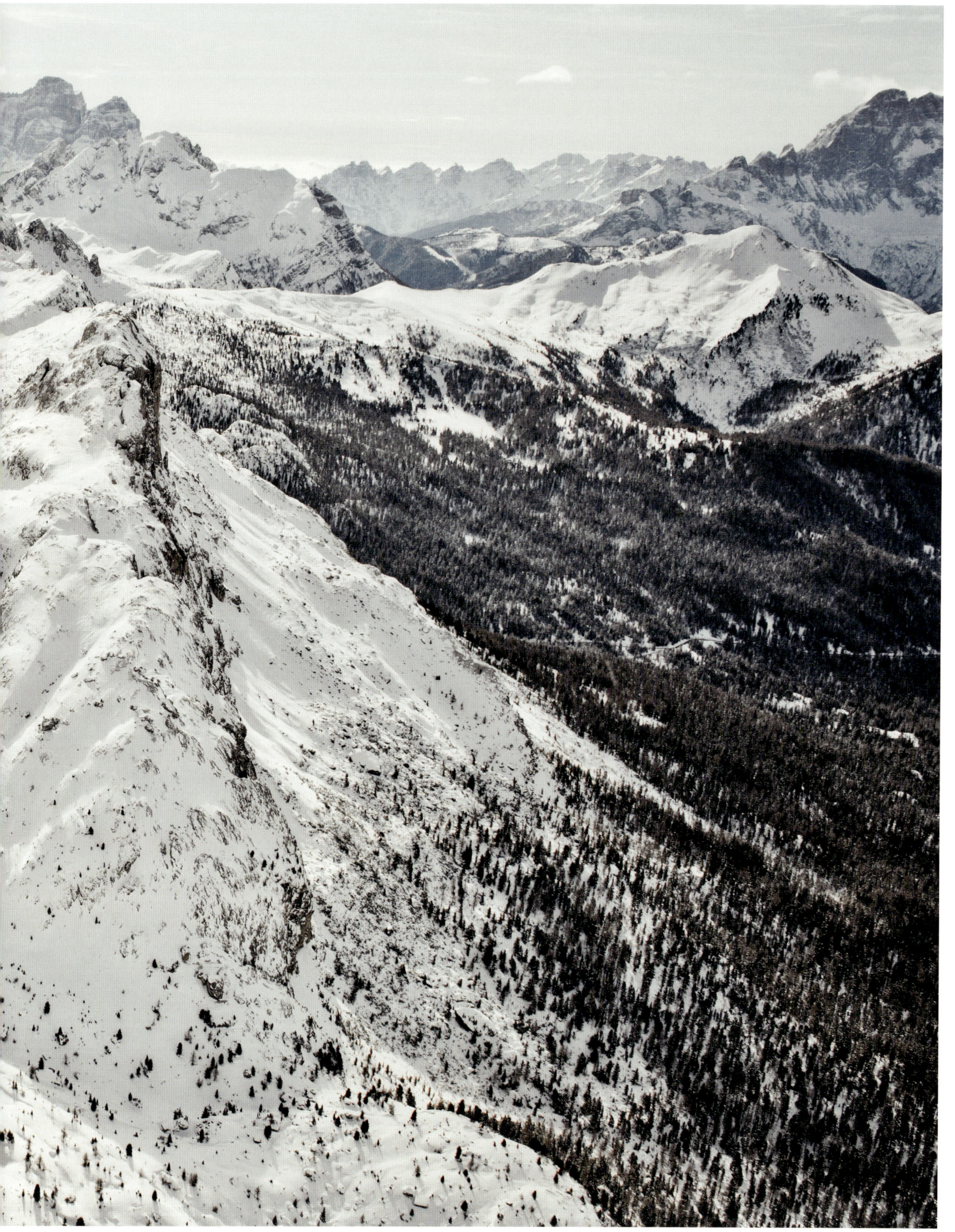

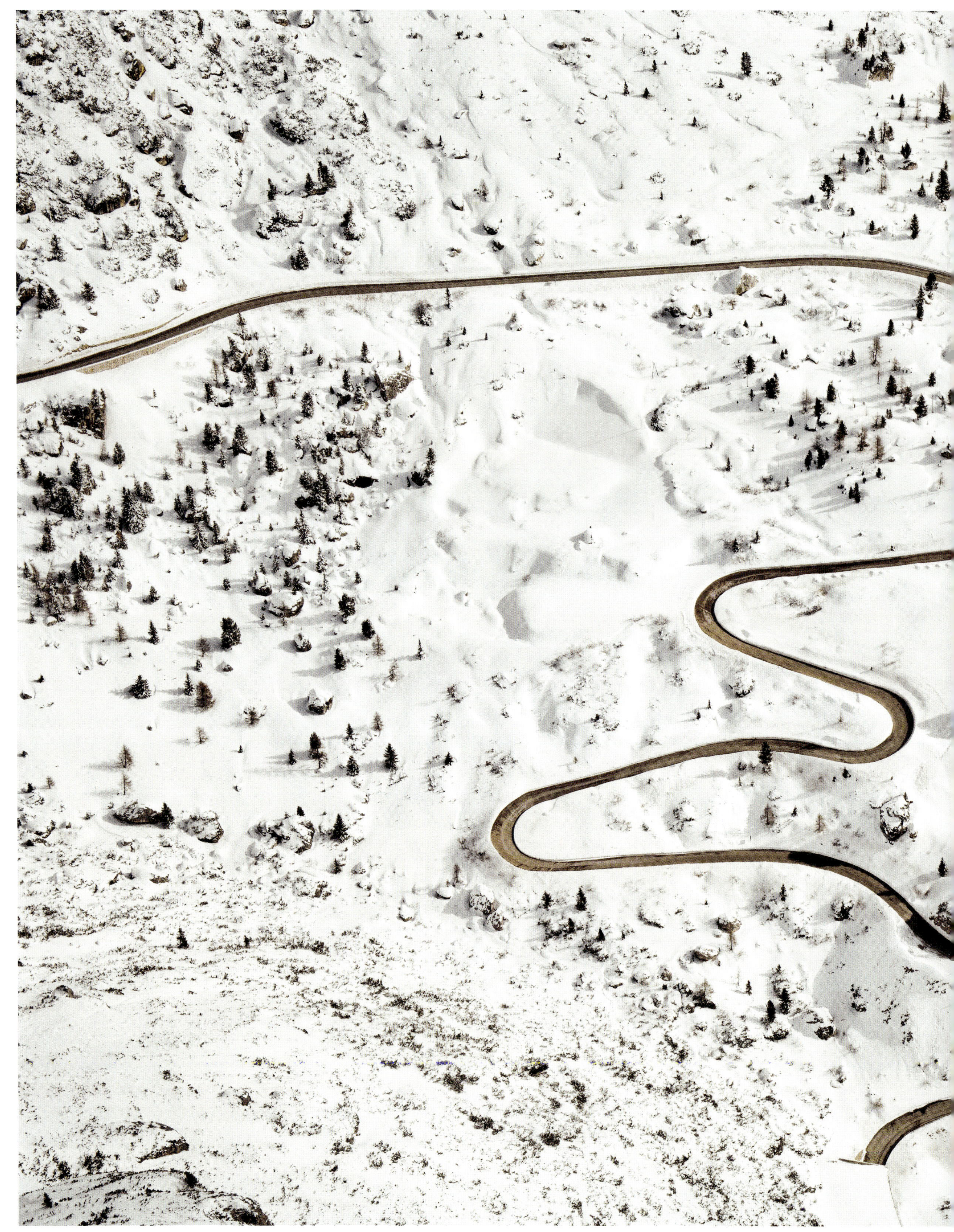

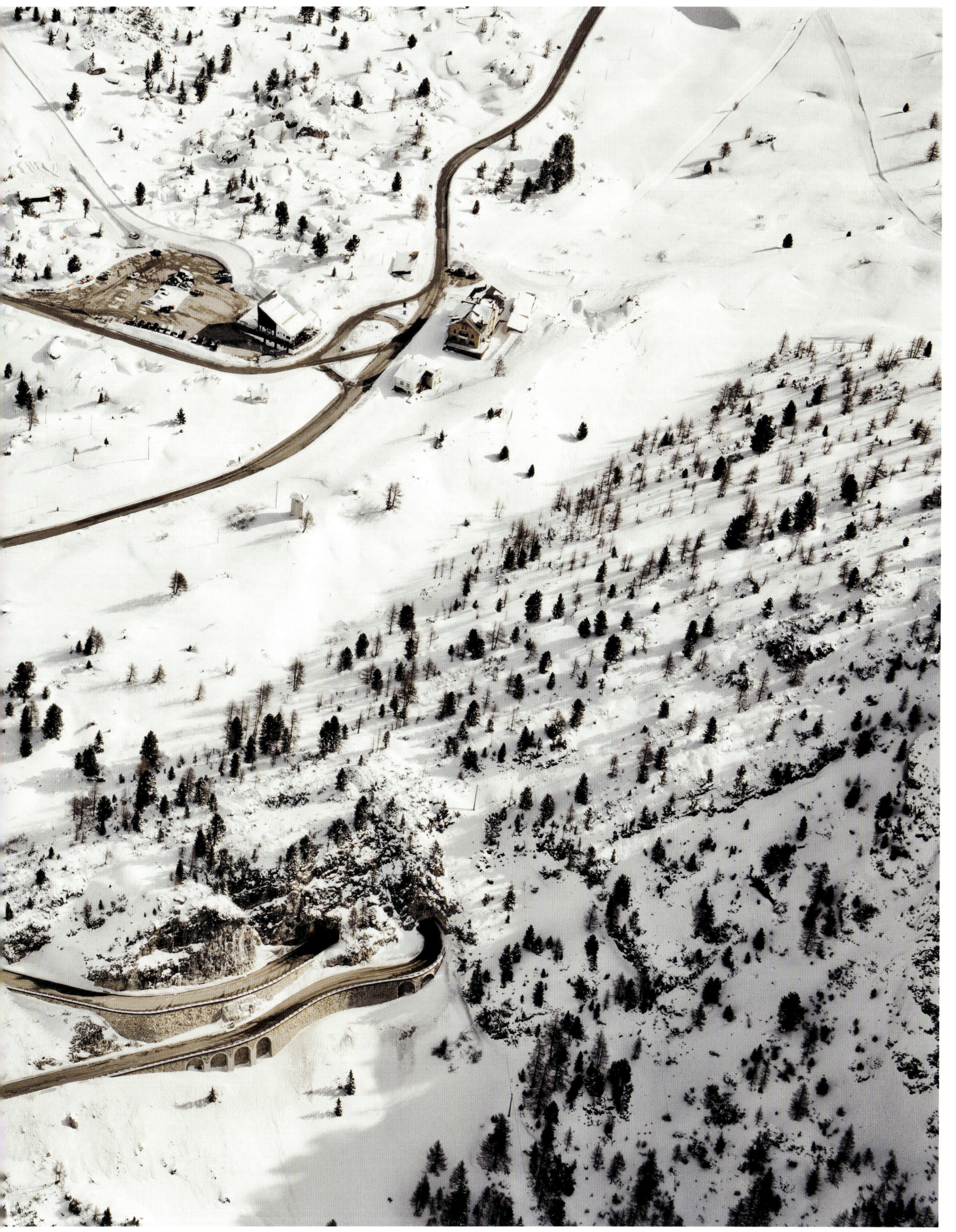

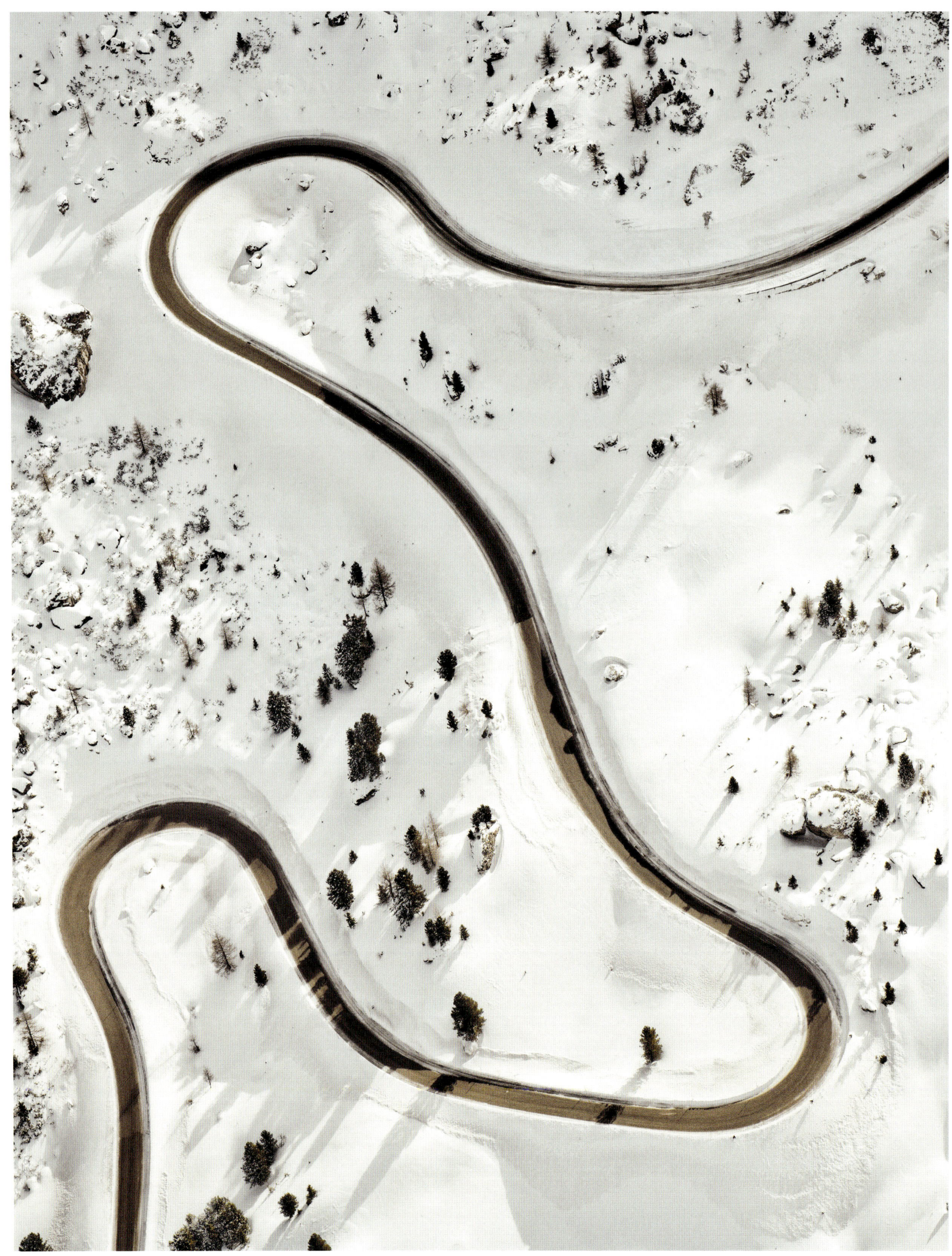

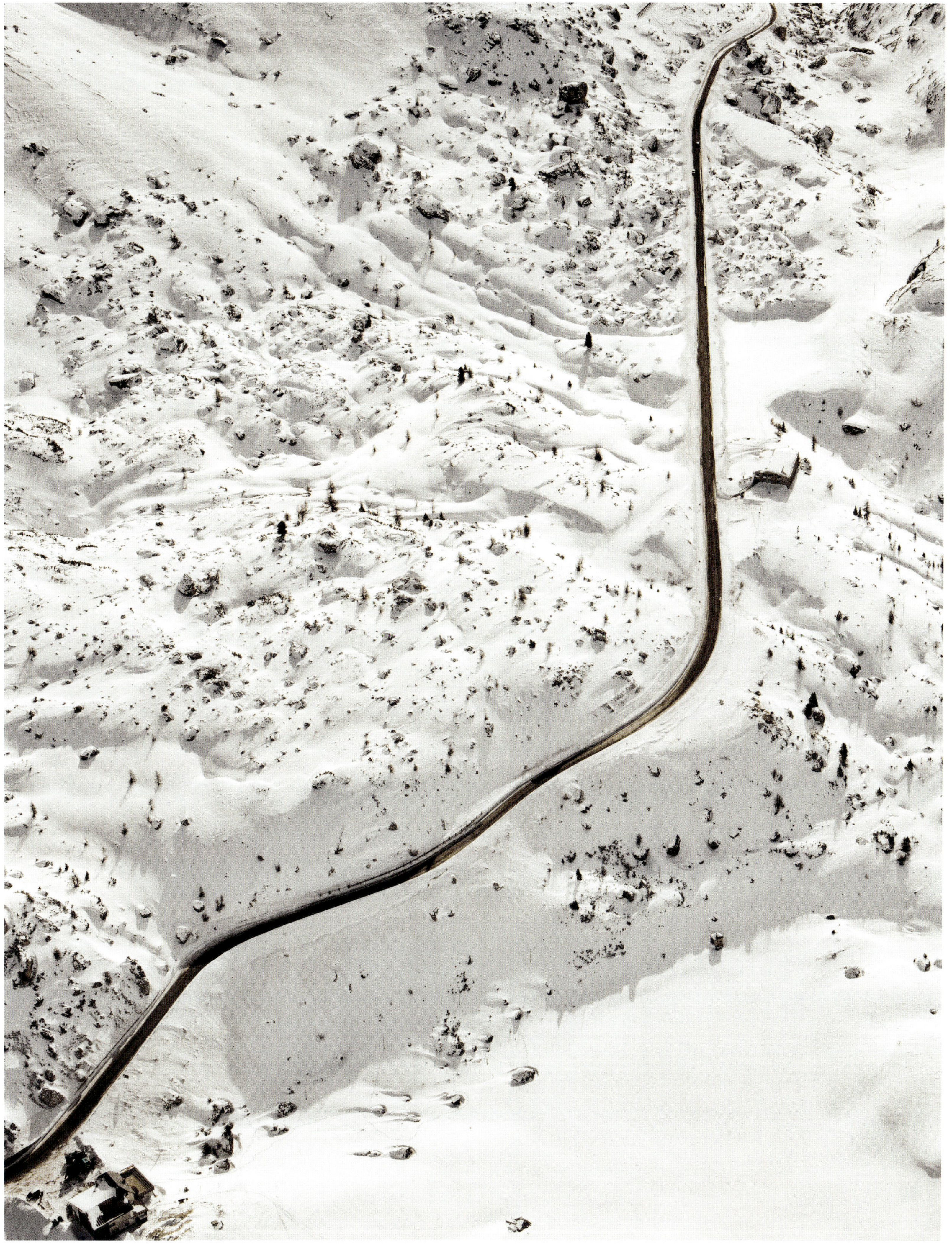

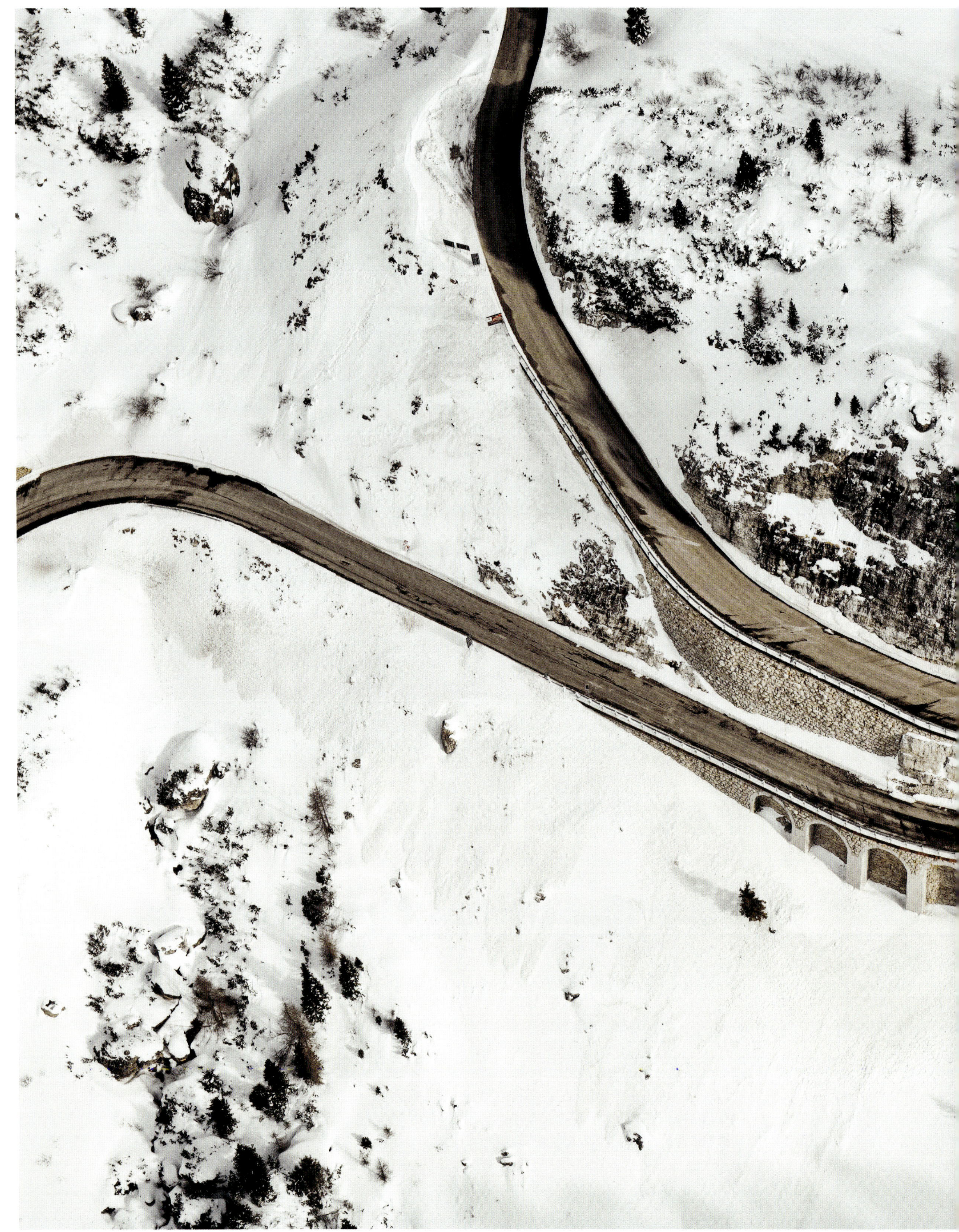

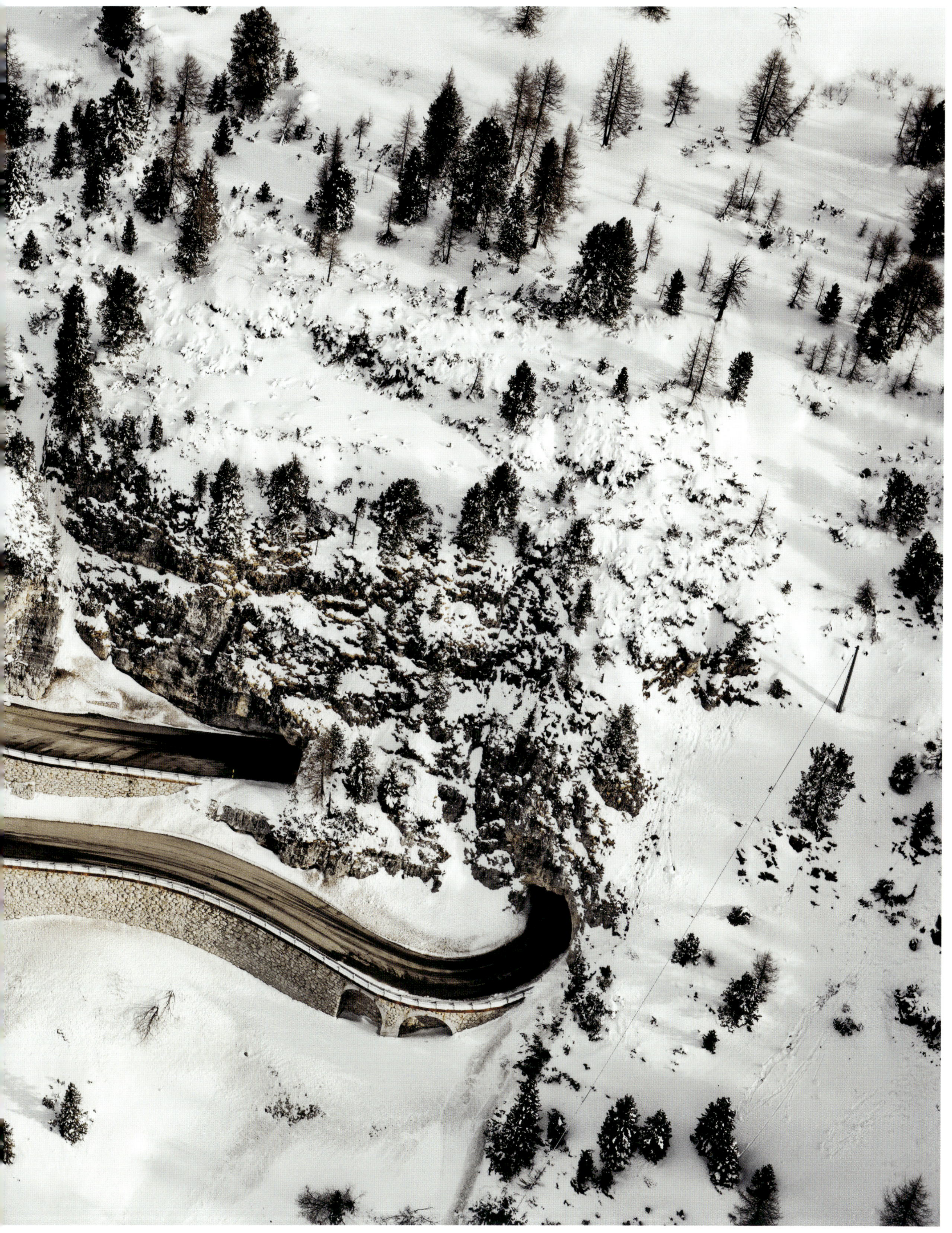

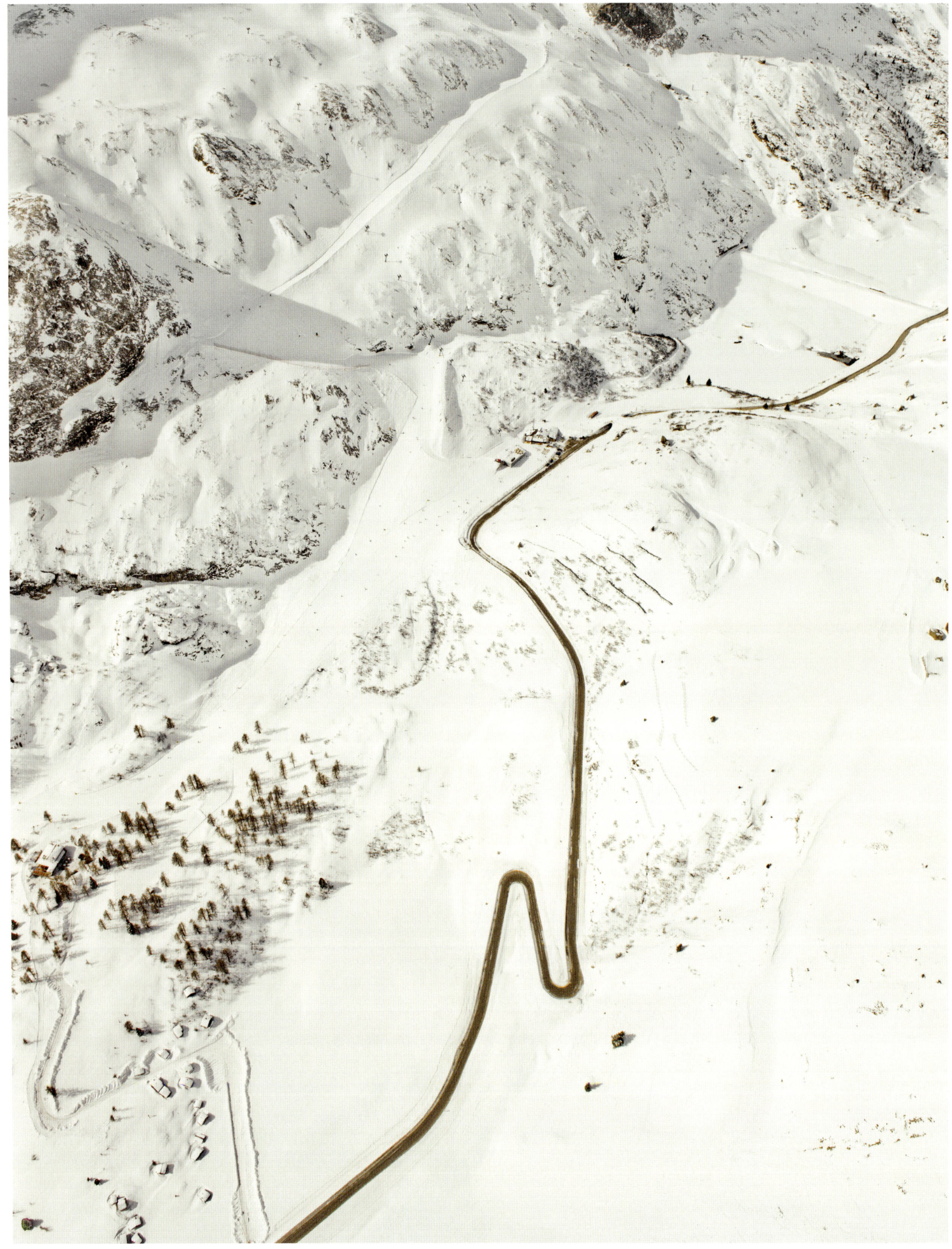

Passo Fedaia

— 2057 Meter
Meist ganzjährig geöffnet

Auch wenn der malerische Stausee unterhalb der Passhöhe im Winter meist vereist, zugeschneit und nur beim zweiten Hinsehen zu entdecken ist, bietet der gut ausgebaute Passo Fedaia eine durchaus interessante Alternative zur Fahrt über das Pordoijoch. Vor allem der Blick auf den mächtigen Marmolatagletscher ist unvergesslich. Wem es bei diesem Anblick in den Skizehen juckt, der kann in Malga Ciapela spontan die „Brettl" vom Dach schnallen, die Seilbahn hinauf auf 3269 Meter nehmen und über den Gletscher nochmals hinab zur Passhöhe sausen.

— 2,057 Meter
Usually open all year round

Although in winter the picturesque lake below the summit is usually frozen over, snowed in and virtually only recognisable at a second glance, the well-built Passo Fedaia offers an attractive alternative to the passage over the Pordoi pass. The view to the mighty Marmolada Glacier in particular is exceptional. If this sight sets the hearts of skiers aflutter, they can simply unfasten the skis from the roof, catch the cable car in Malga Ciapela up to 3,269 metres, and whizz down again over the glacier to the pass below.

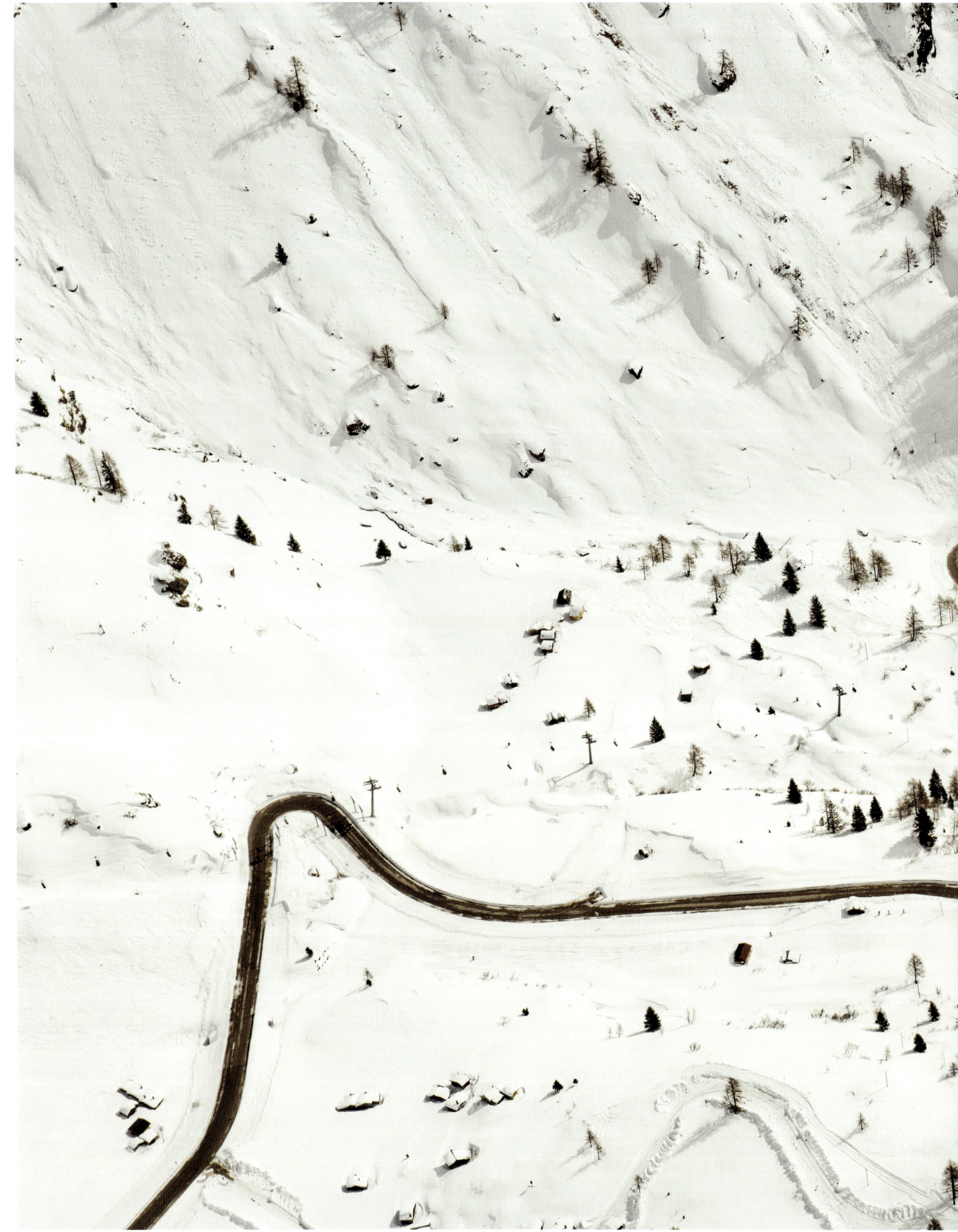

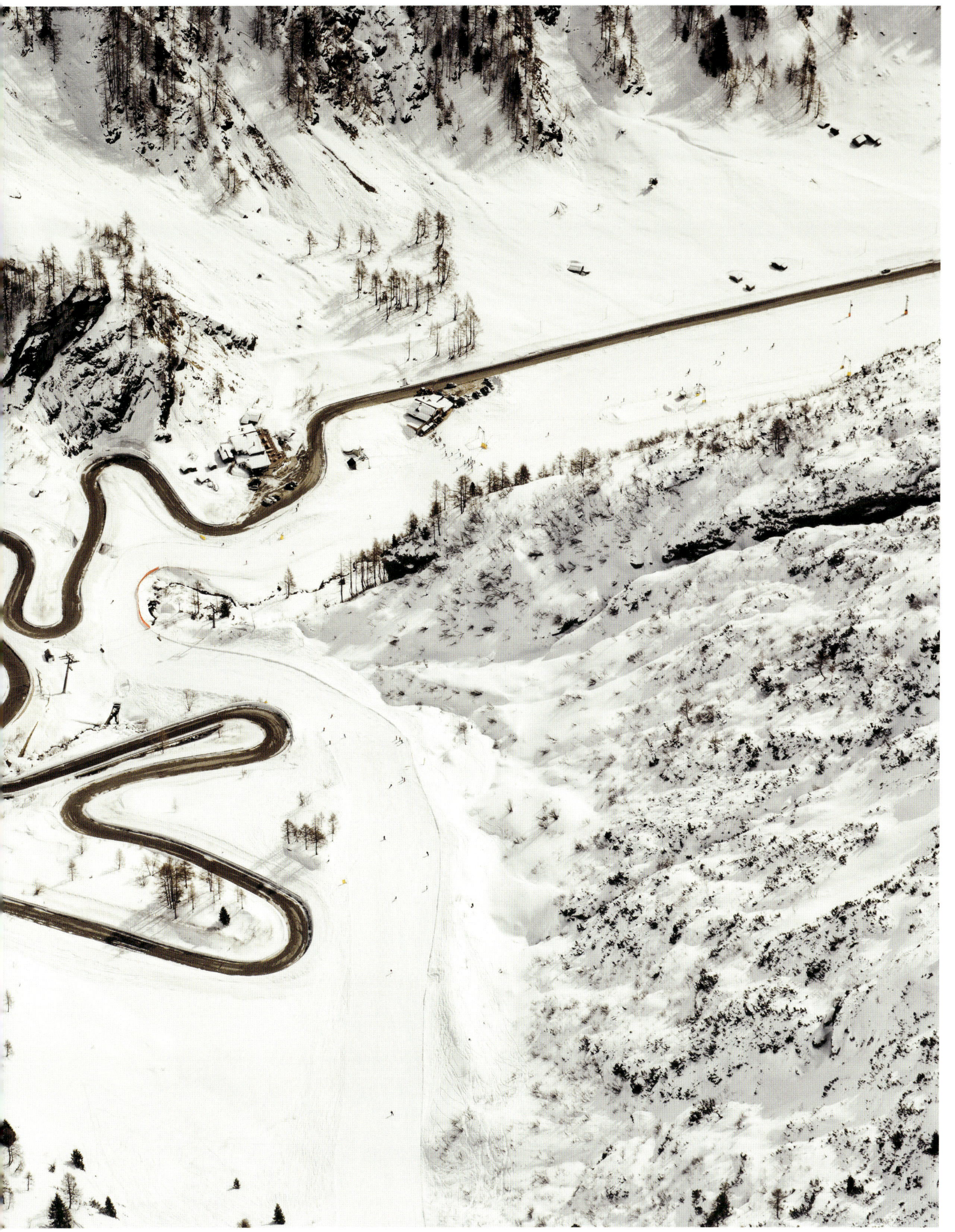

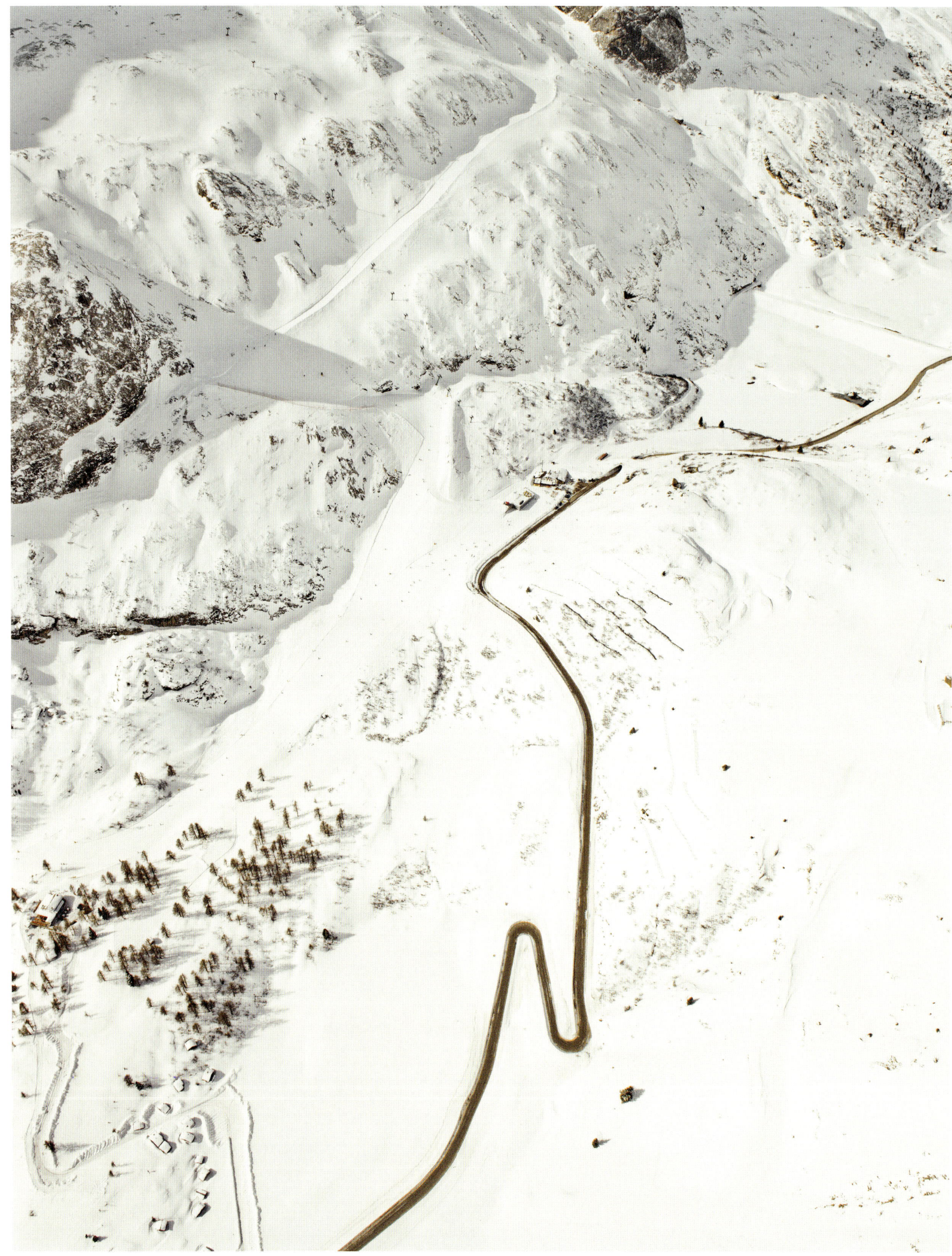

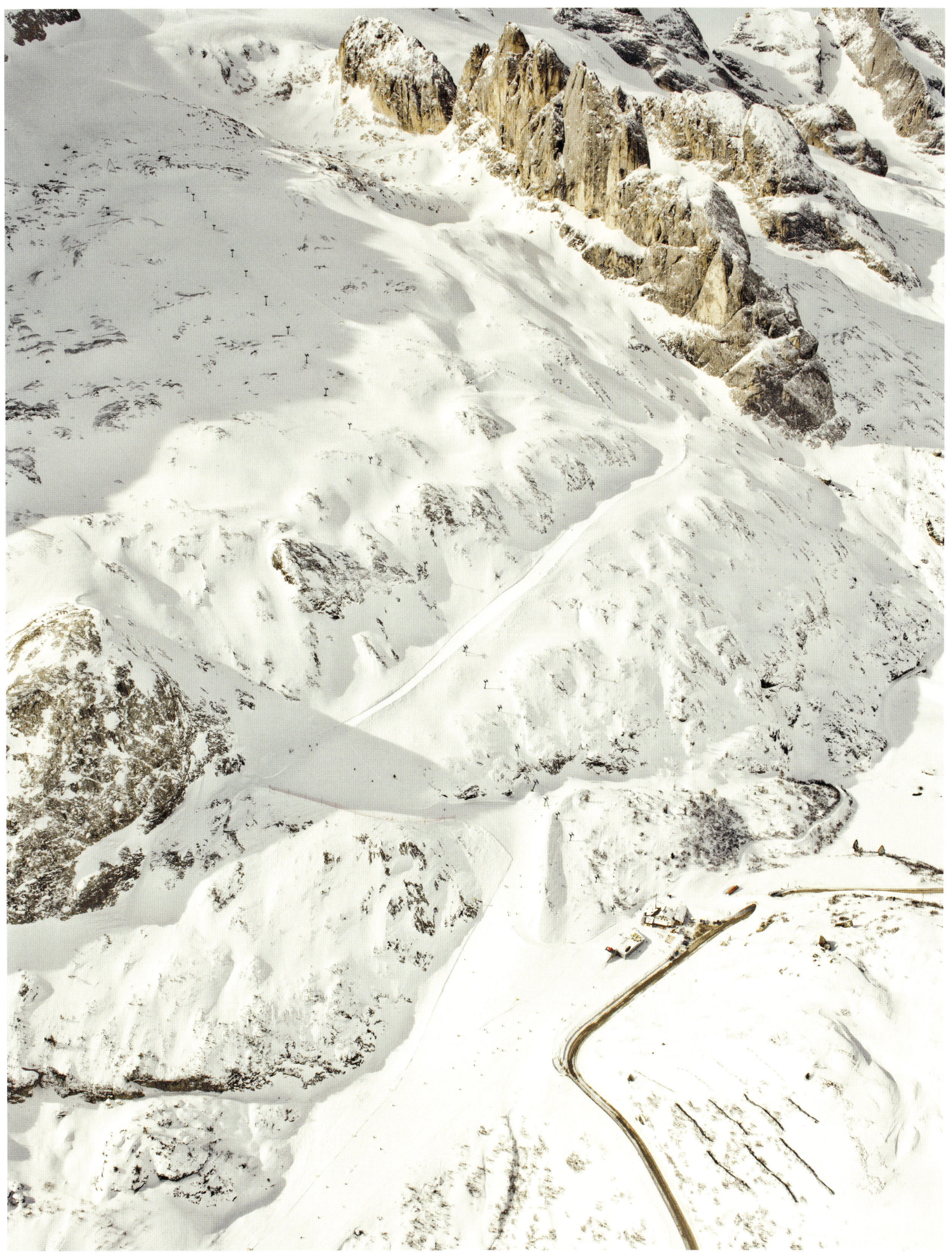

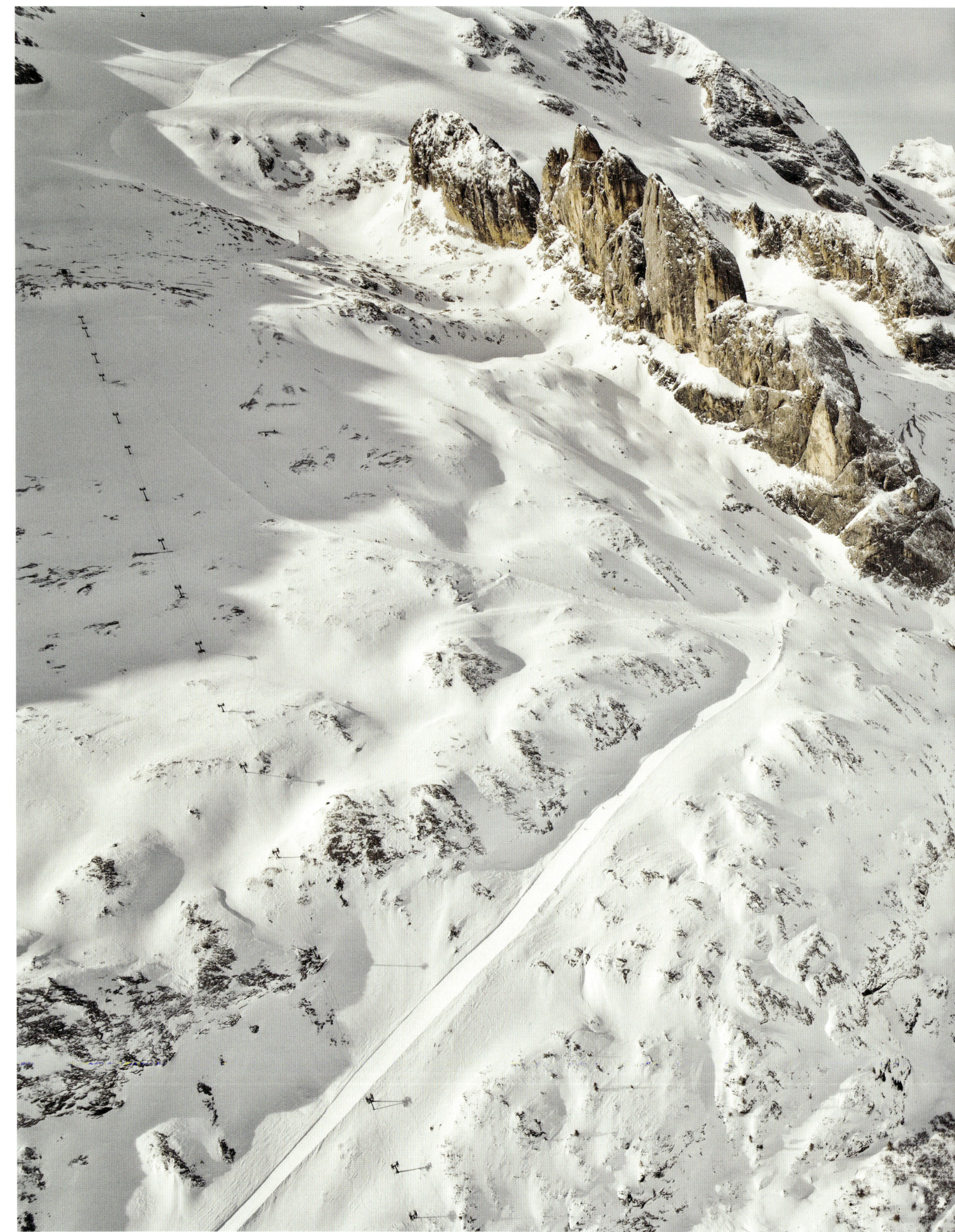

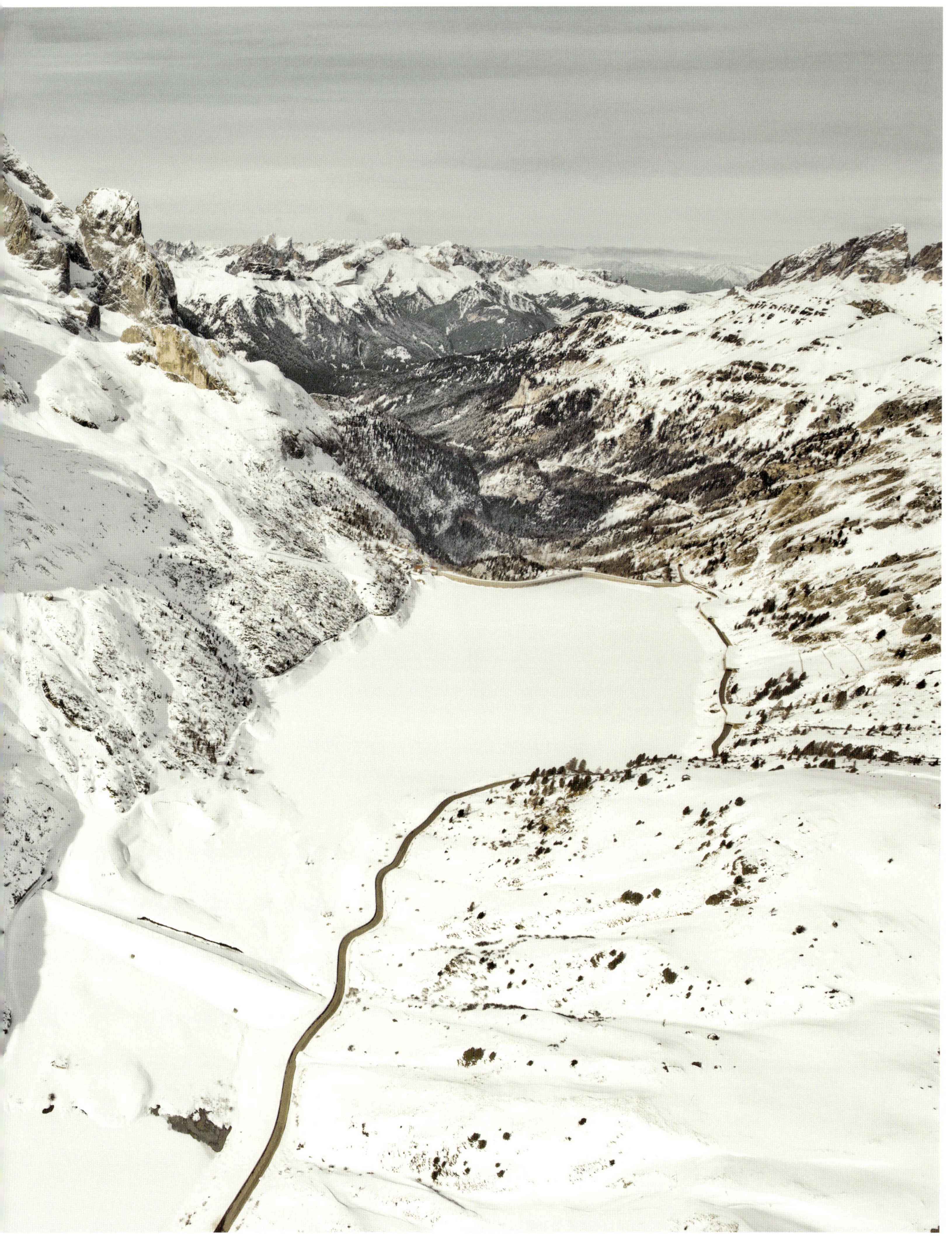

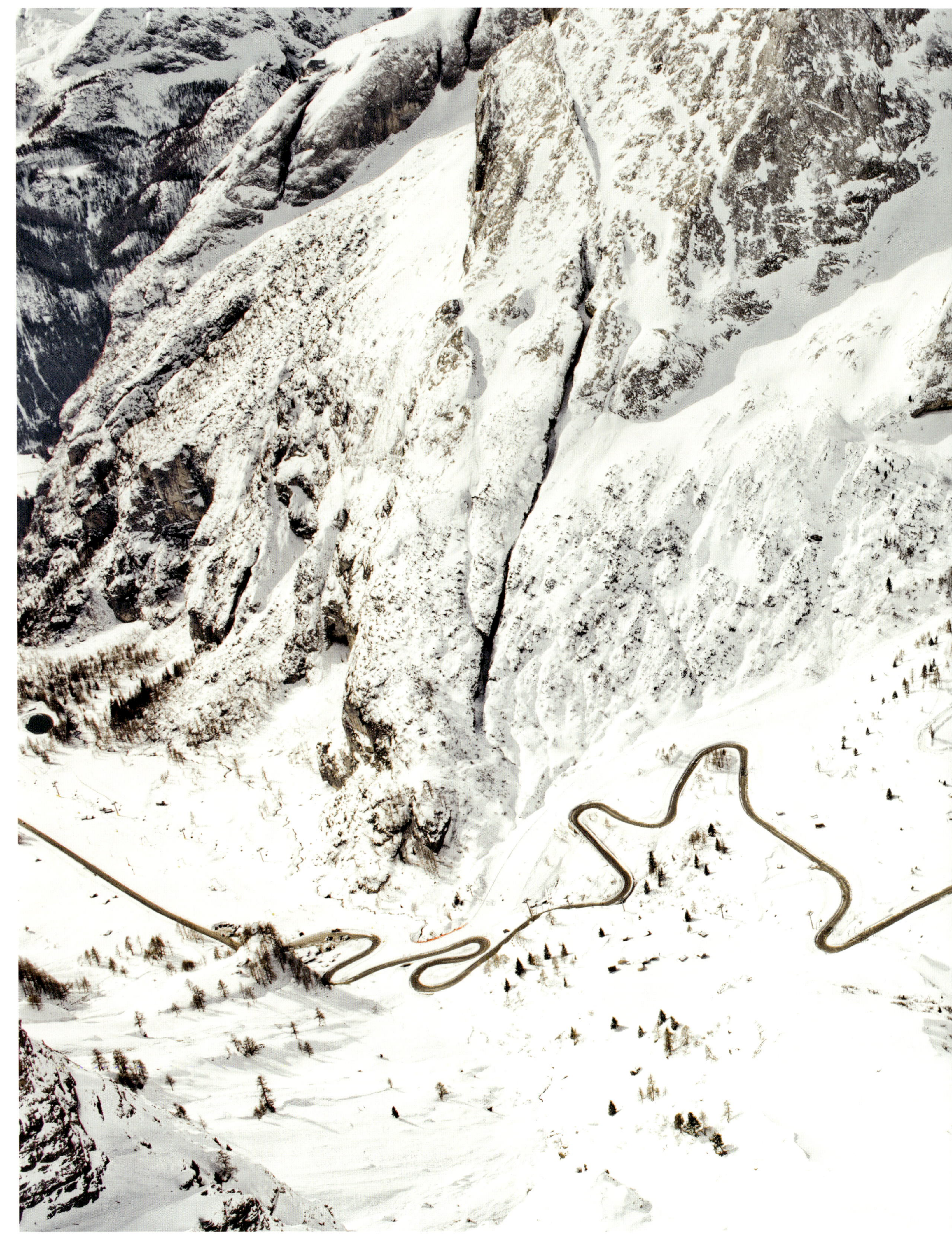

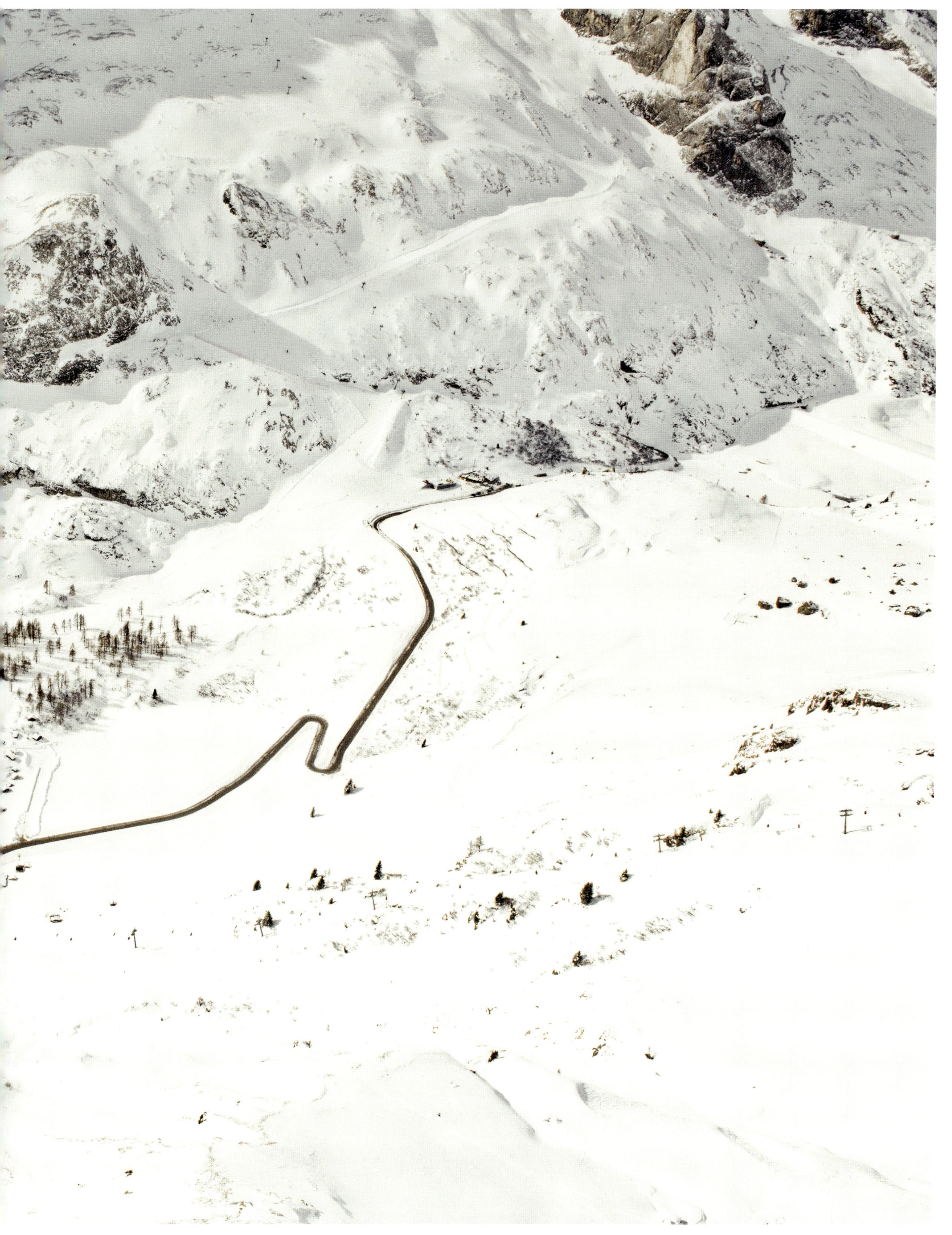

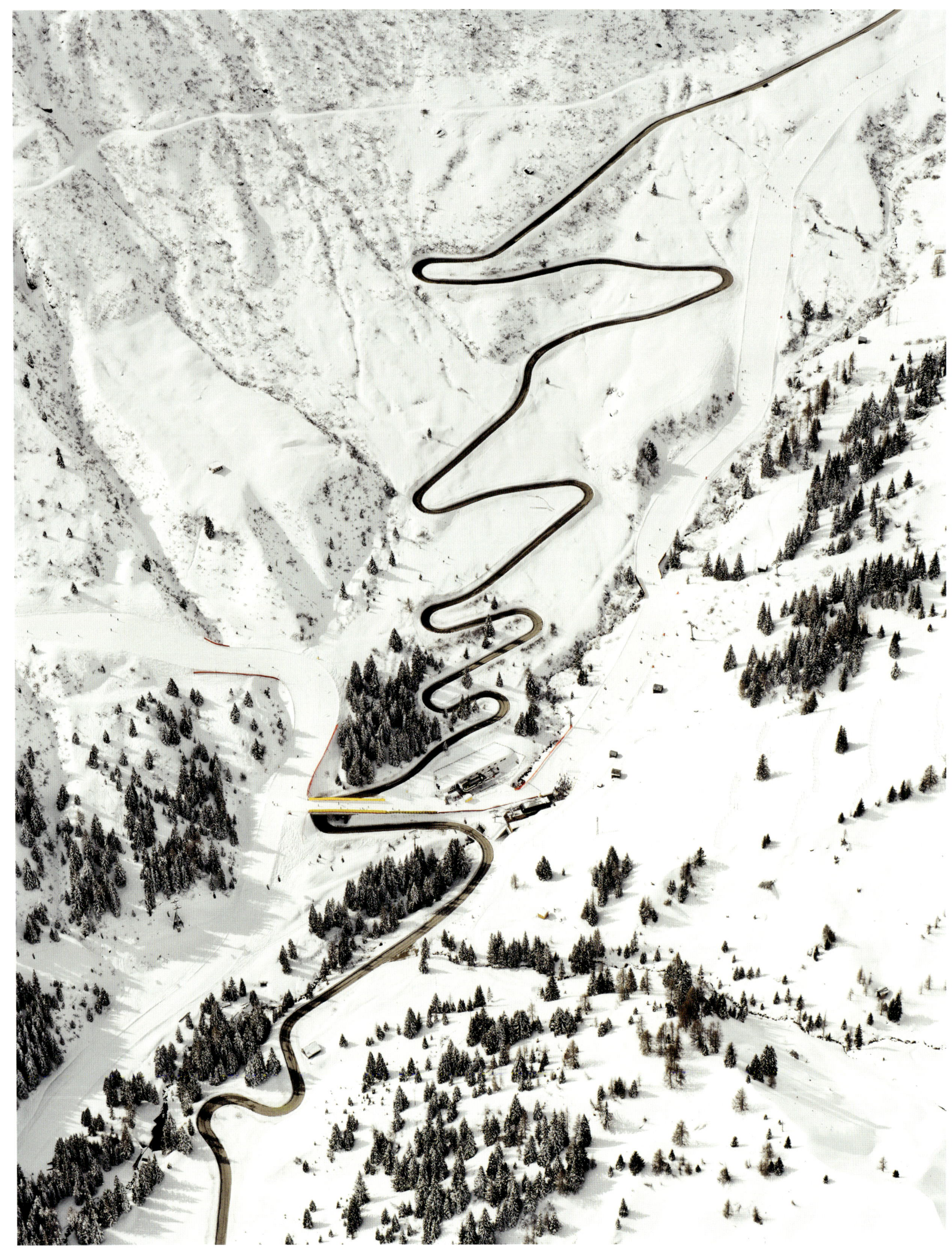

Passo Pordoi

— 2239 Meter
Meist ganzjährig geöffnet

Spätestens mit der Eröffnung der Großen Dolomitenstraße zwischen Bozen und Cortina d'Ampezzo im Jahr 1909 rückte die schroffe Bergwelt auf die Agenda des internationalen Tourismus. Der Passo Pordoi (Pordoijoch) ist mit 2239 Meter der höchstgelegene Punkt der Alpenstraße – entsprechend lohnenswert ist das schroffe Panorama der winterlichen Dolomitengipfel im Osten. Geübte Freerider steigen an der Passhöhe übrigens in die Seilbahn hinauf zum Sass Pordoi, wo einige legendäre, aber durchaus anspruchsvolle Tiefschneetouren beginnen.

— 2,239 metres
Usually open all year round

The 1909 opening of the Great Dolomite Road, linking Bolzano and Cortina d'Ampezzo, turned the rugged mountainous world of the Dolomites into an international tourist destination. At 2,239 metres, the Passo Pordoi (Pordoi Pass) is the highest point of the Alpine road, and from here the jagged panorama of the wintery Dolomite peaks to the east is particularly stunning. Seasoned and ambitious freeriders can climb into the cable car at the top of the pass and catch a ride up to the Sass Pordoi, the starting point of some legendary deep-powder slopes.

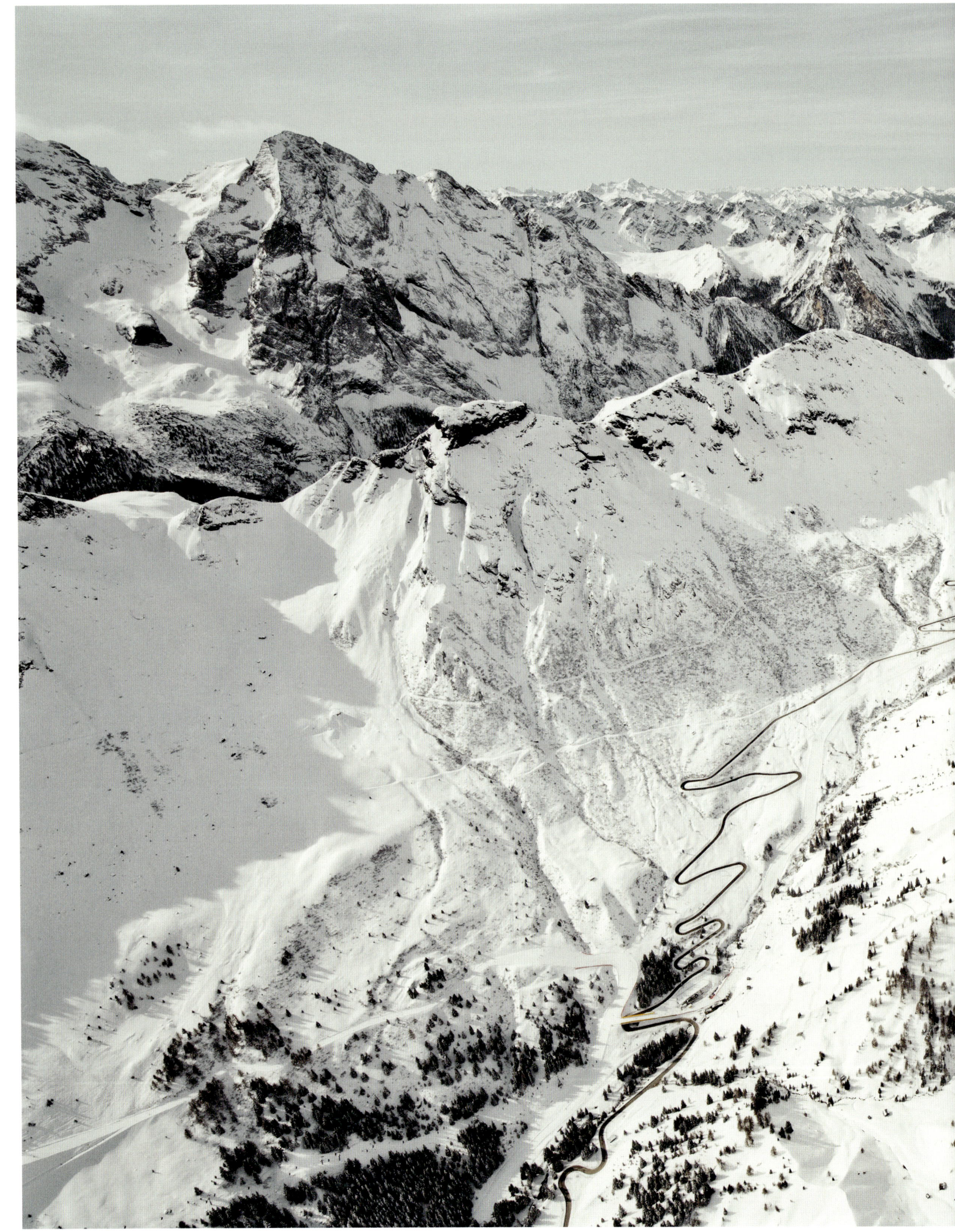

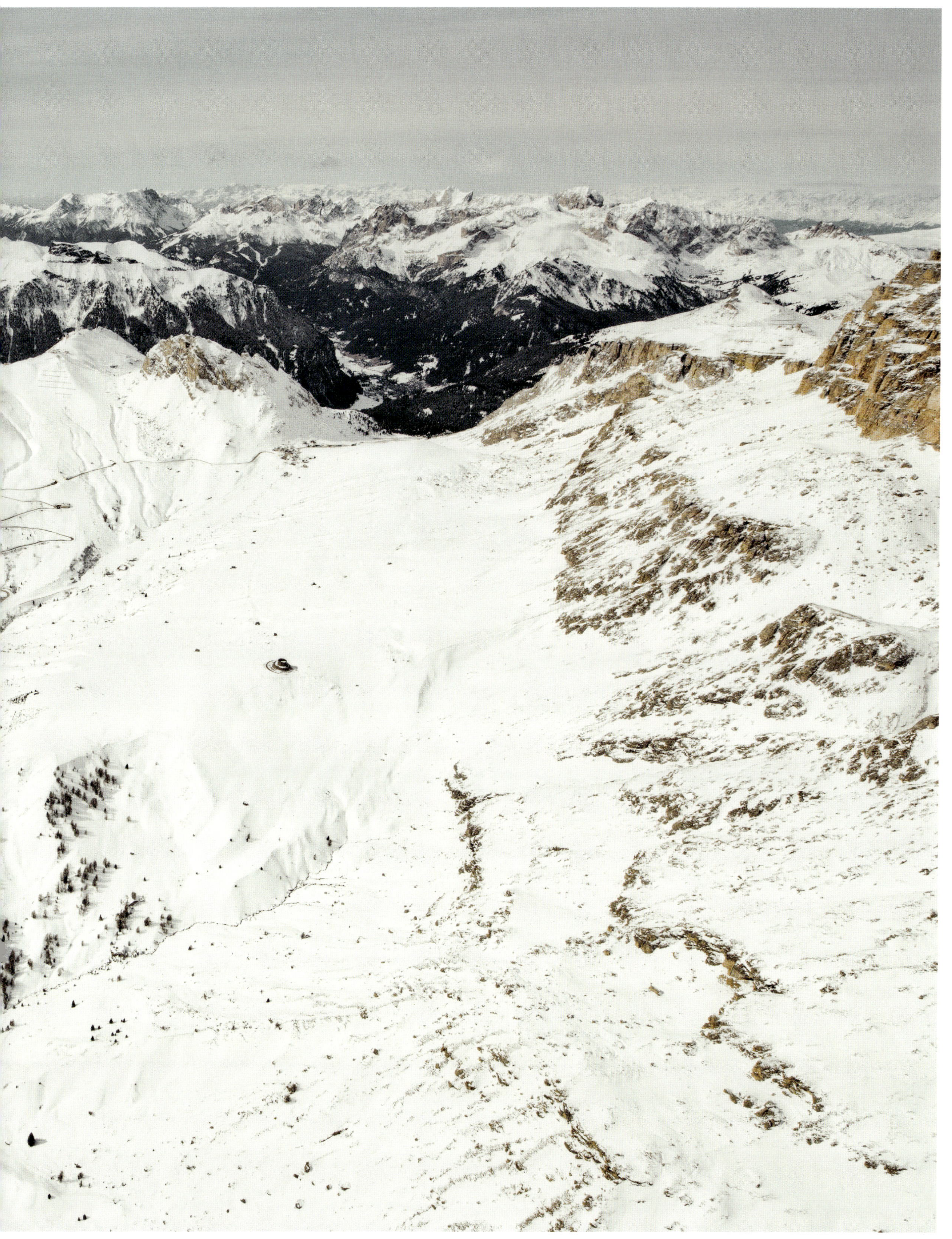

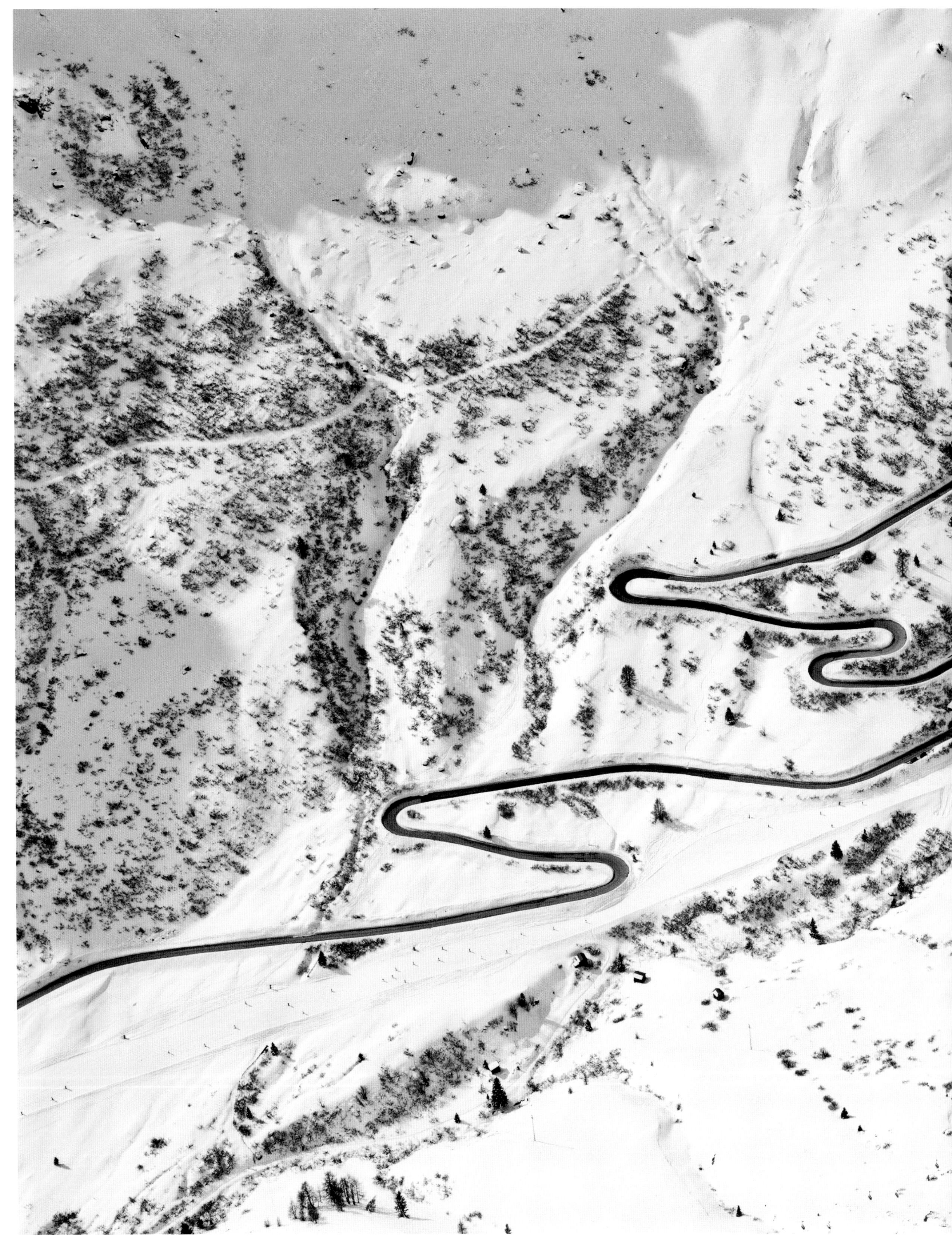

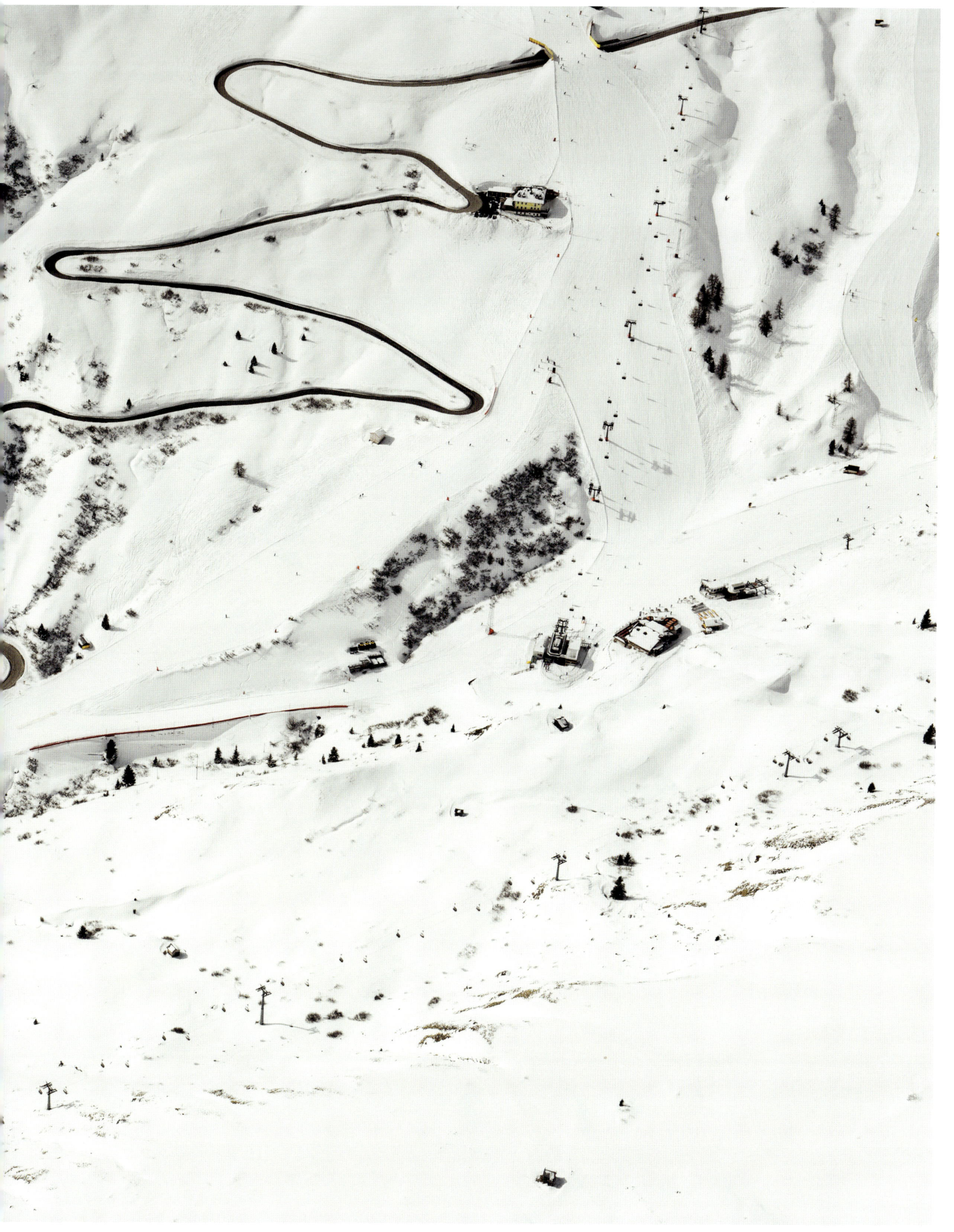

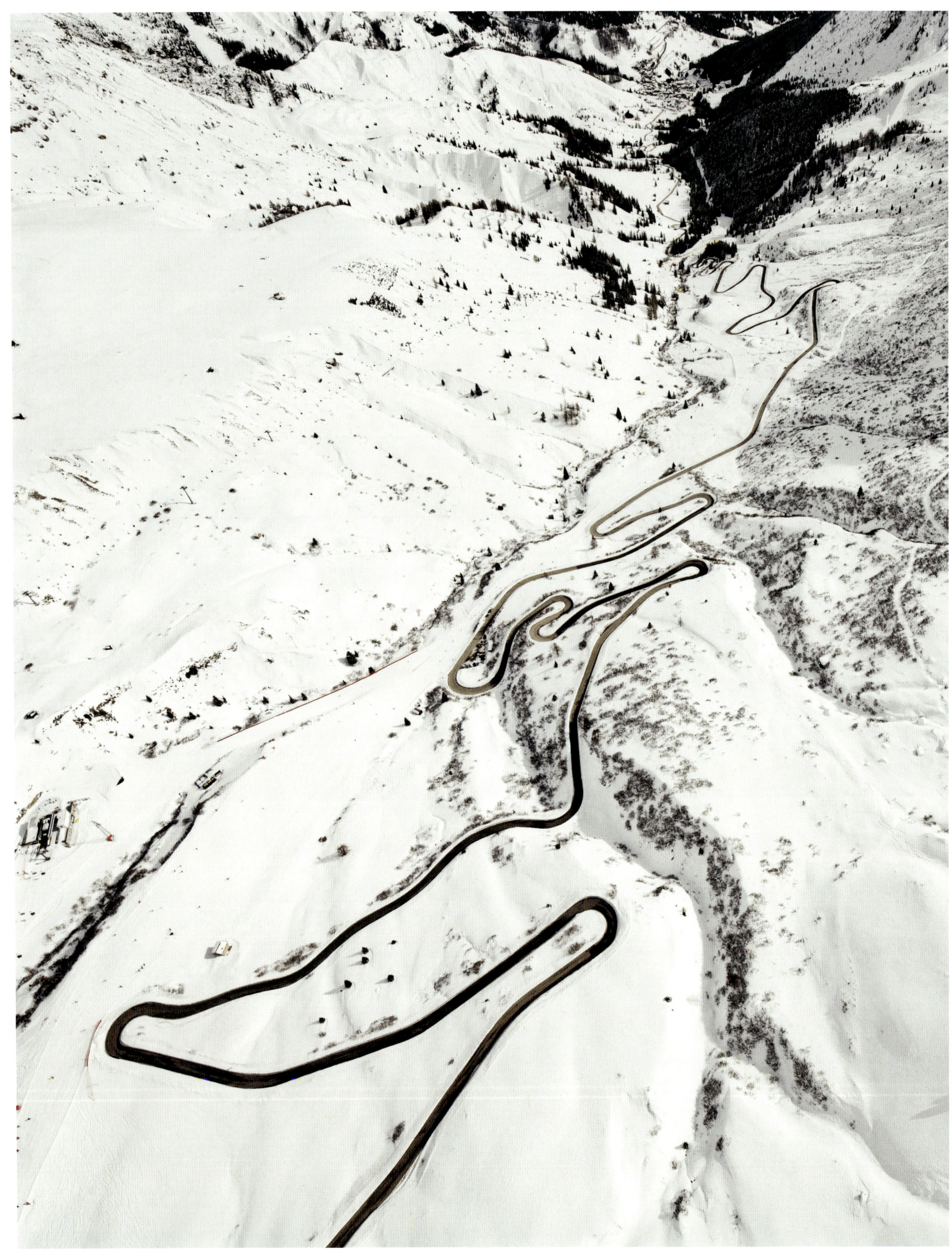

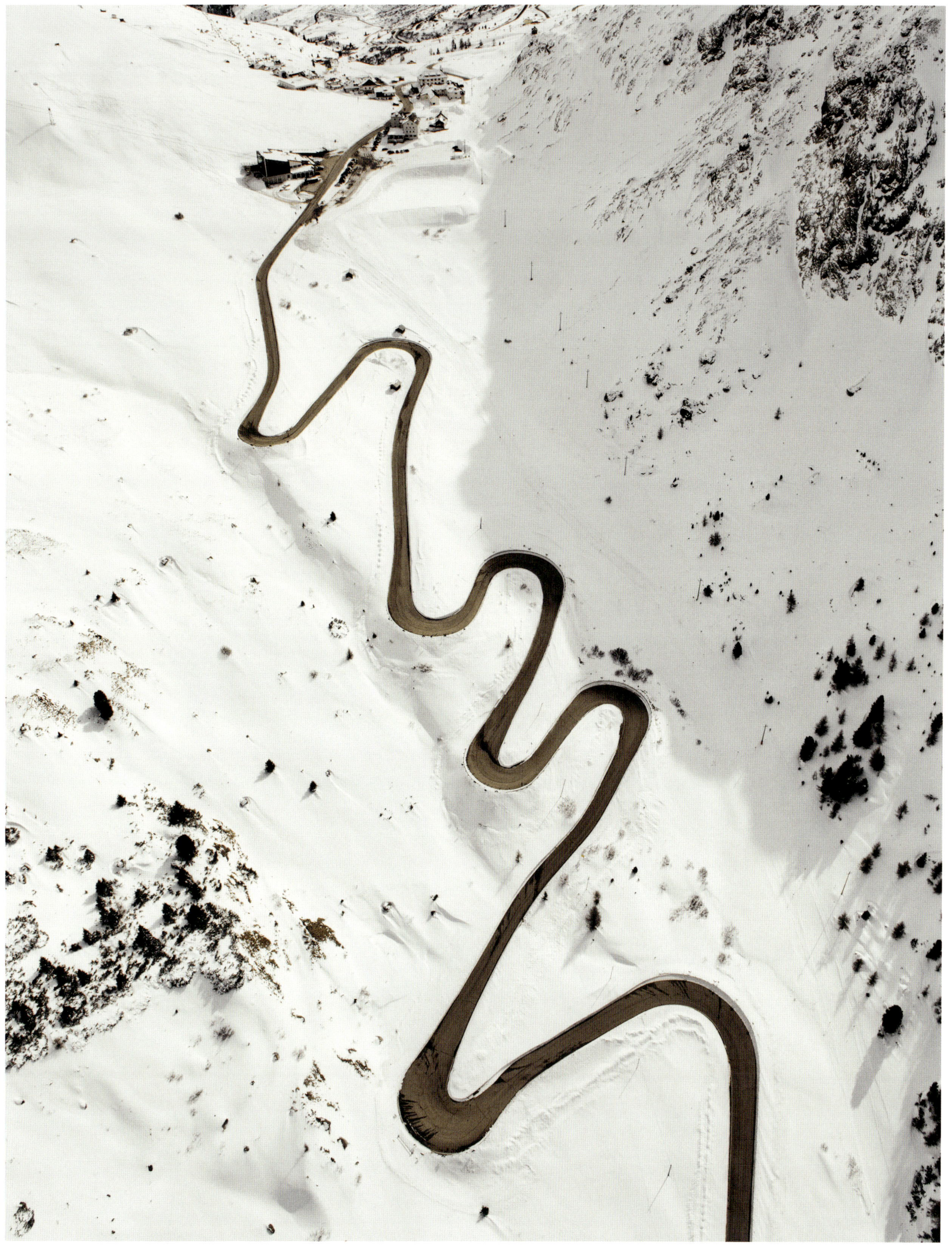

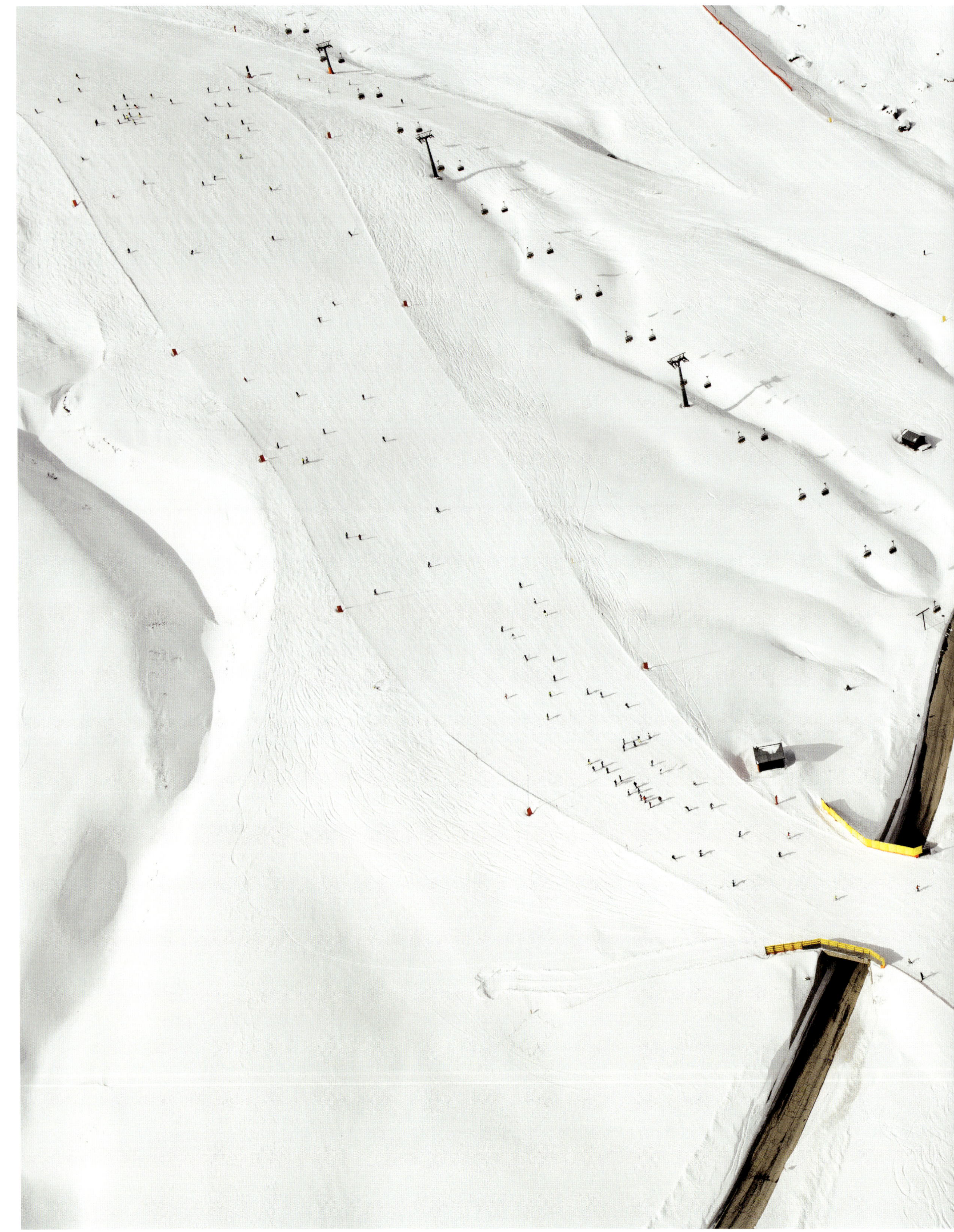

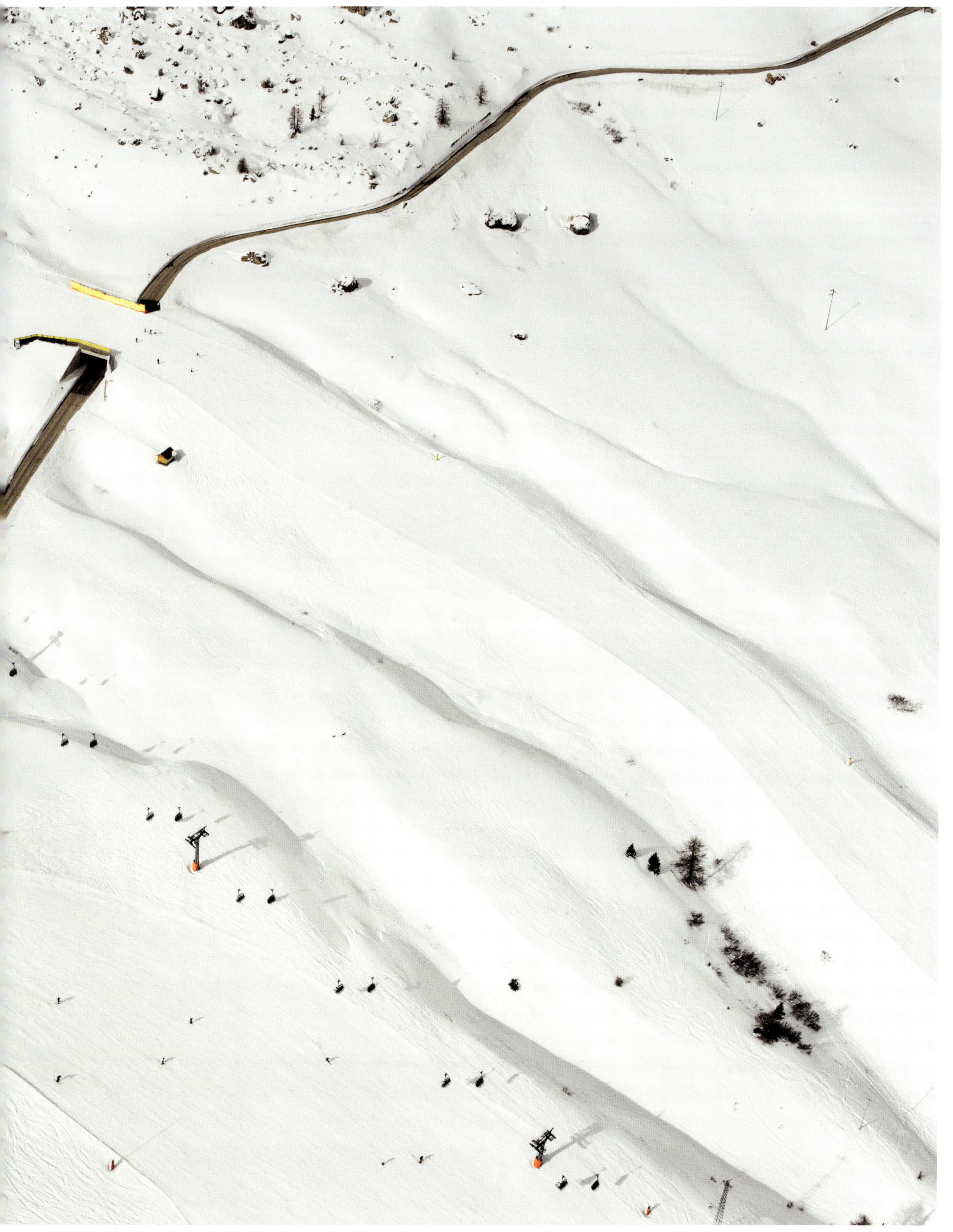

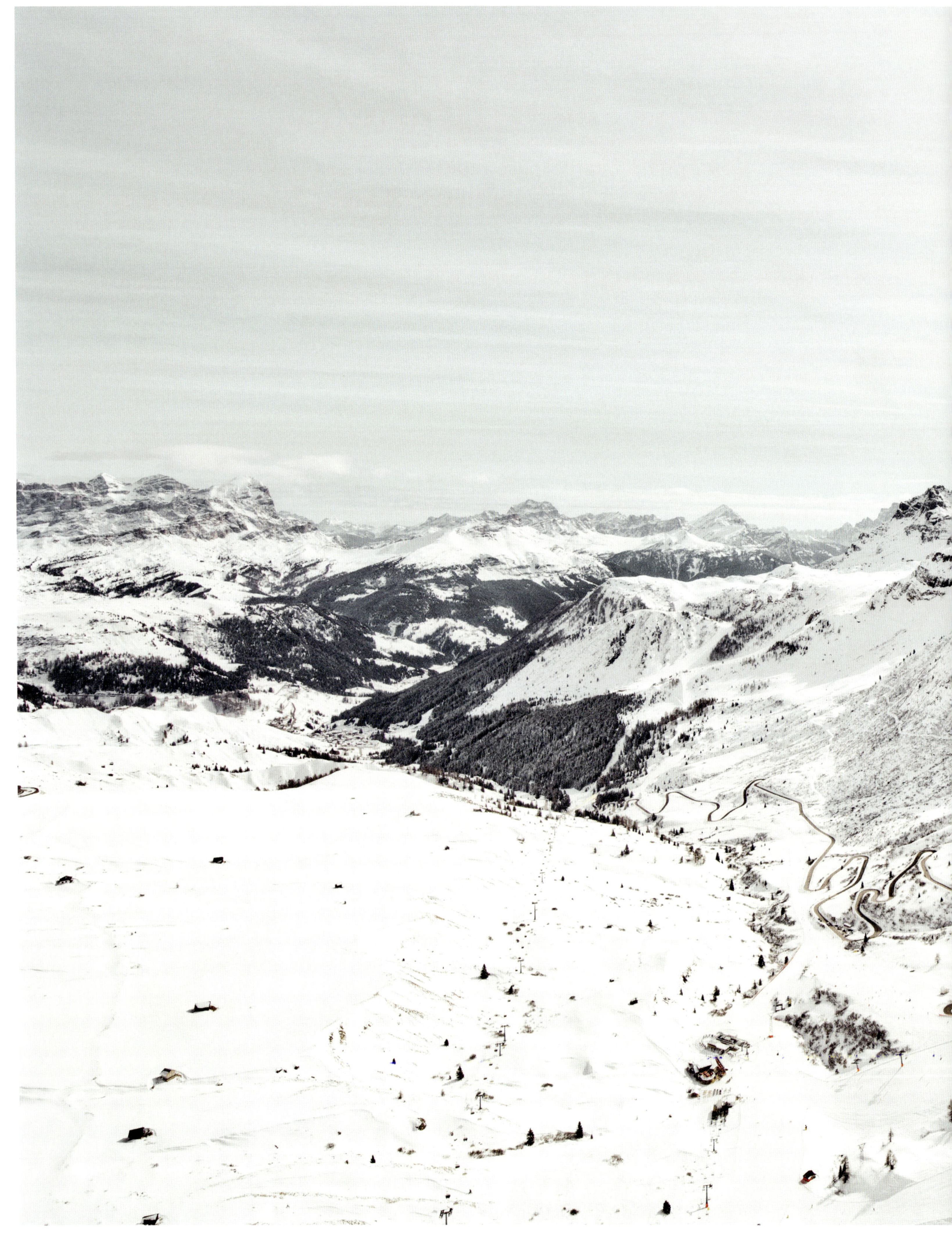

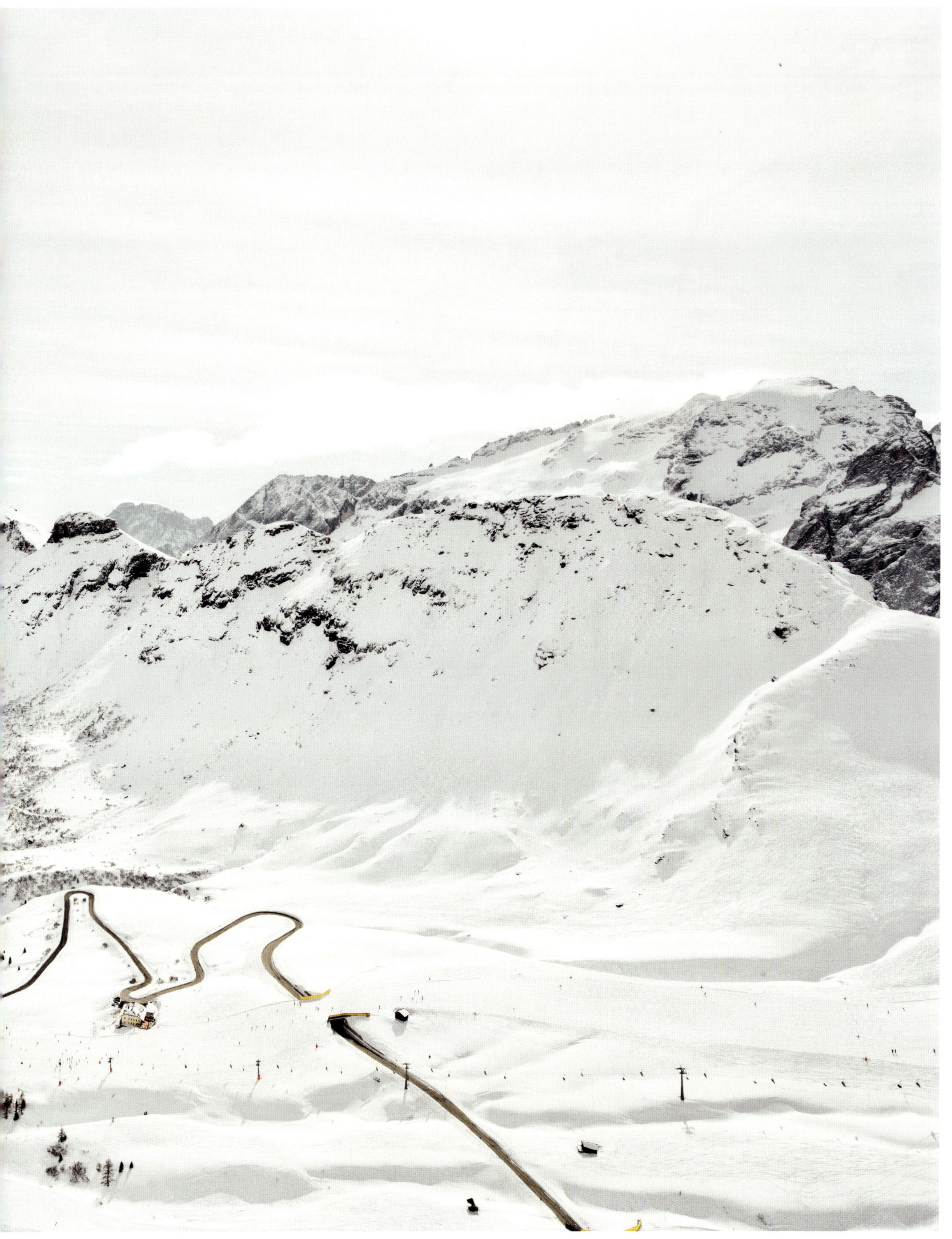

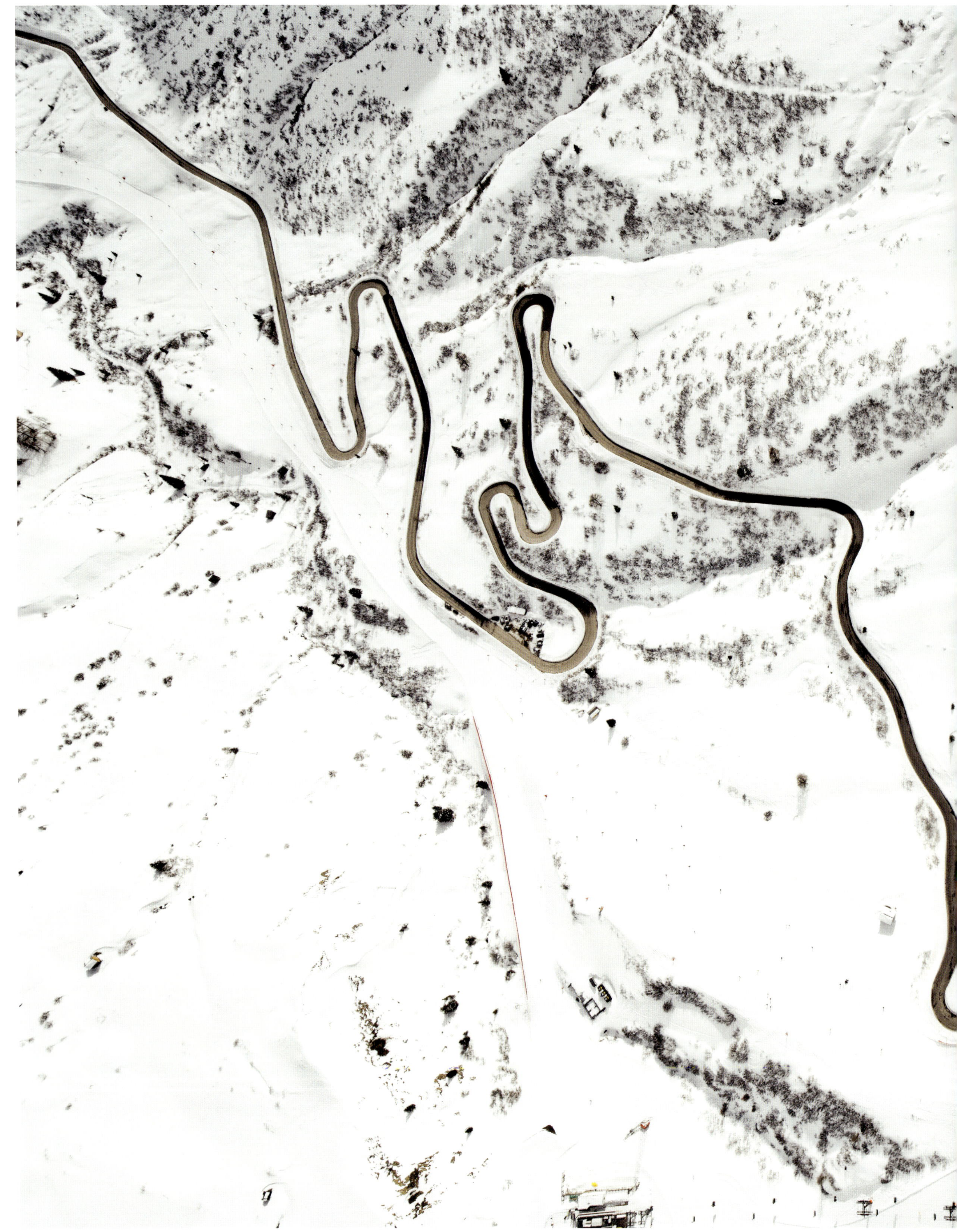

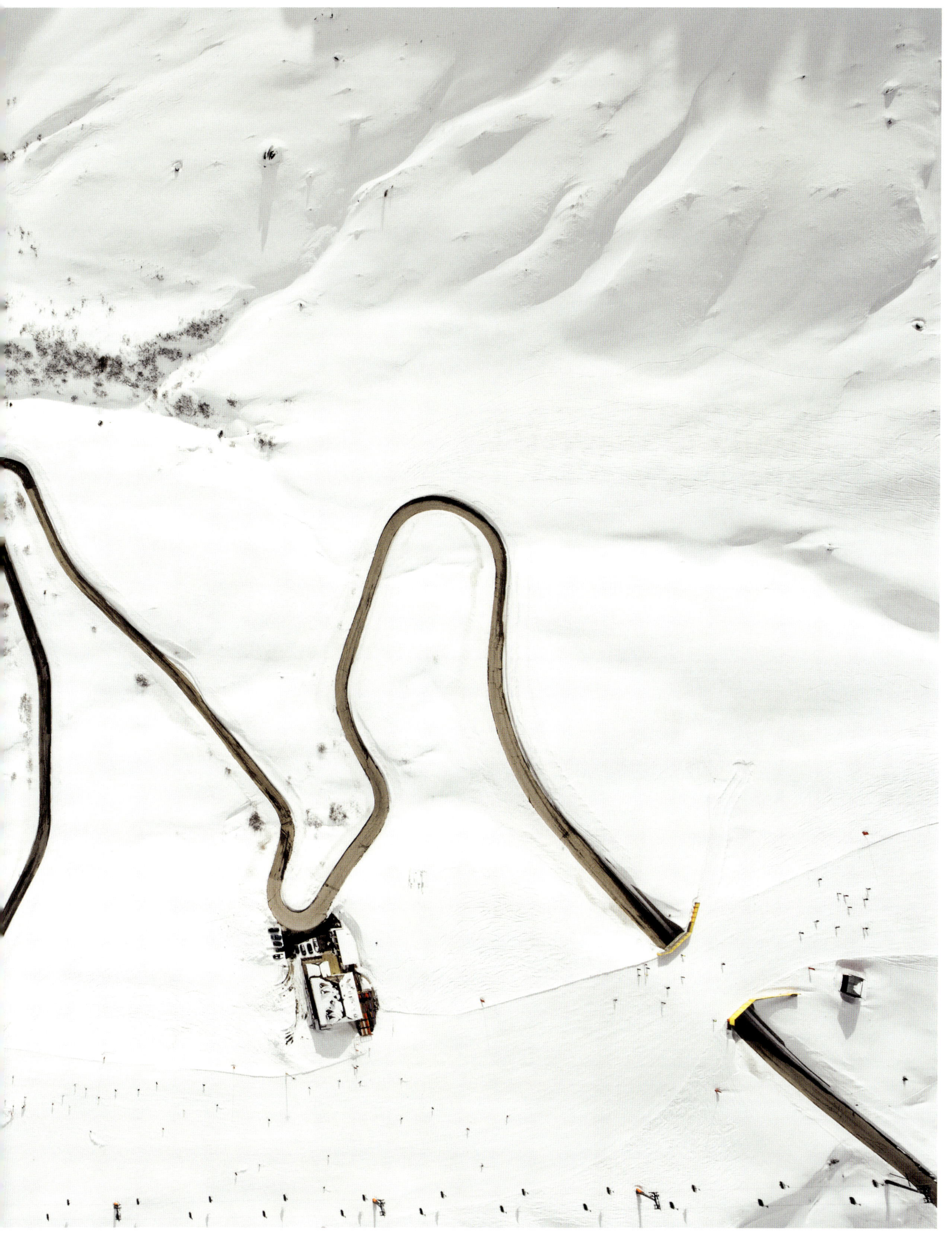

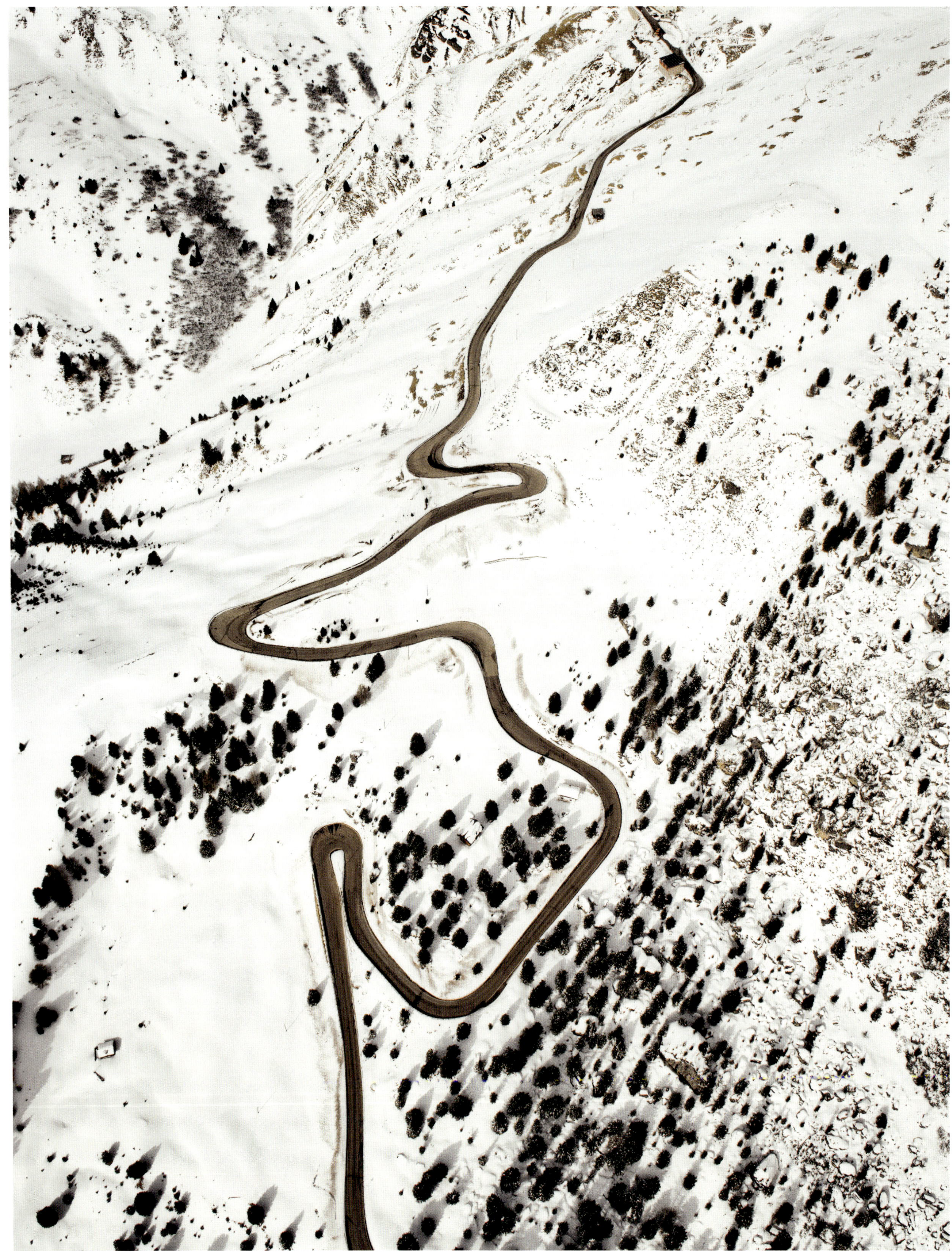

Passo Sella

— 2244 Meter
Meist ganzjährig geöffnet

Der Passo Sella (das Sellajoch) verbindet Wolkenstein in Südtirol mit dem Fassatal im Trentino. Während man die berühmte „Sella Ronda" im Winter am besten auf Ski zurücklegen kann, ist der späte Frühling kurz nach der Schneeschmelze die beste Zeit, die Vierpässetour um den Sellastock mit dem Rennrad, Motorrad oder Auto in Angriff zu nehmen. Die bereits 1872 eröffnete Sella-Passstraße gehört übrigens nicht nur zu den ältsten, sondern auch den eindrucksvollsten Bergstraßen der Dolomiten – die scharf geschnittenen Kehren unter den mächtigen Felswänden des Sellastocks sind unter Kurvenkennern berühmt, erfordern aber eine ruhige Hand am Volant.

— 2,244 metres
Usually open all year round

The Passo Sella (Sella Pass) links Selva Val Gardena in South Tyrol with the Fassa Valley in Trentino. While in winter the famous "Sella Ronda" is best negotiated on skis, the prime time to experience the four pass trail around the Sella Group is in late spring, shortly after the snow melts, either on a racing bicycle, motorbike or in a car. Opened in 1872, the Sella Pass road is regarded not only as the oldest but also one of the most impressive mountain roads in the Dolomites – the tight zigzags under the majestic cliffs of the Sella Group are revered amongst cornering aficionados, how-ever a steady hand is needed at the wheel.

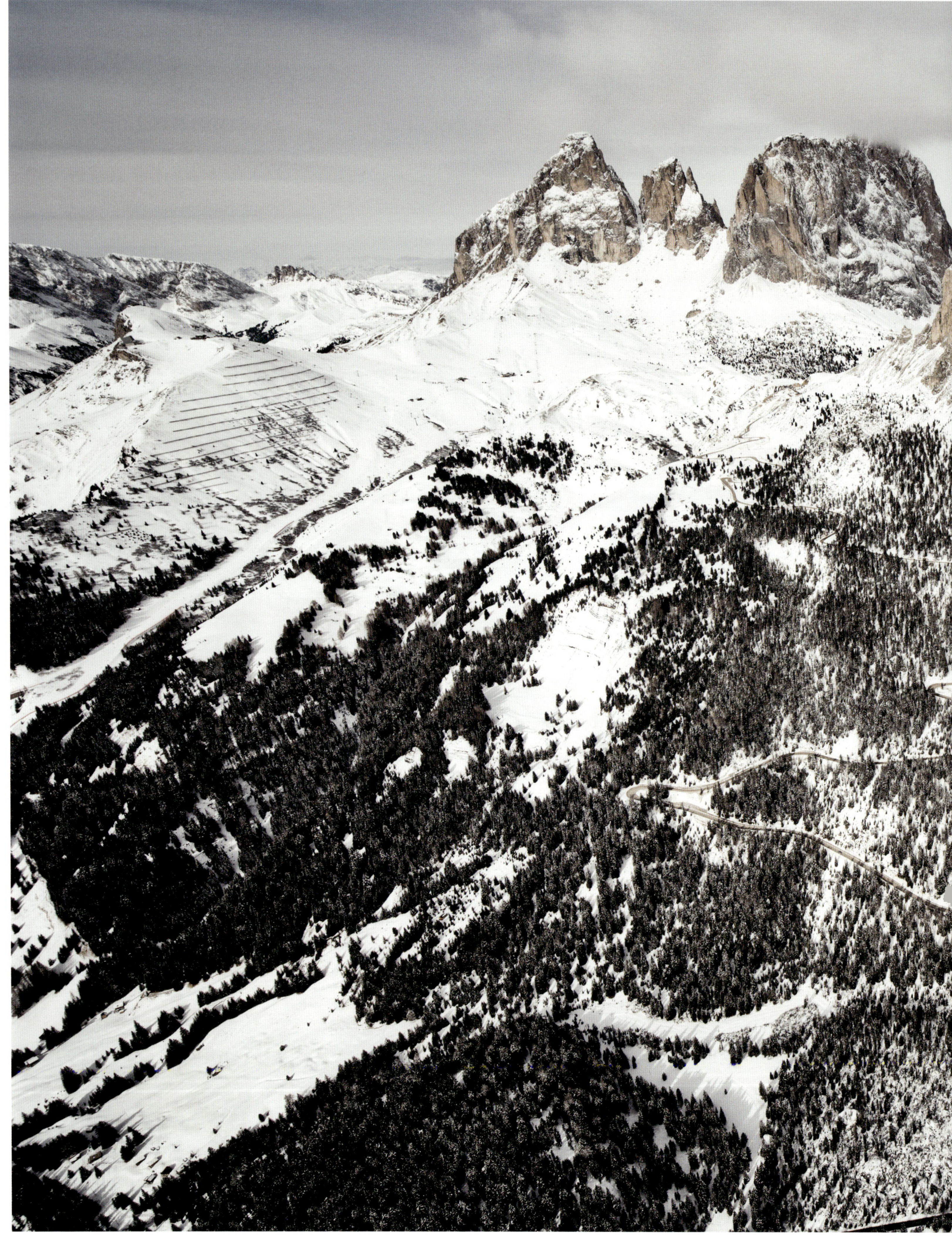

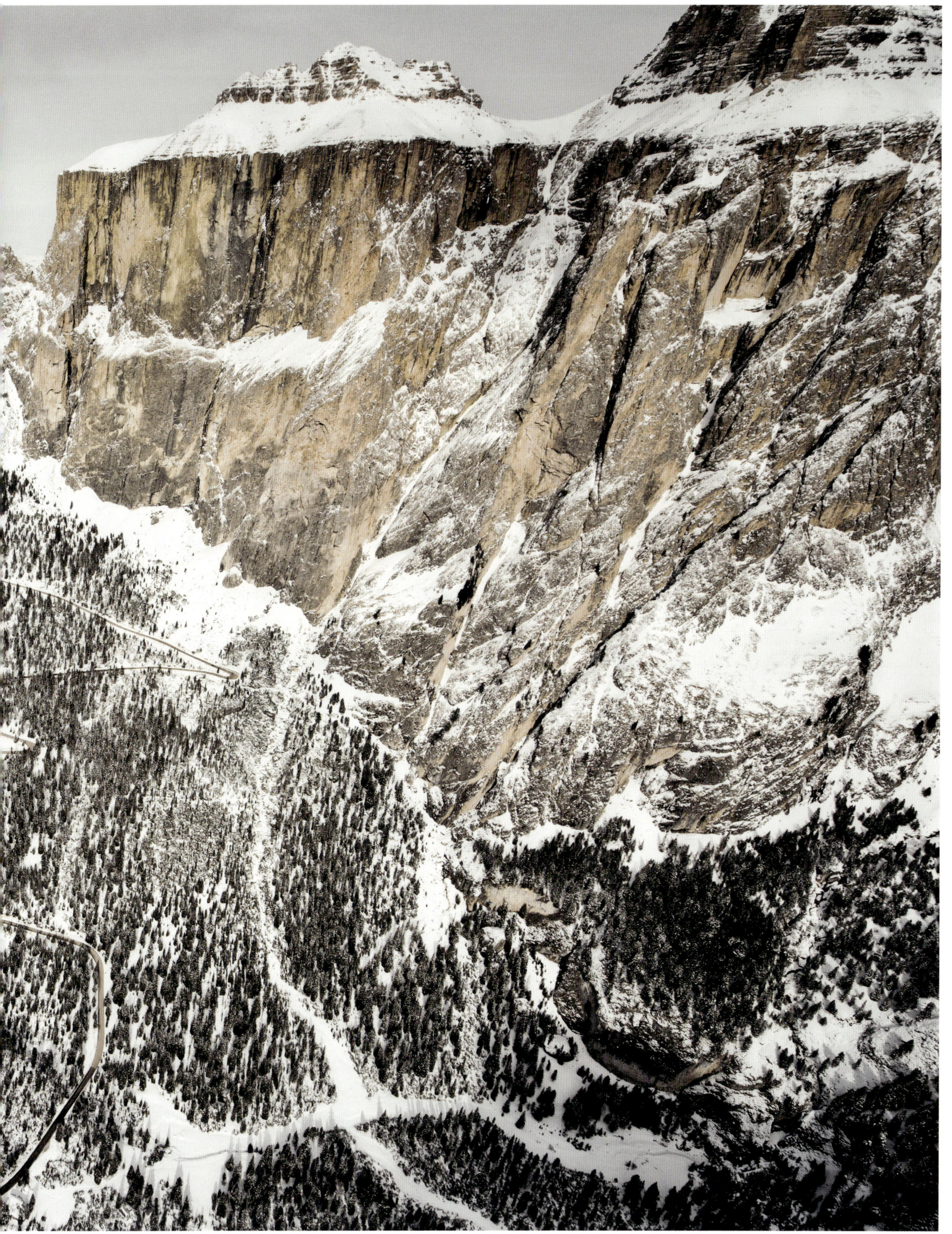

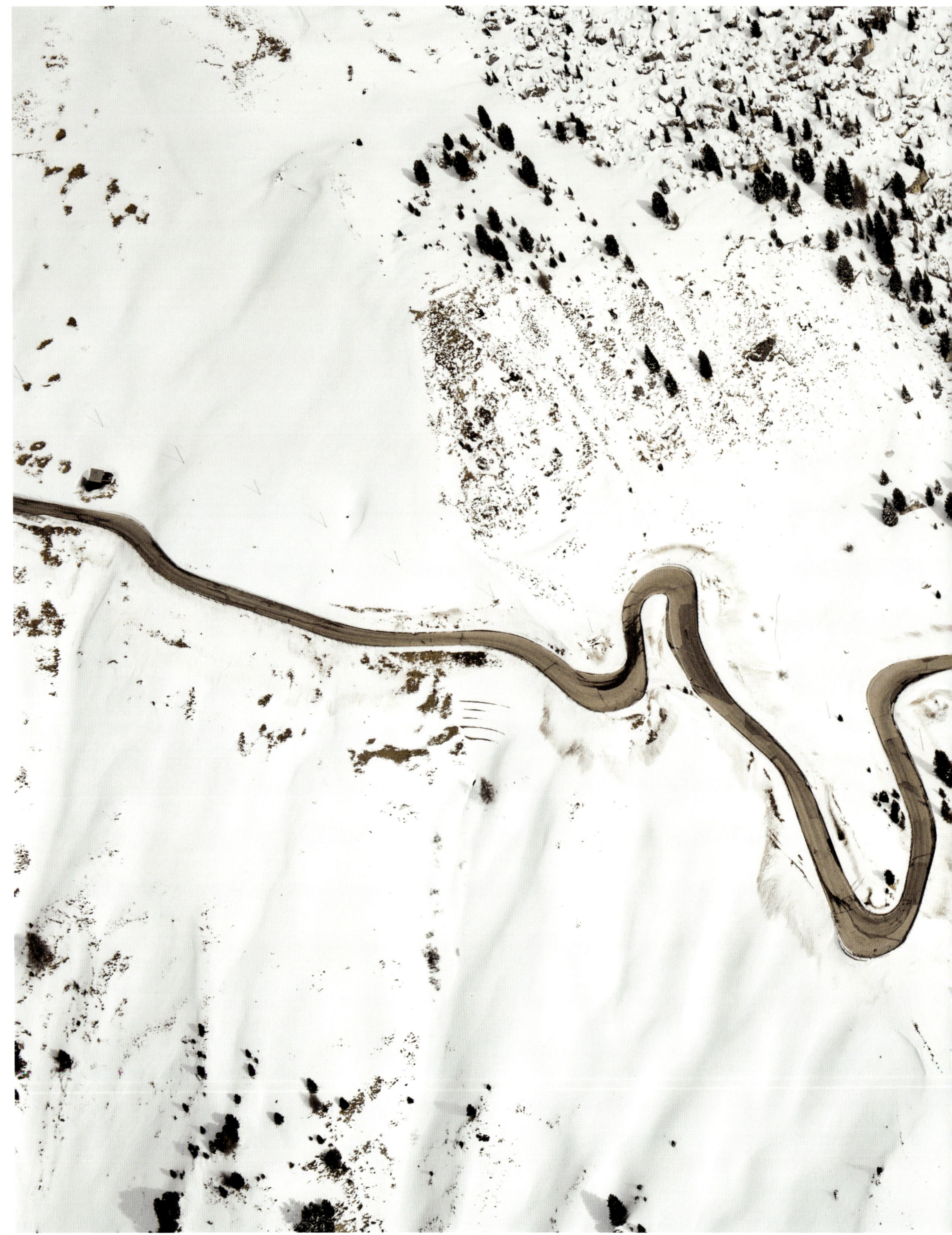

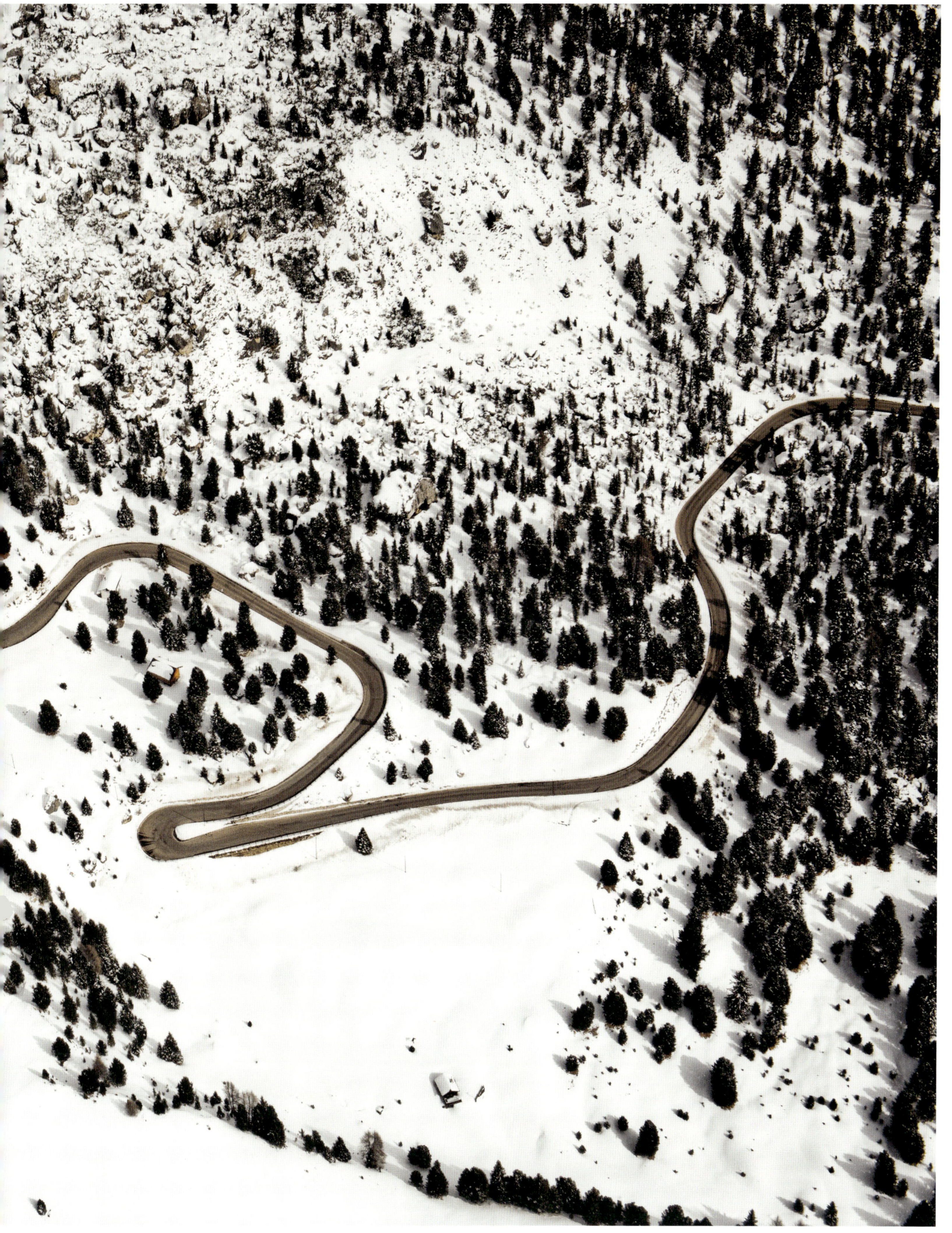

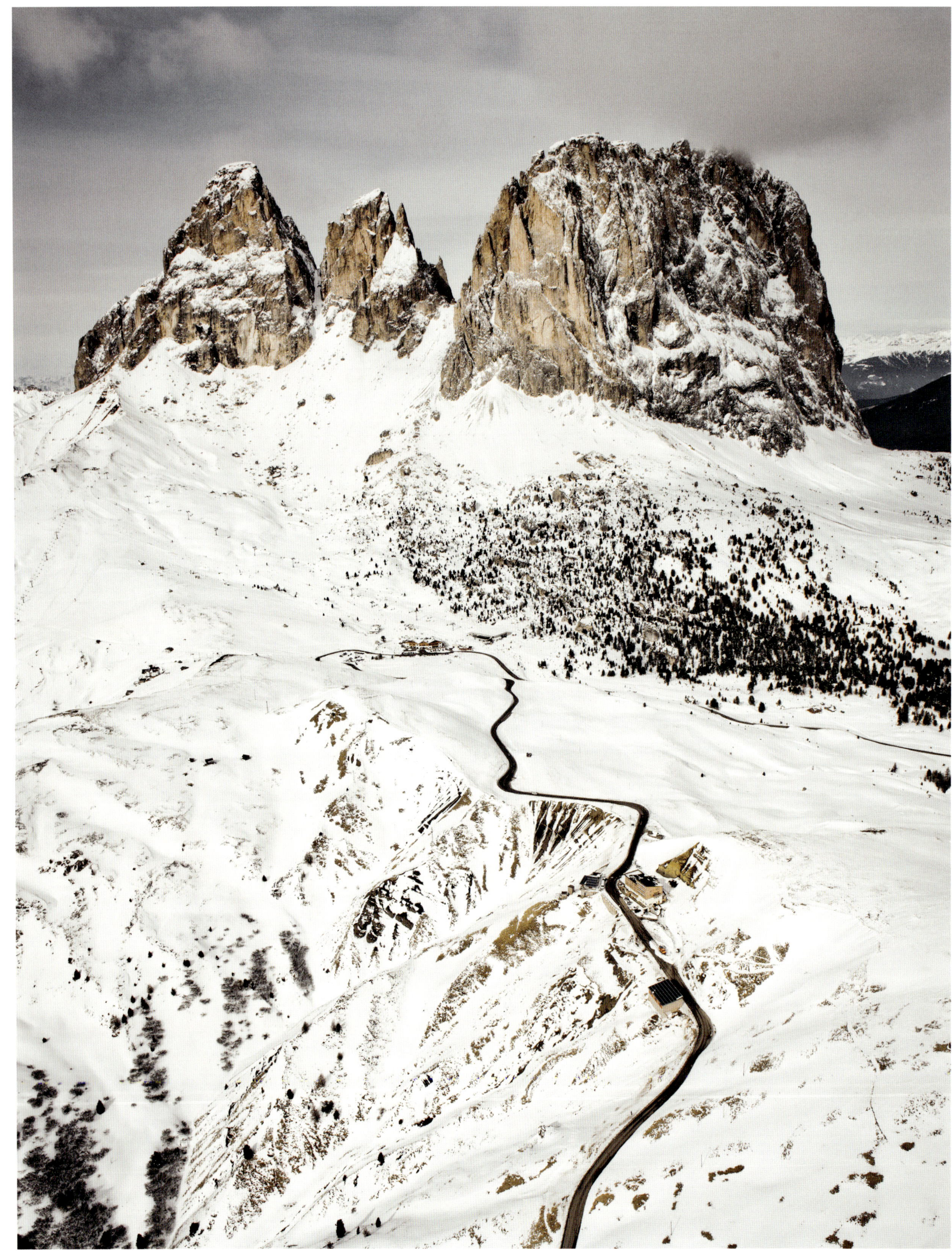

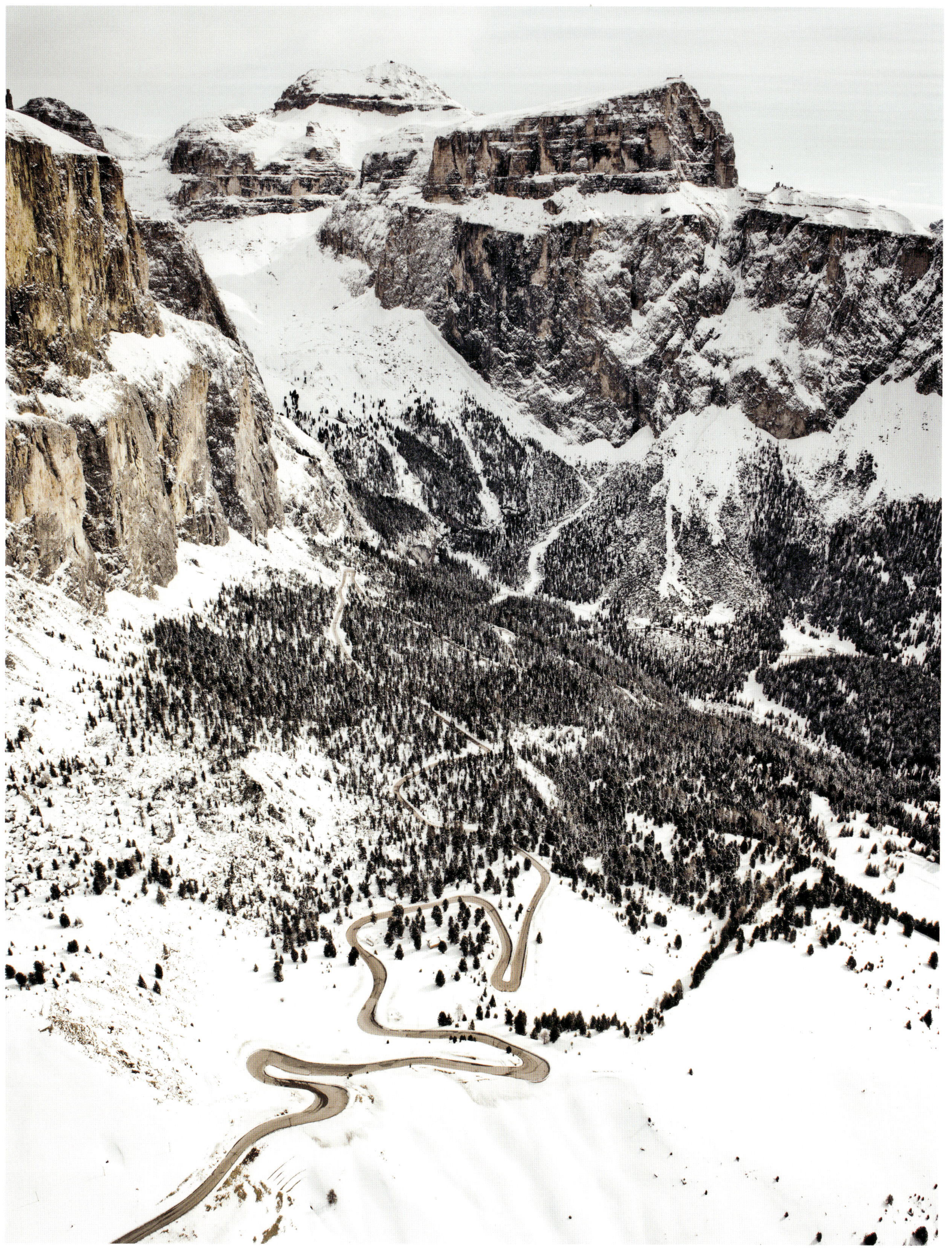

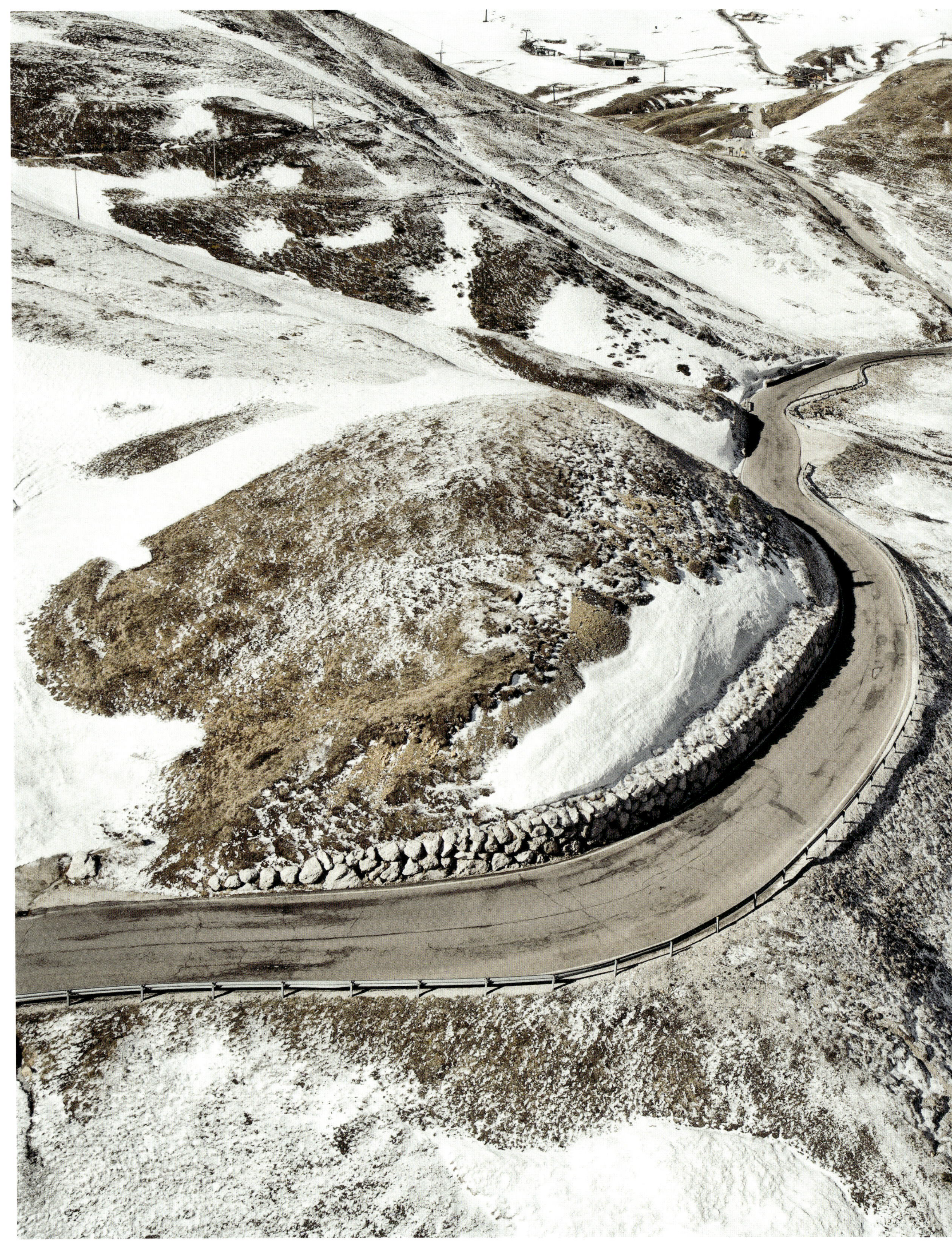

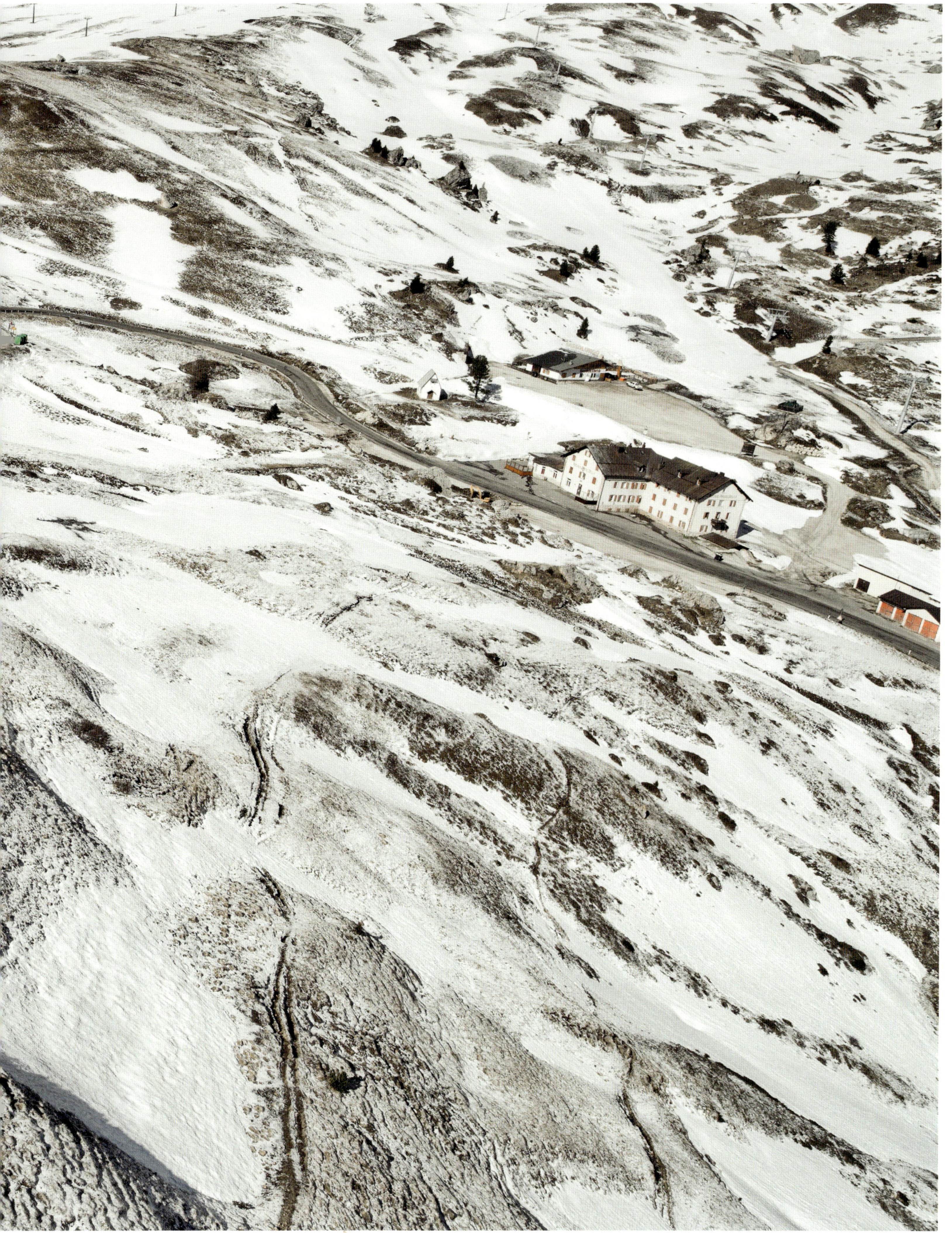

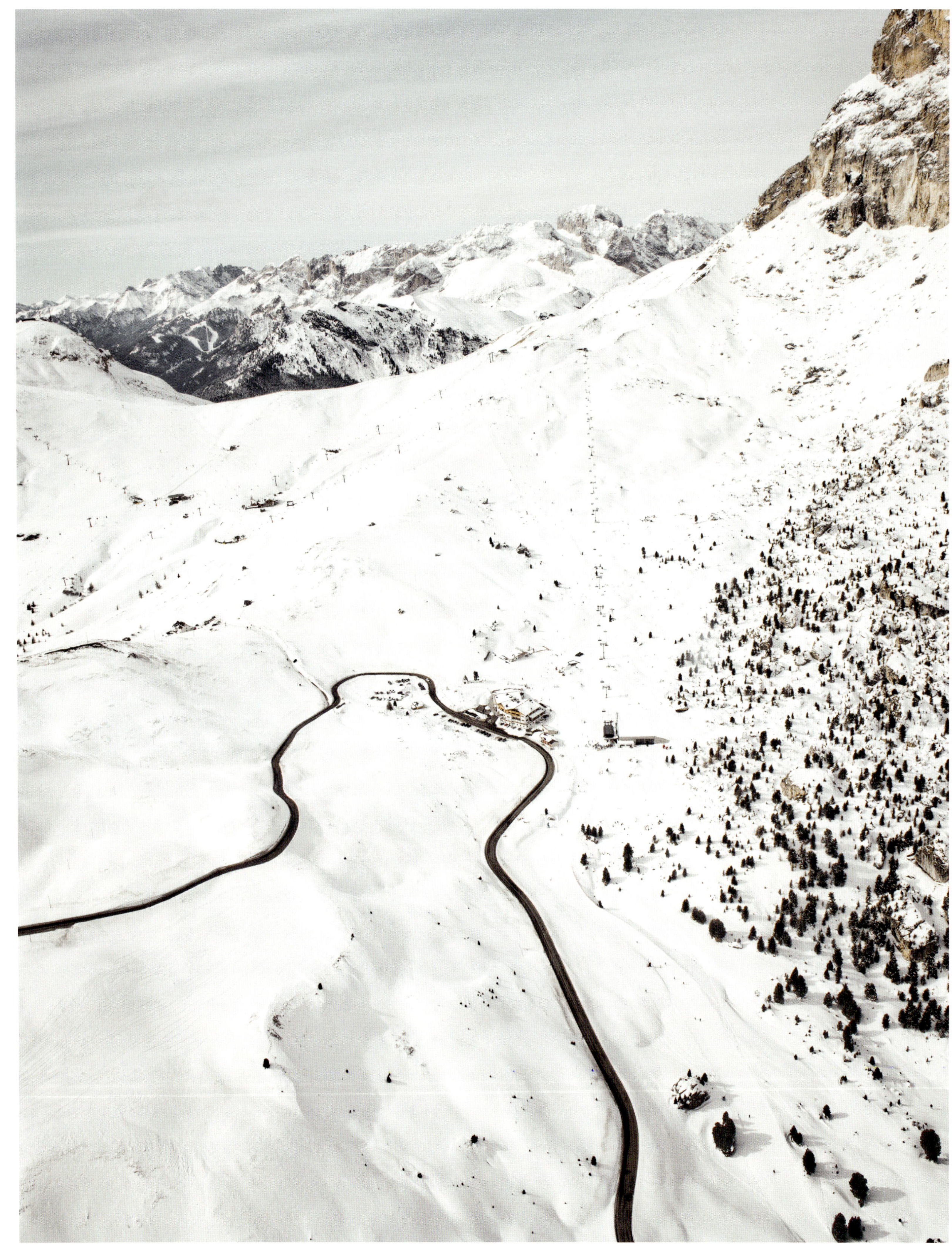

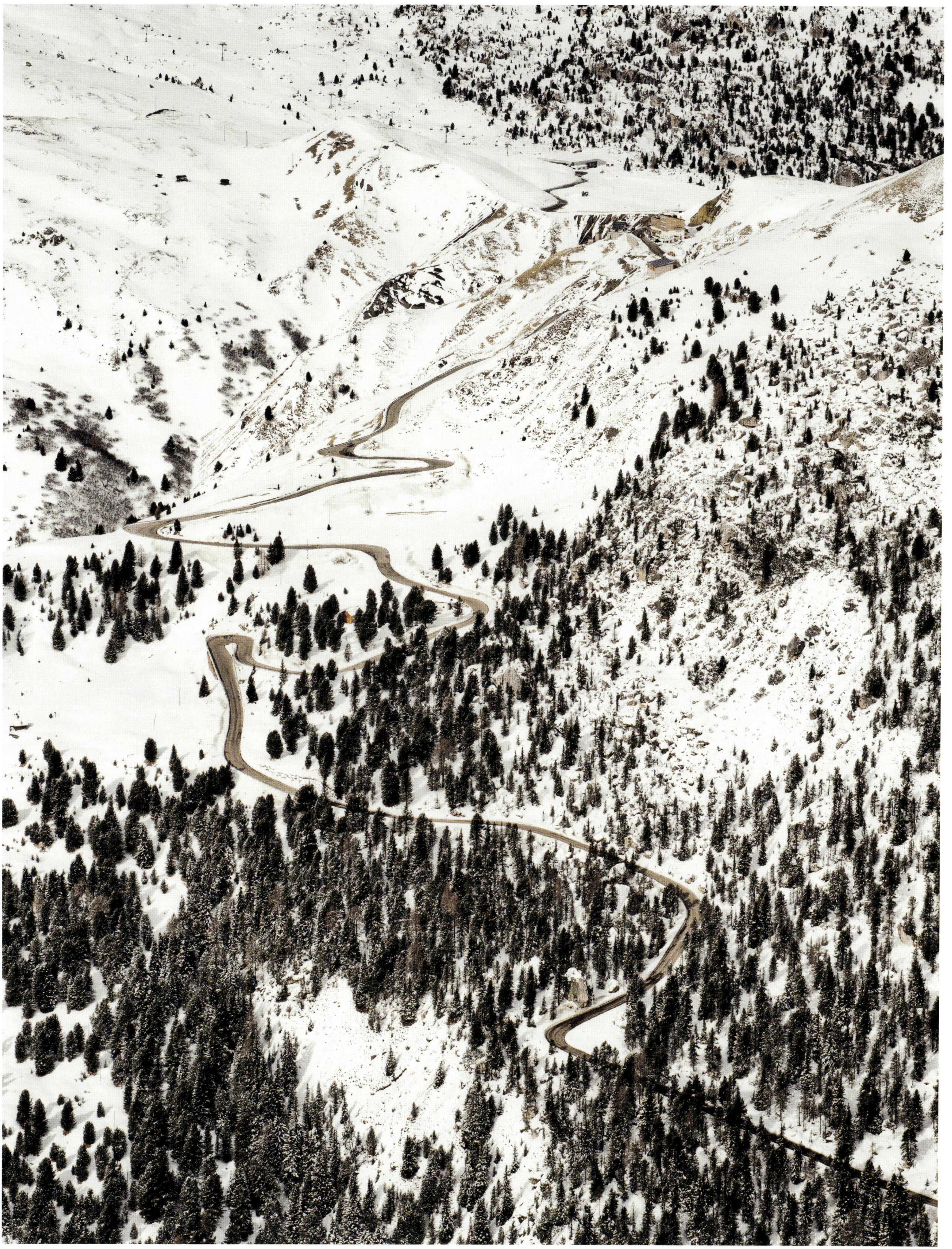

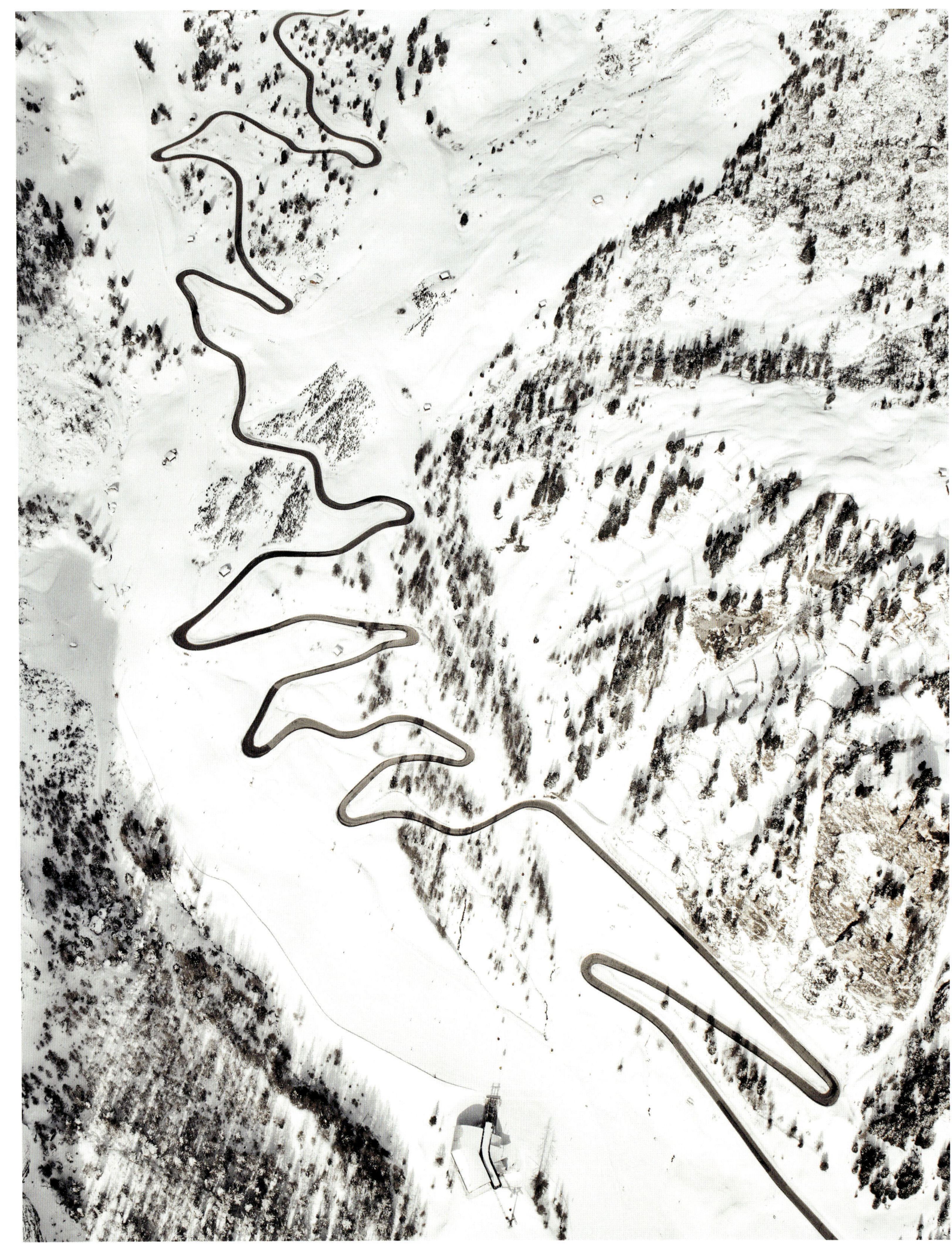

Passo Gardena

— 2121 Meter
Meist ganzjährig geöffnet

Wie viele Straßen und Bauwerke Südtirols geht auch die Straße zum Passo Gardena, dem Grödner Joch, auf die Stellungskämpfe während des Ersten Weltkriegs zurück, die das Gesicht der Dolomiten nachhaltig gezeichnet haben. Die Fahrt von Gröden bei Wolkenstein hinauf zur Passhöhe auf 2121 Meter ist spektakulär: Nirgends sonst auf der „Sellaronda" kommt man den Felswänden des Sellastockes so nah. Auch der Blick vom Joch auf die mächtigen, weiß bepuderten Gipfel im Osten ist an klaren Wintertagen äußerst eindrucksvoll.

— 2,121 metres
Usually open all year round

Like many roads and buildings in South Tyrol, the road to Passo Gardena (Gardena Pass) goes back to the battles during the First World War, which have left deep scars on the face of the Dolomites. The drive from Selva Val Gardena up to the pass at 2,121 metres is spectacular. Nowhere else on the Sella Ronde can you get as close to the rock walls of the Sella Group as here. And on a clear winter's day, the outlook from the summit to the mighty, white-powdered peak in the East is absolutely magnificent.

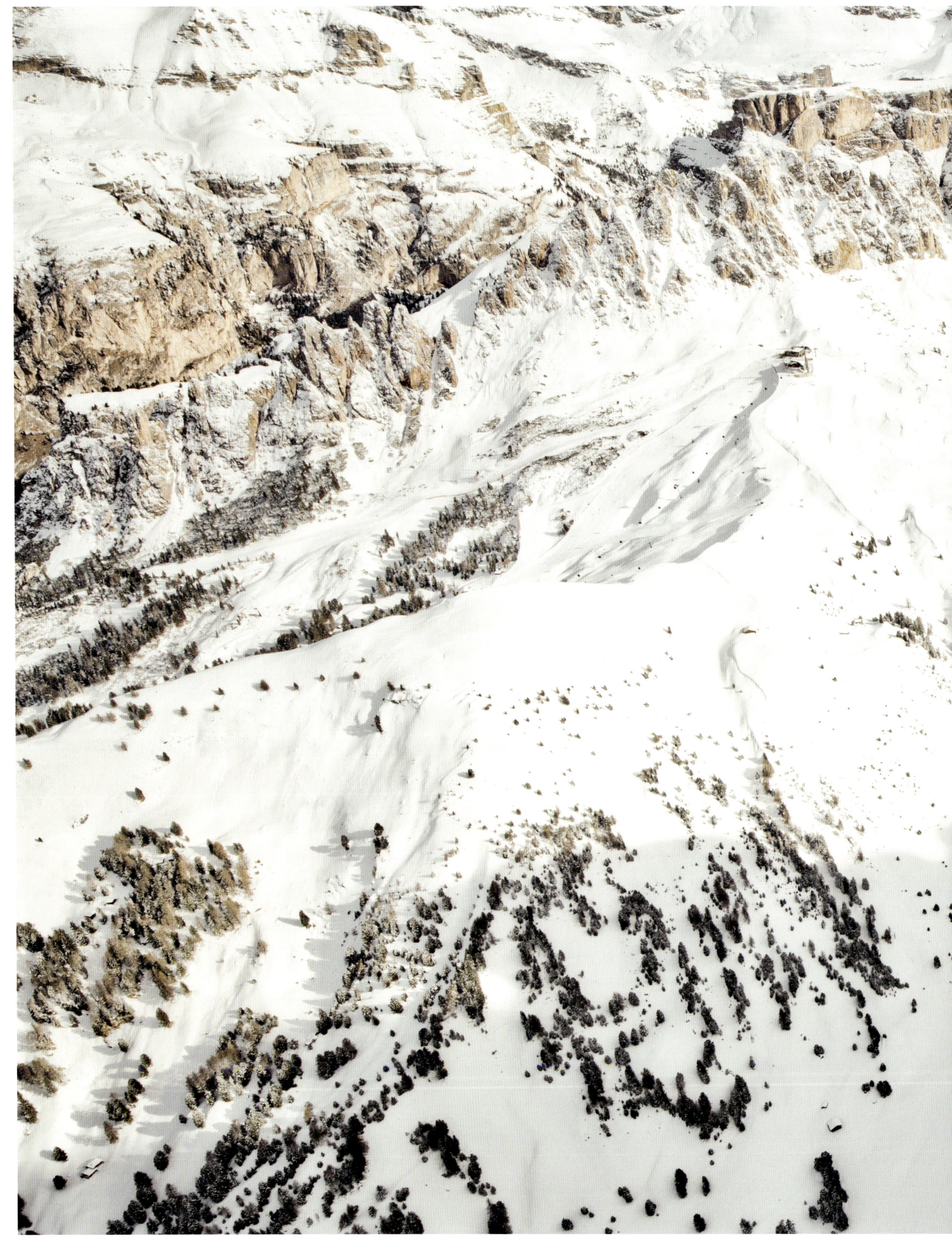

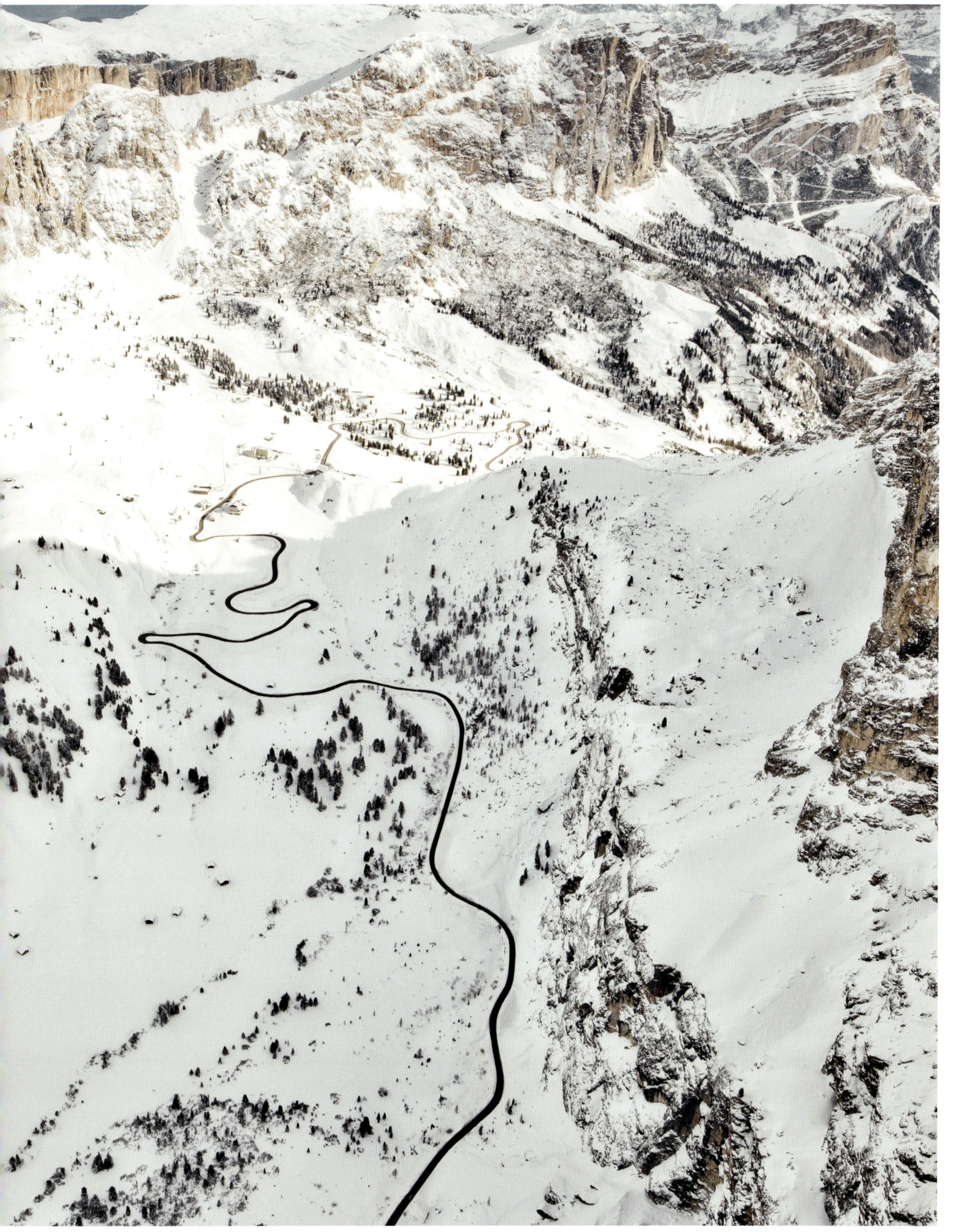

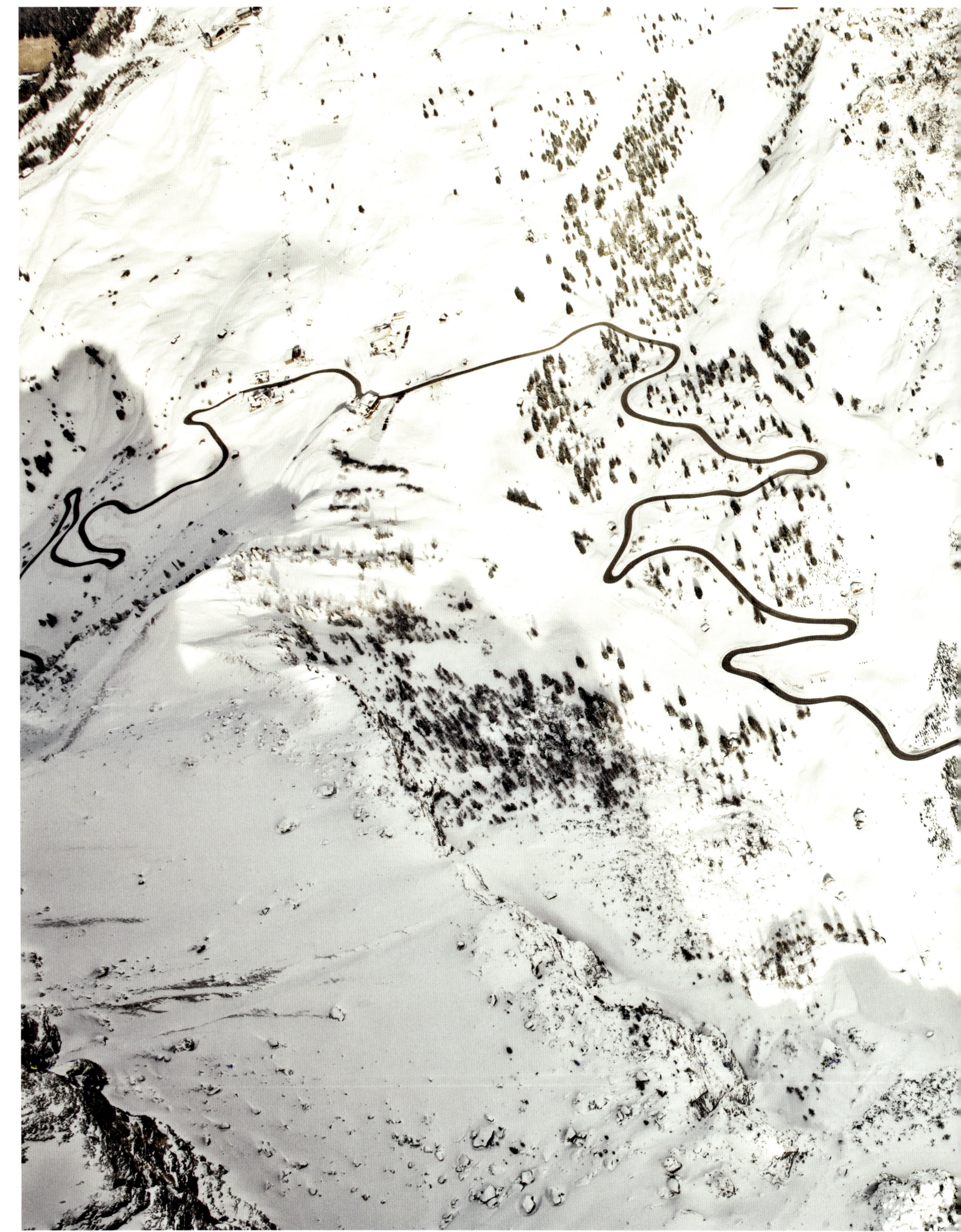

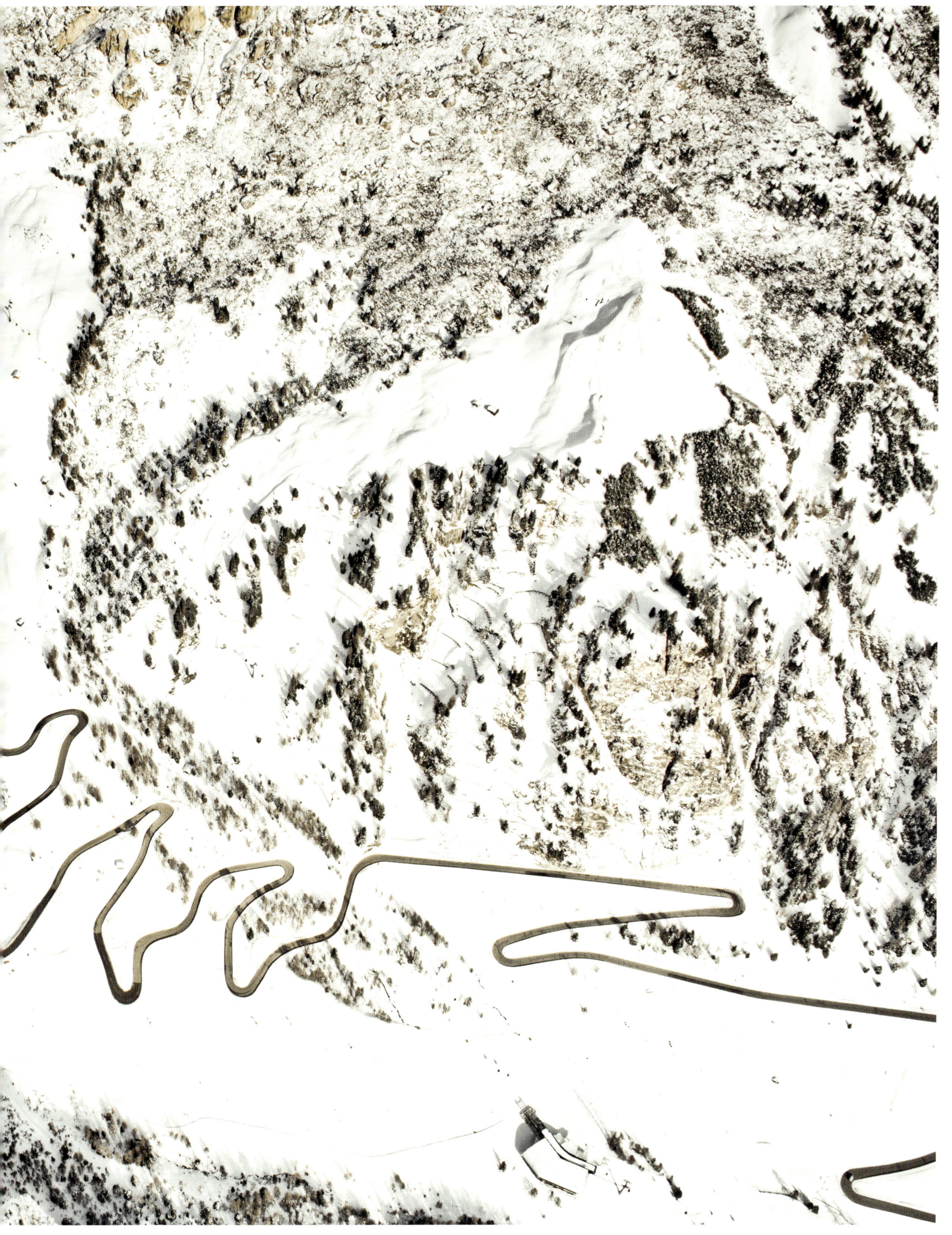

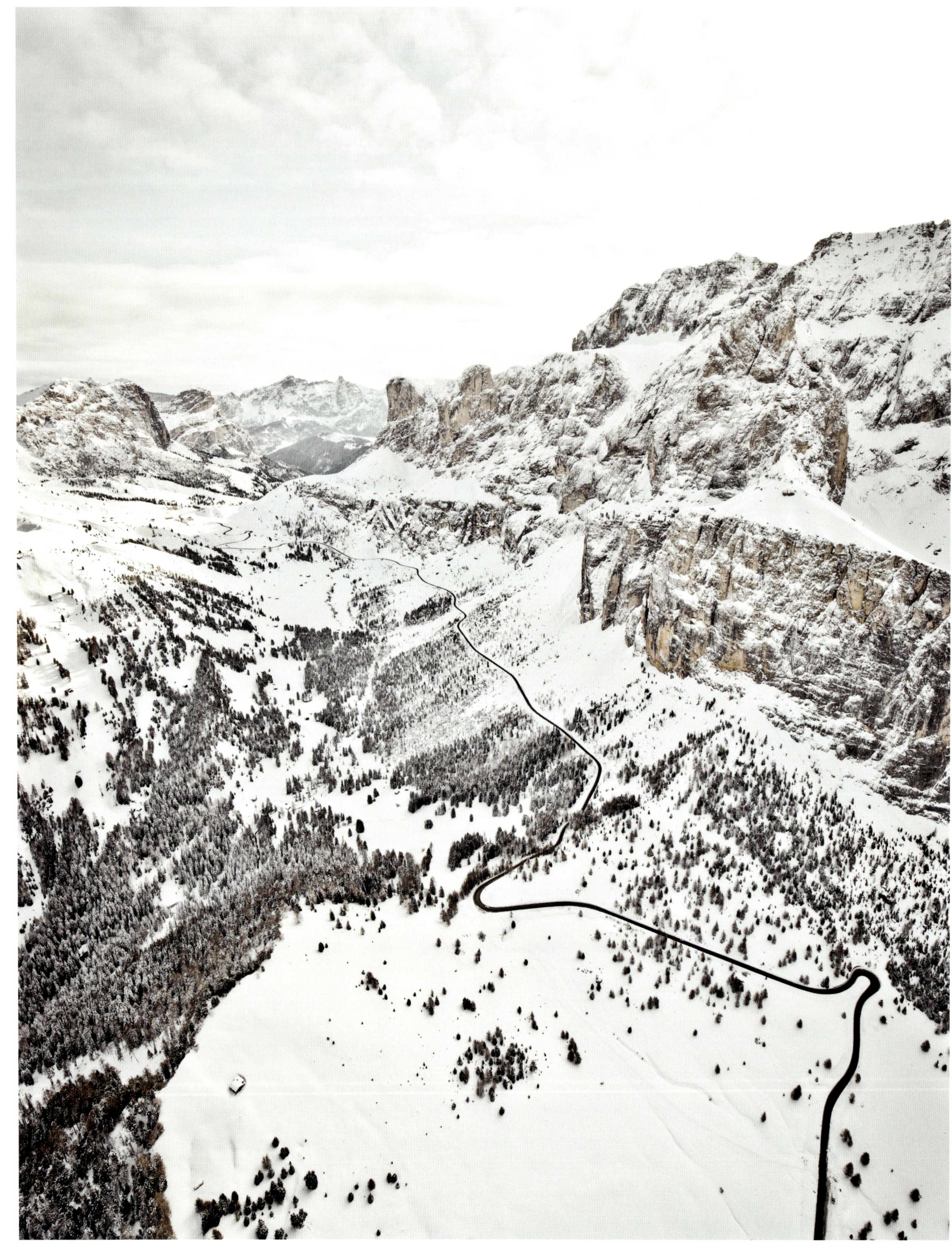

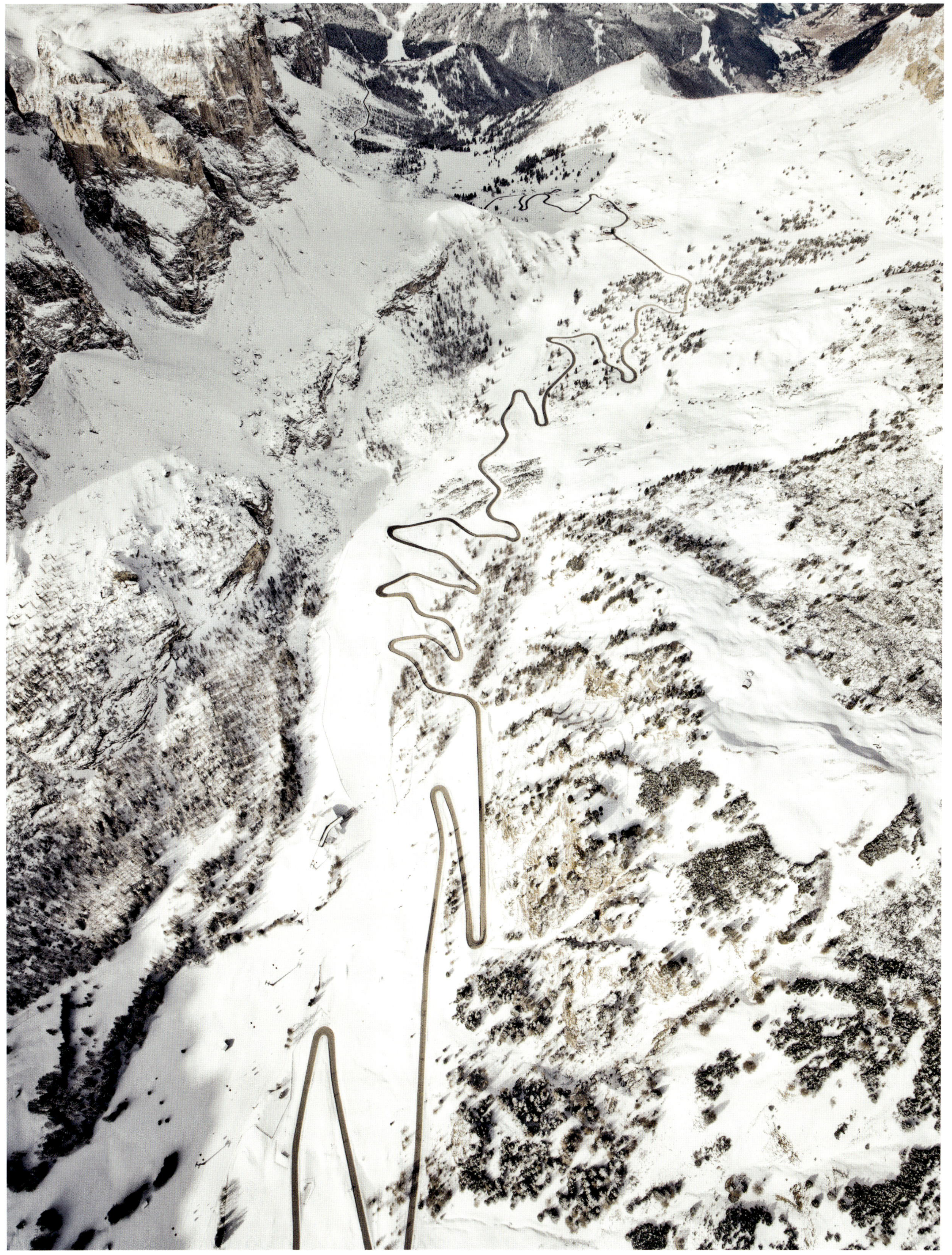

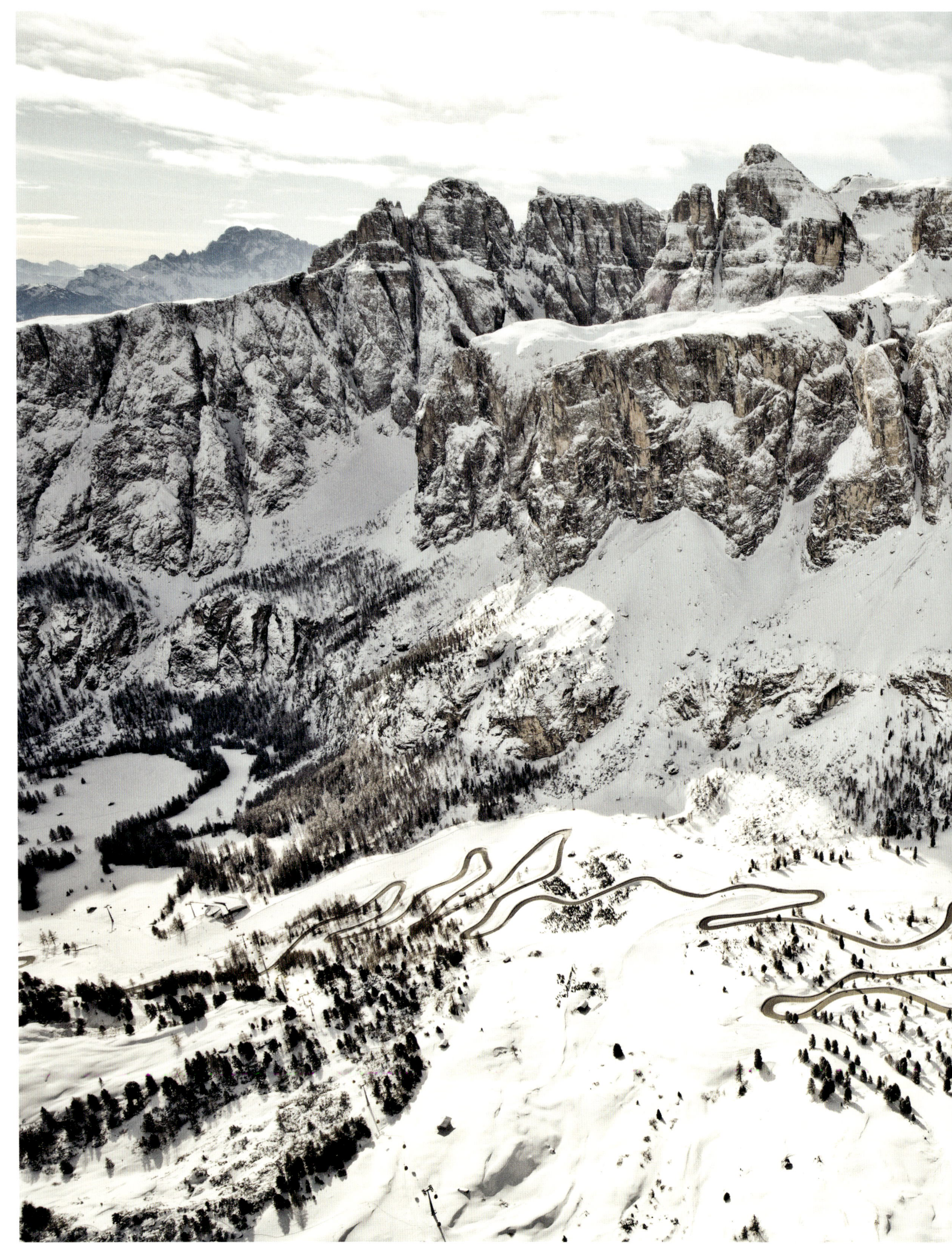

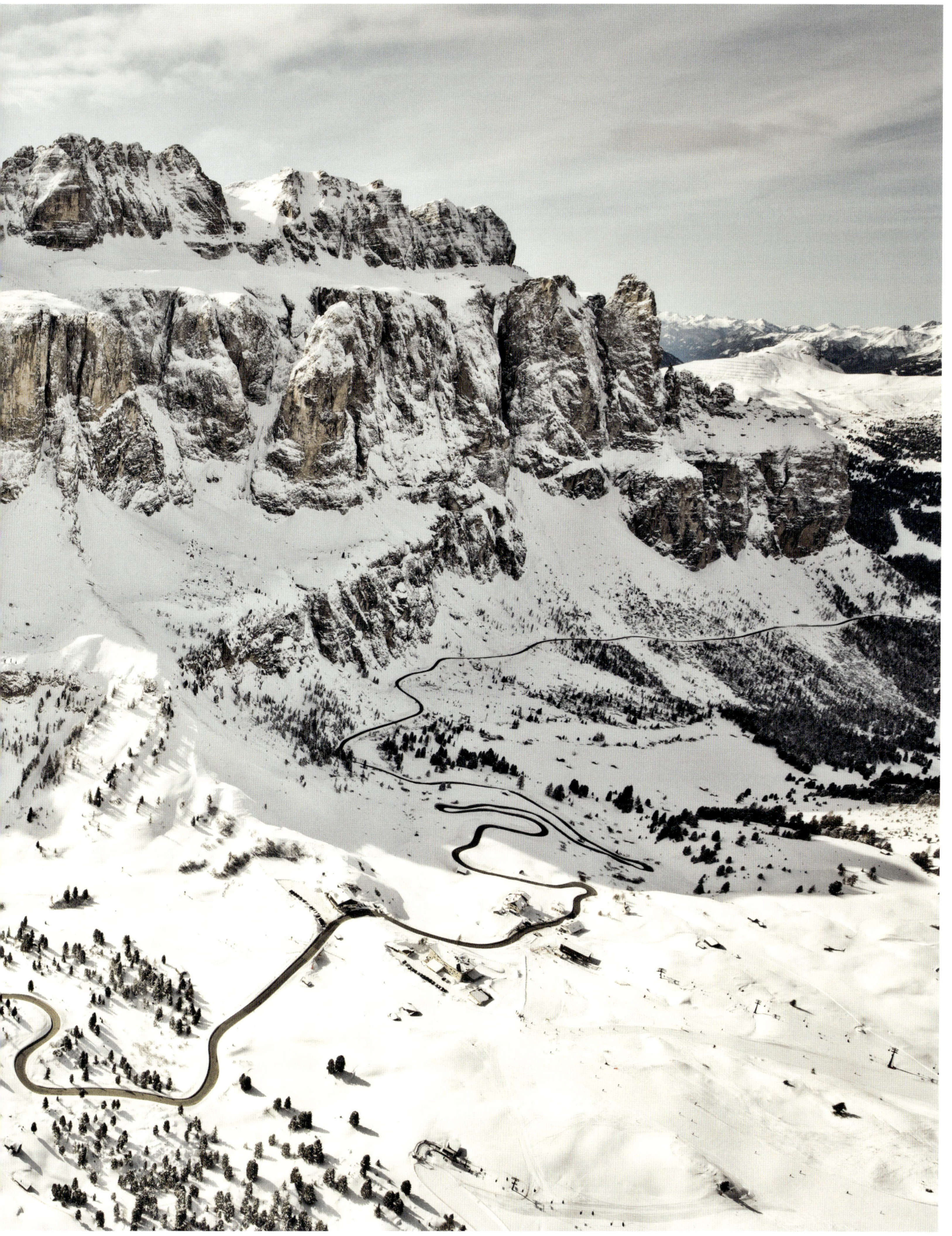

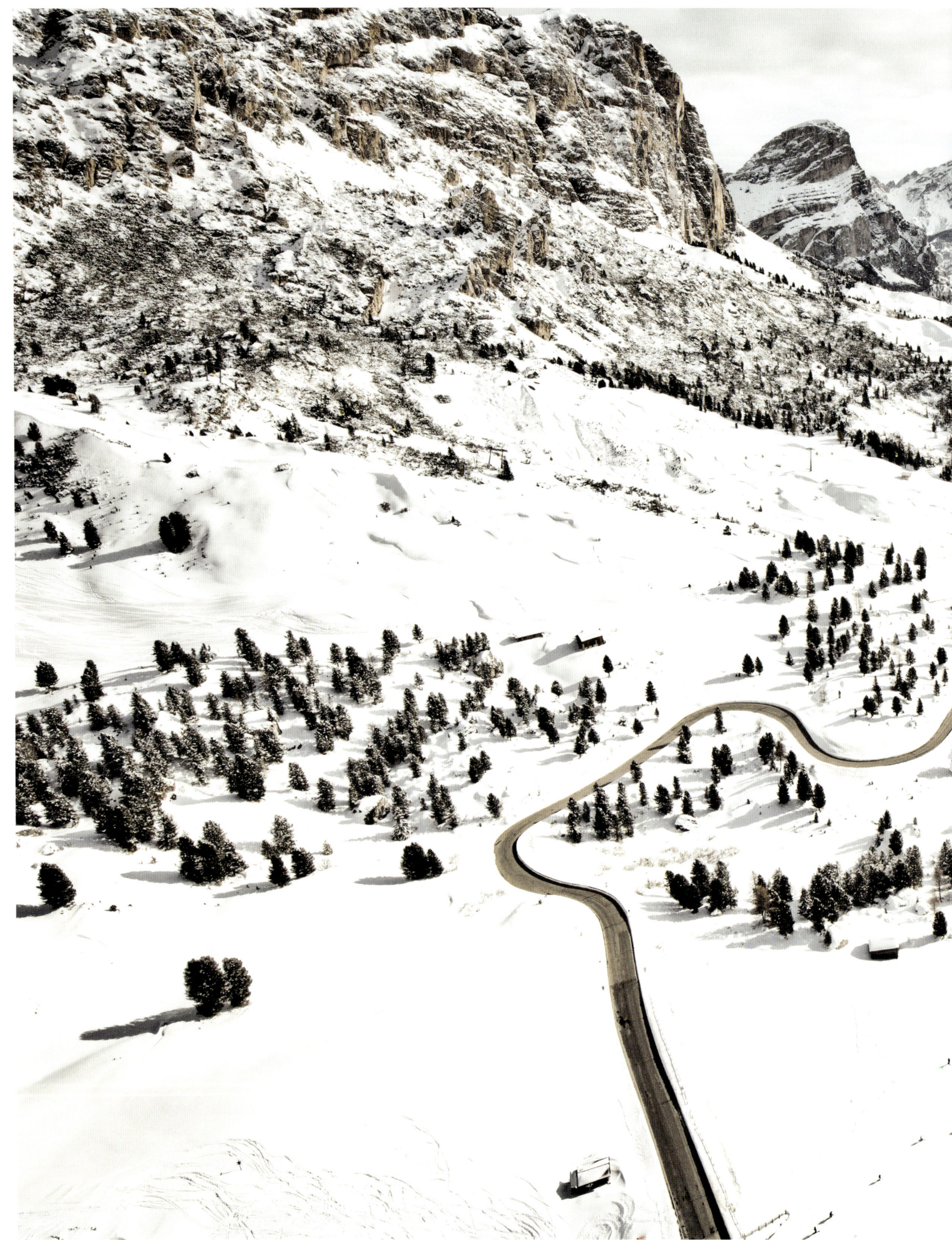

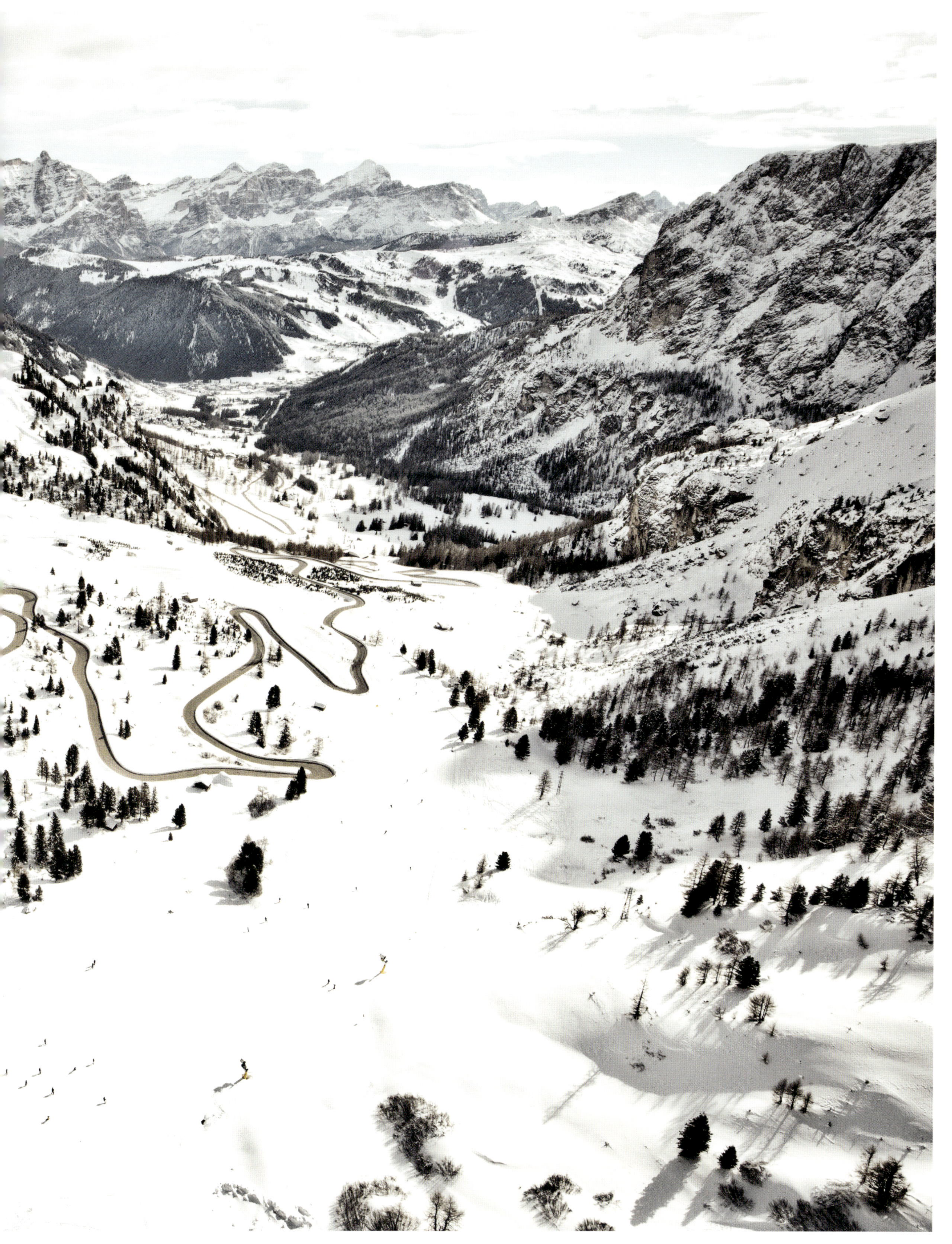

Passo dello Stelvio

— 2757 Meter
Wintersperre meist von November bis Juni

Der Passo dello Stelvio (das Stilfser Joch) verbindet das Vinschgau mit dem Veltlin und gilt als eine der kühnsten und kurvenreichsten Passstraßen der Alpen. Die 48 schwindelerregenden Serpentinen der Nordostrampe gehören zu den eindrucksvollsten Bauten des Österreichischen Kaiserreiches – und ziehen bis heute besonders sportliche Gipfelstürmer in ihren Bann. Da die Straße im Winter unmöglich freizuhalten ist, hat das hochgelegene Skigebiet am Stilfser Joch übrigens nur im Sommer und Herbst geöffnet. Dort kann man nach erfolgreicher Kurvenjagd auf 3450 Meter mit den Weltcup-Fahrern um die Wette wedeln.

— 2,757 metres
Usually closed in winter from November to June

The Passo dello Stelvio (Stelvio Pass) connects Vinschgau with Valtellina and is considered one of the most adventurous and twisty of the Alpine passes. The 48 vertiginous serpentines of the north-east ramp, built by the Austrian Empire, are some of the most impressive constructions, and still lure the staunchest of mountain climbers. Because the road is impossible to keep clear in winter, the high-altitude ski area at the Stelvio Pass is open only in summer and autumn. There, at 3,450 metres, you can pit your carving skills on the slopes against some of the world's best.

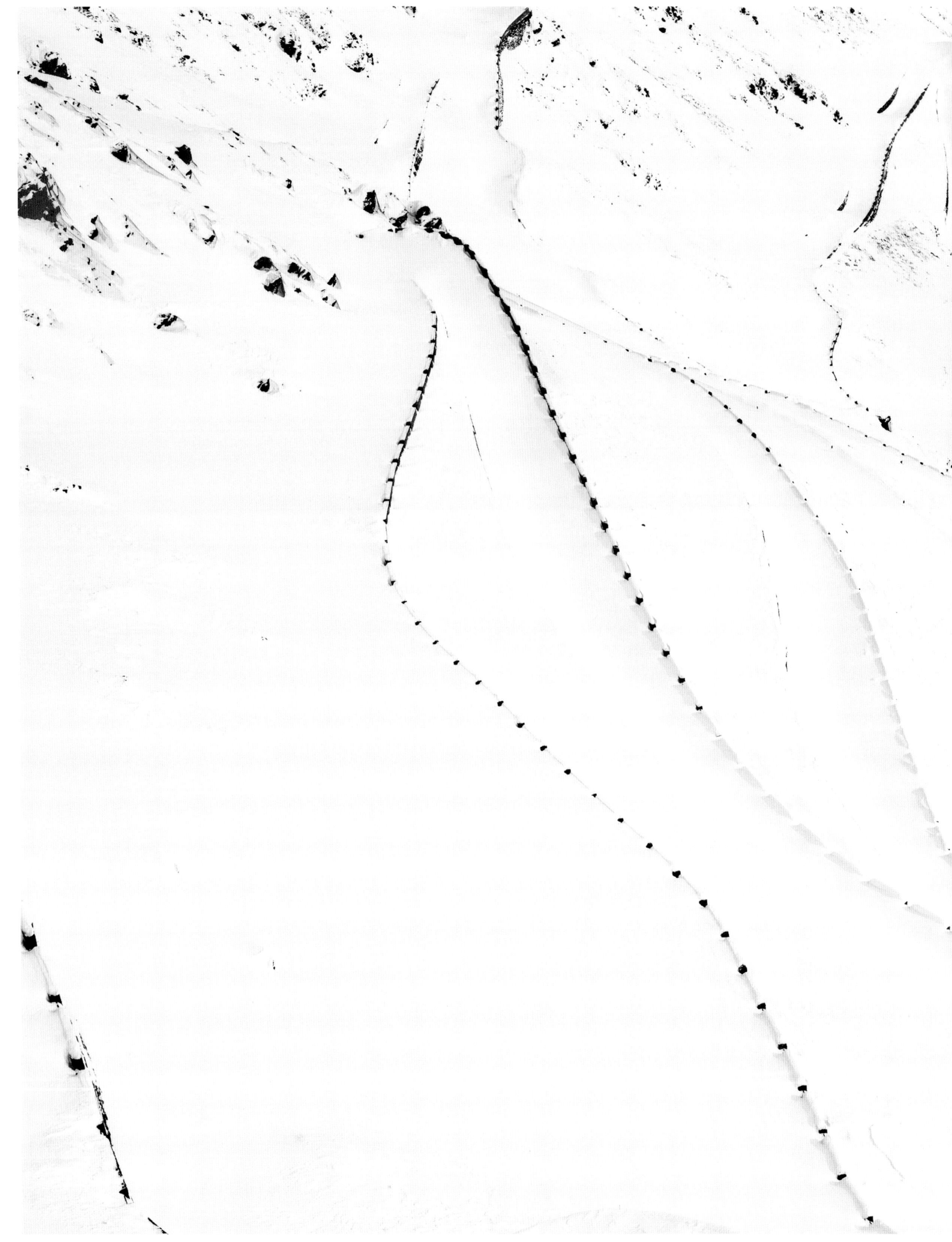

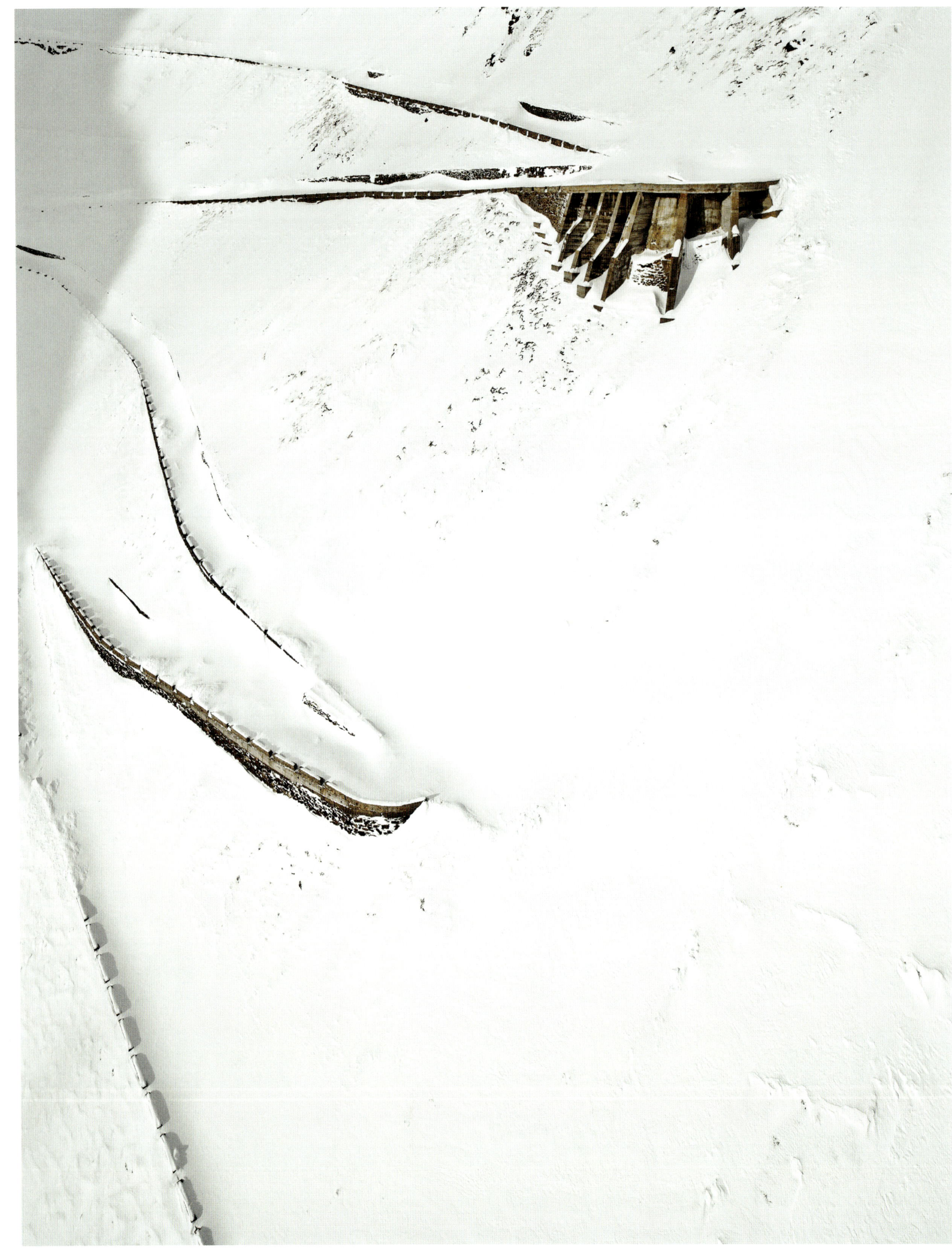

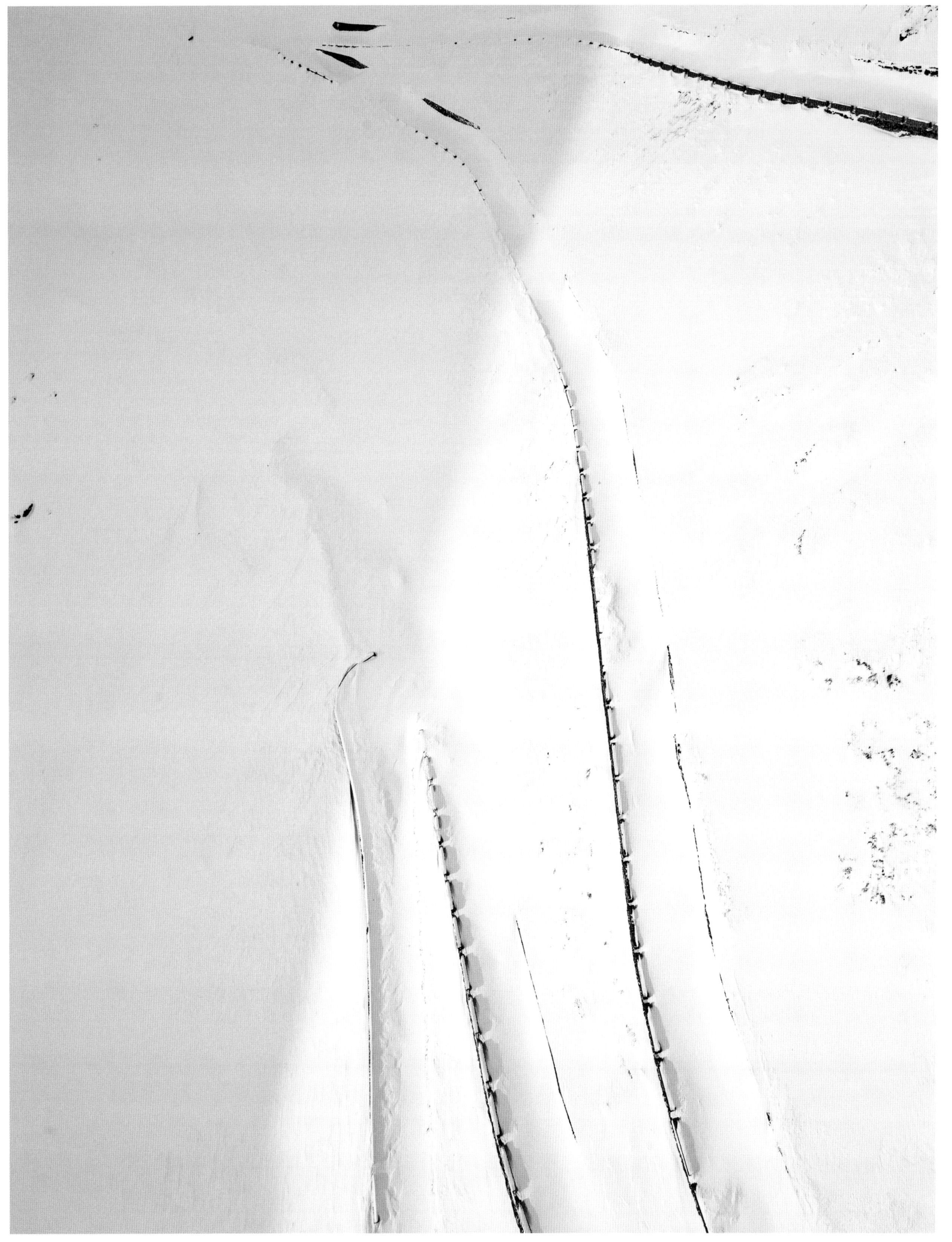

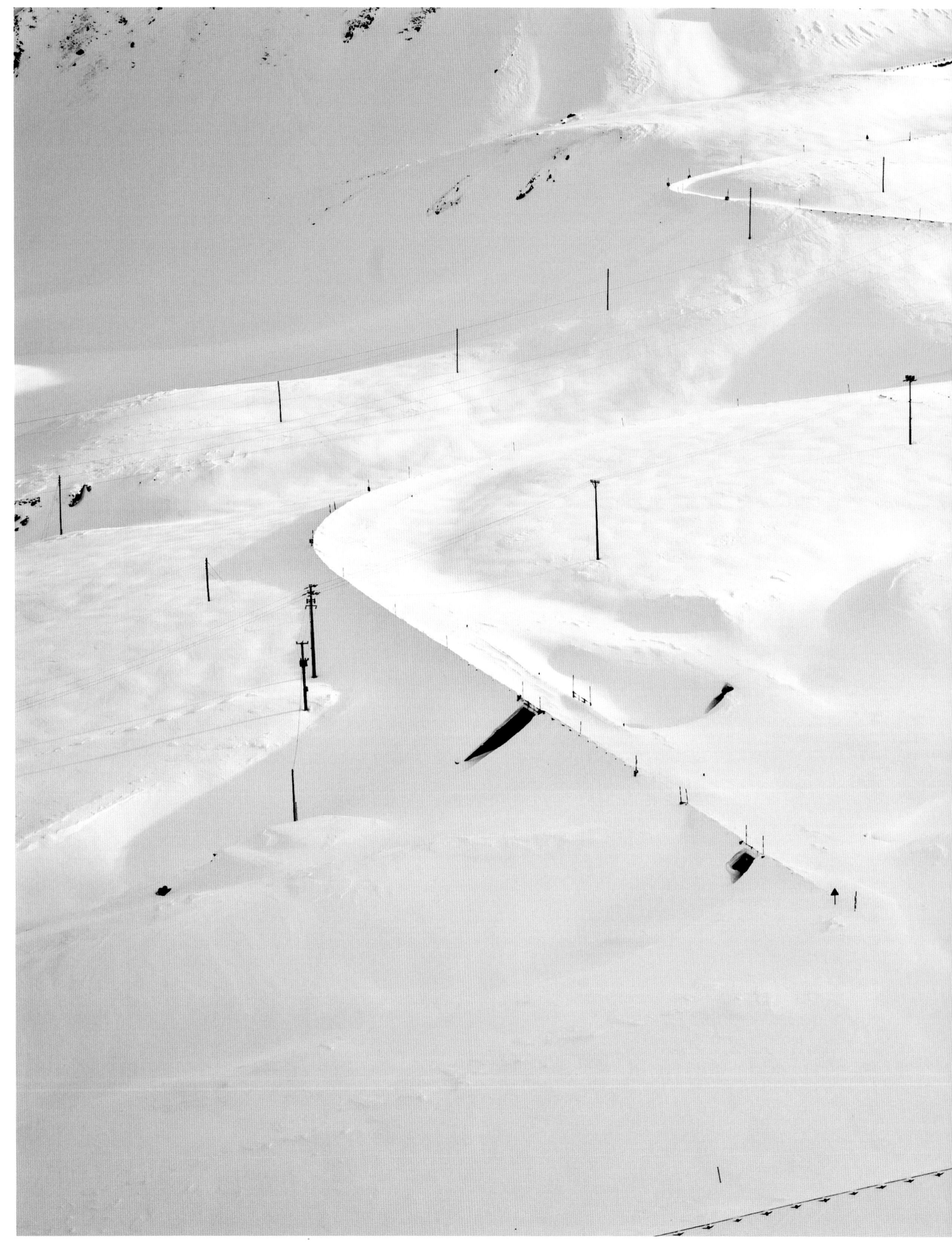

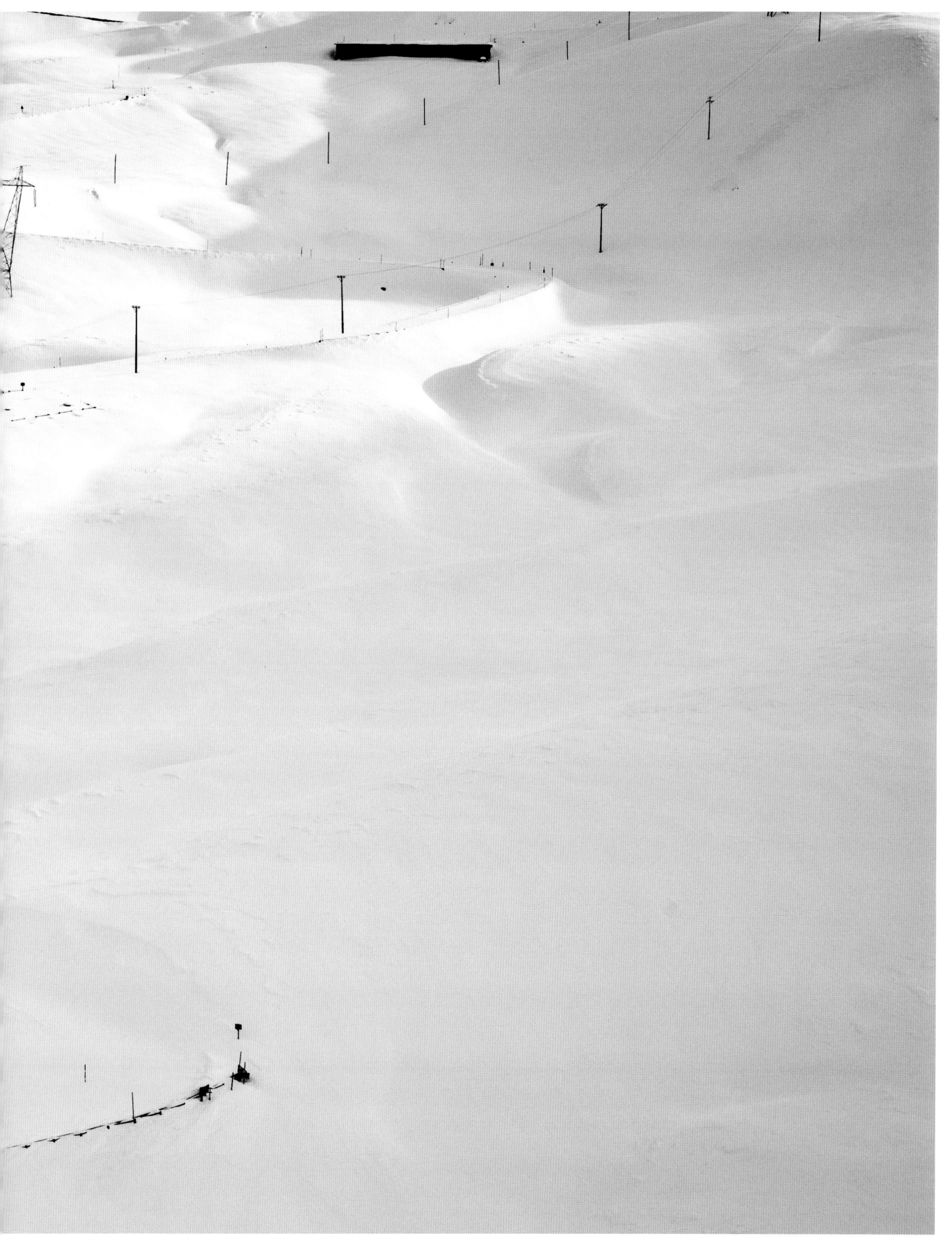

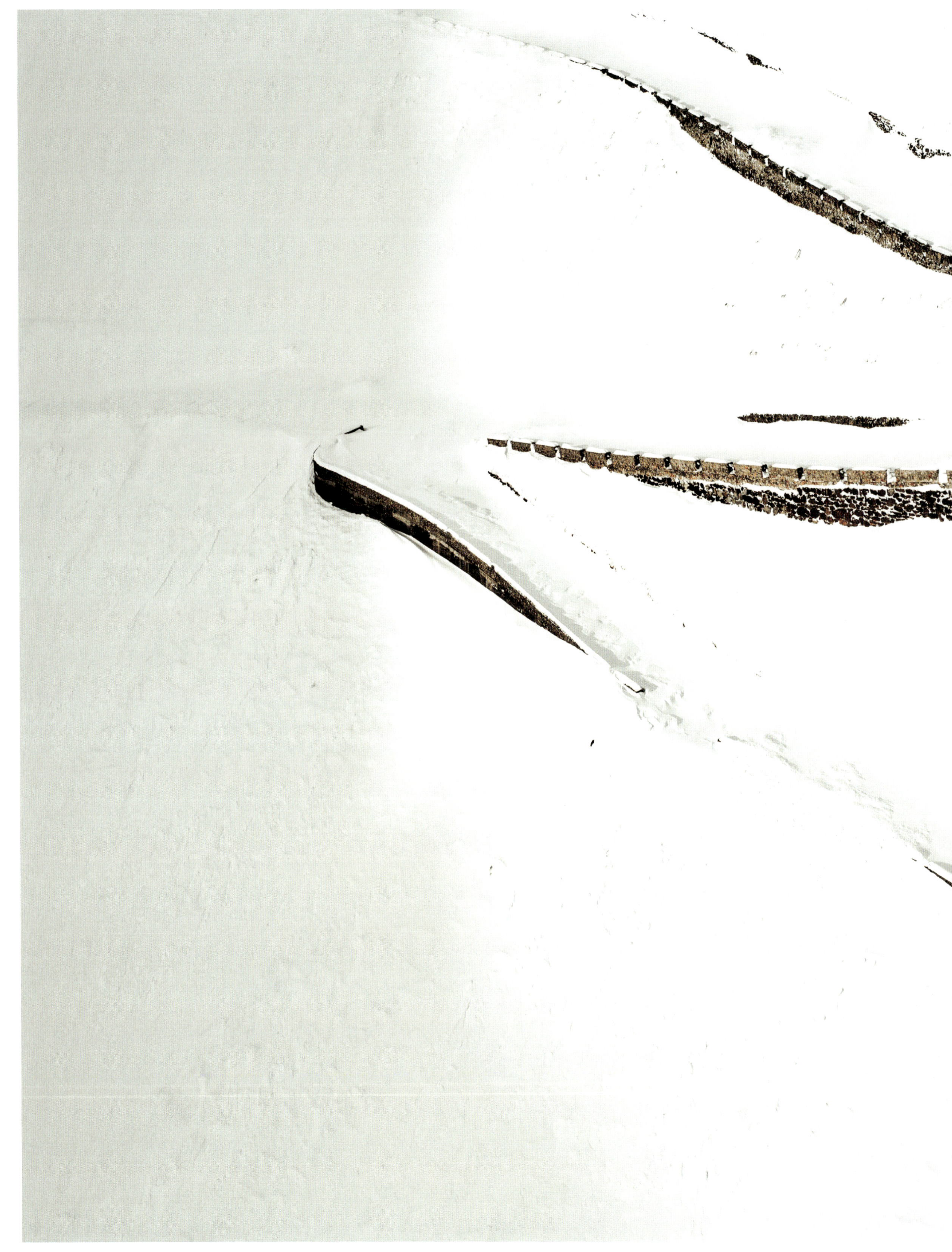

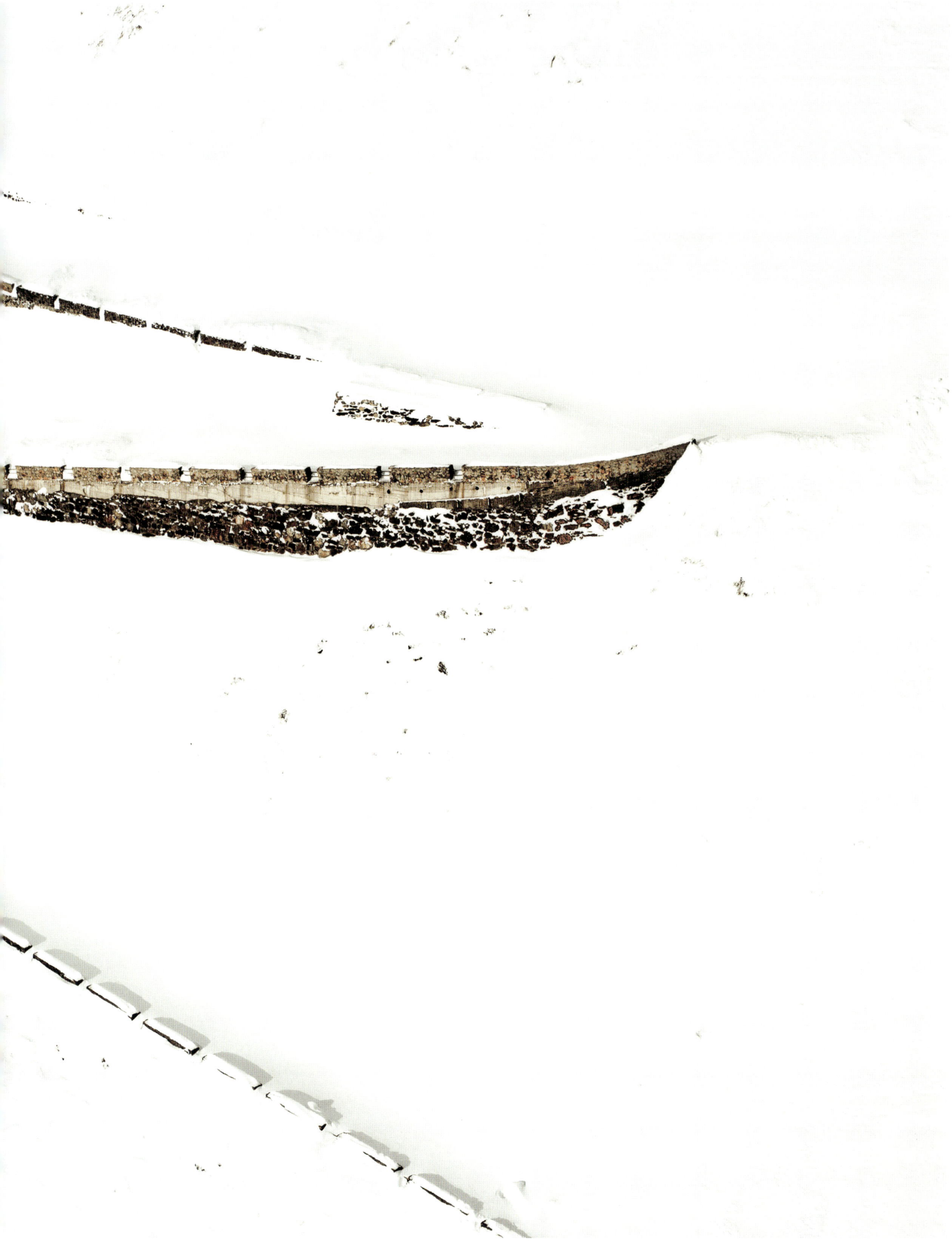

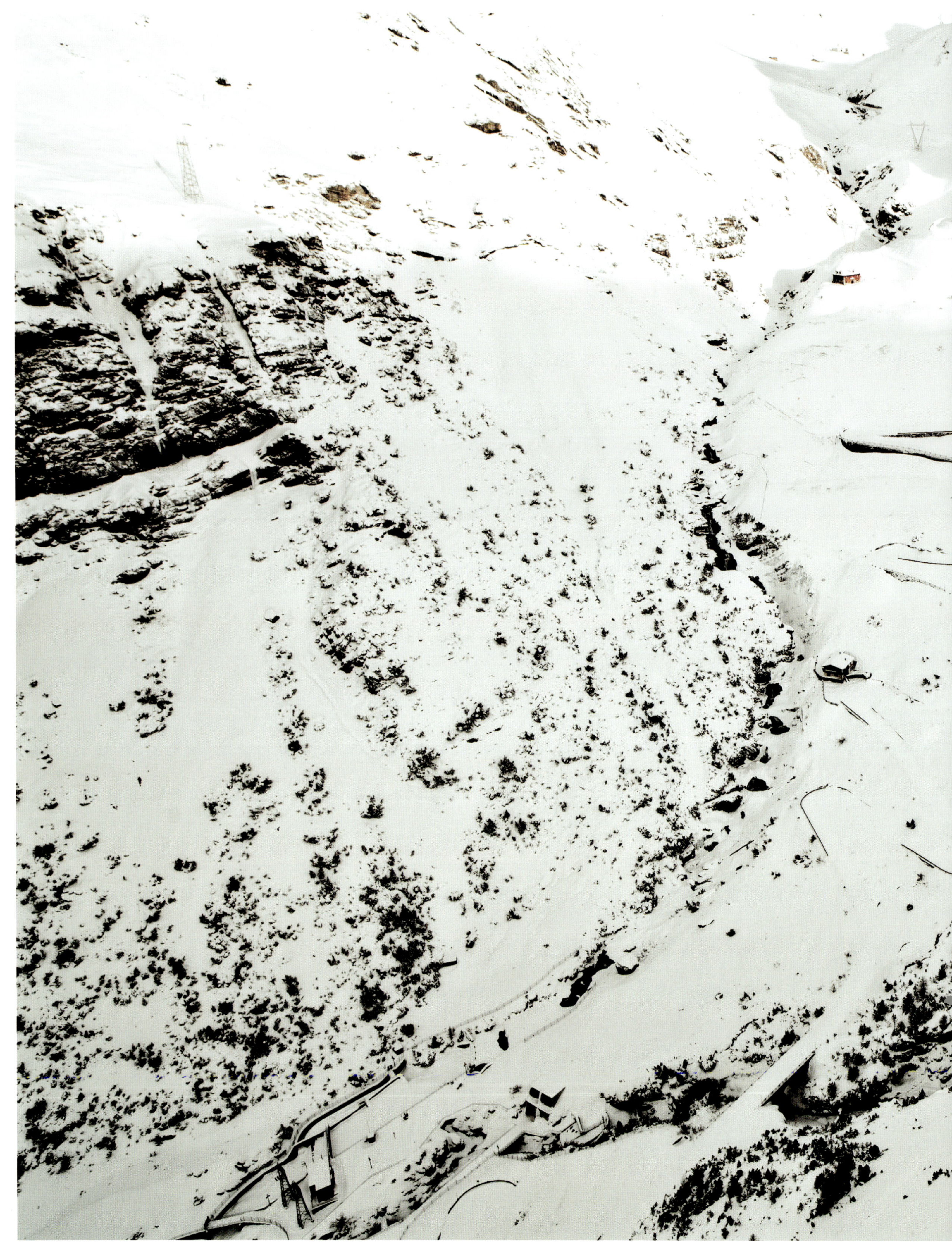

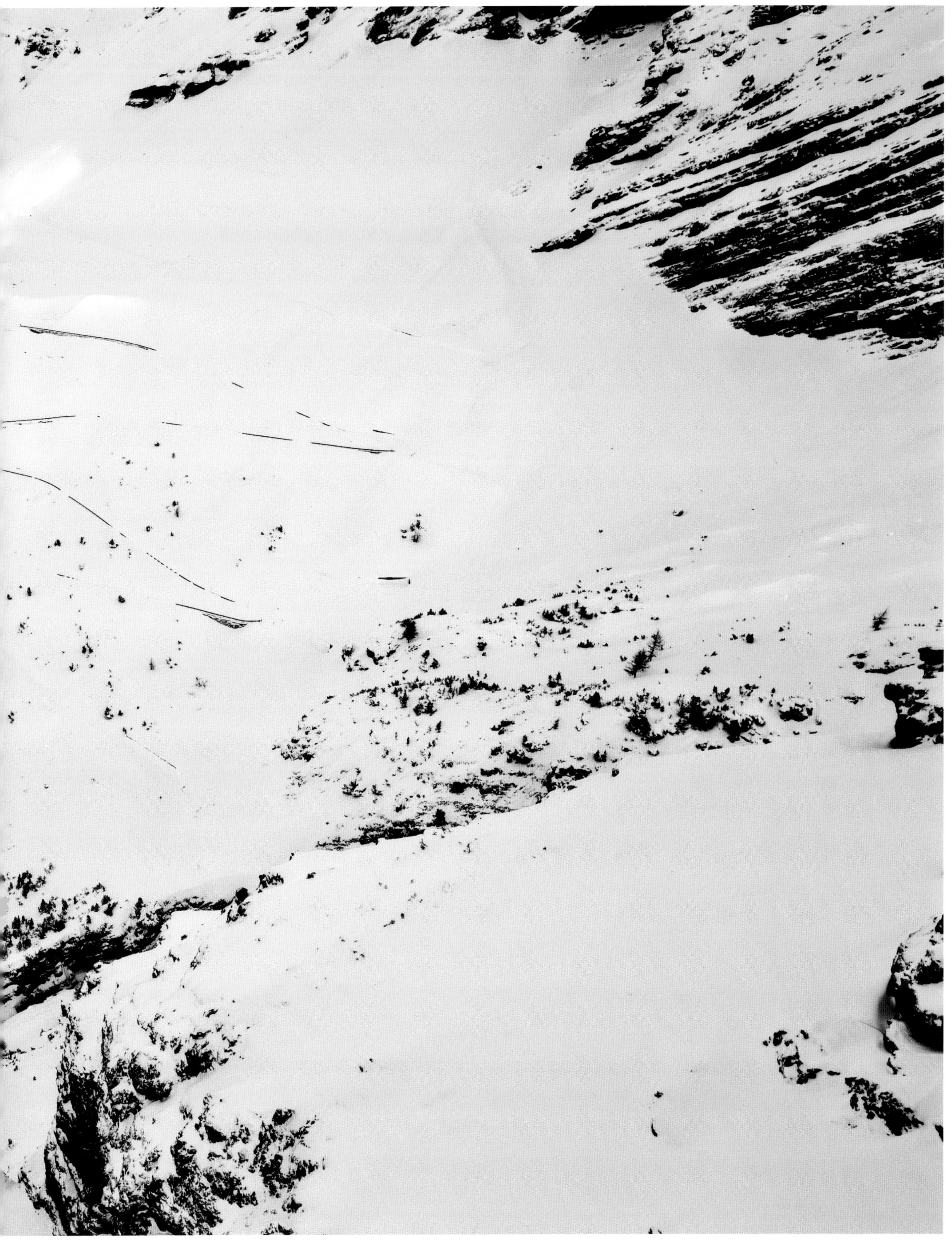

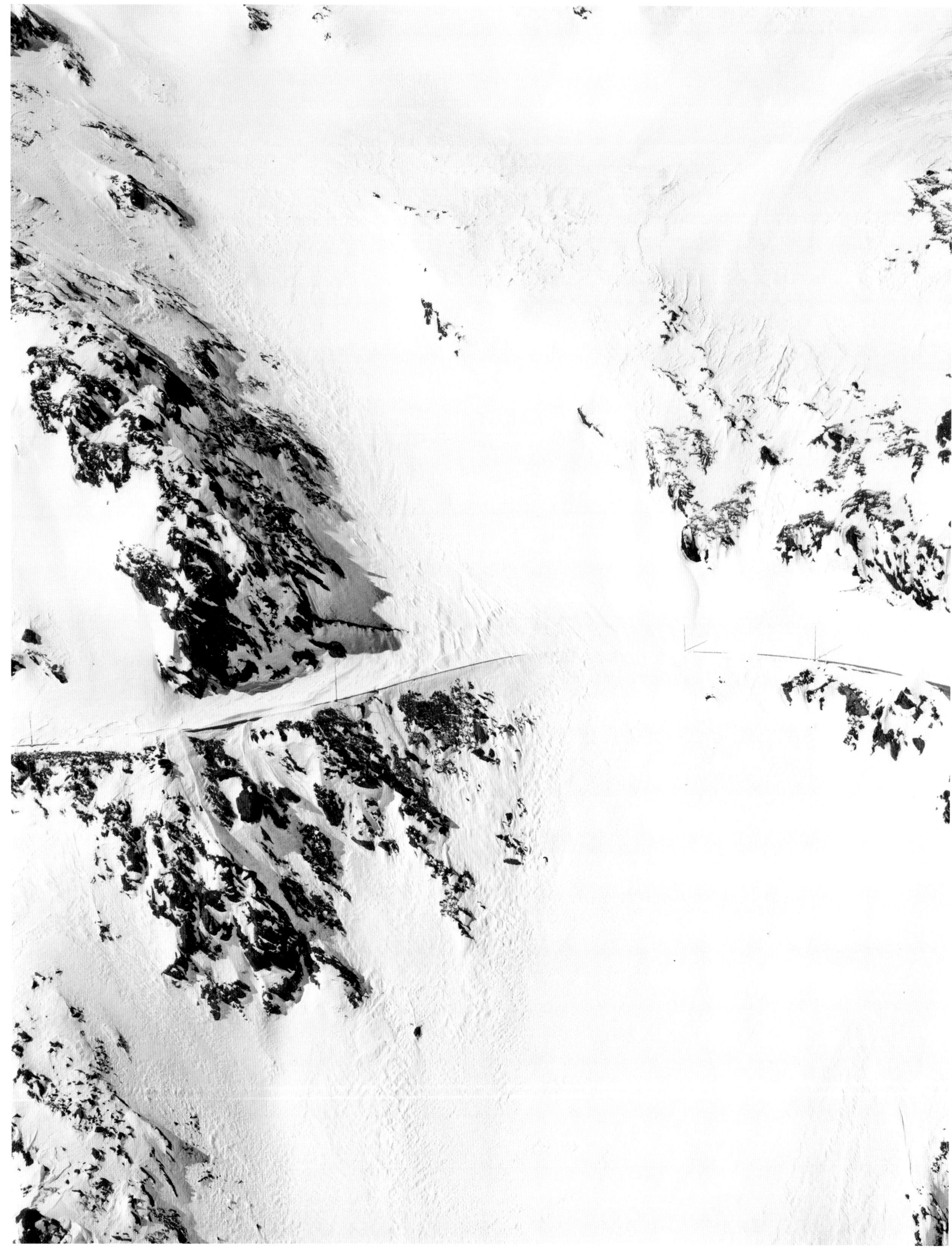

Timmelsjoch-Hochalpenstraße

— 2509 Meter
Wintersperre meist zwischen Oktober und Juni

Die Timmelsjoch-Hochalpenstraße zwischen dem Ötztal im österreichischen Tirol und Passeier im italienischen Südtirol ist mit einer Passhöhe von 2509 Meter die höchstgelegene Straße der Ostalpen. Vor allem für ambitionierte Radfahrer sind die fast 1800 Höhenmeter und mehr als 60 Kurven über das Timmelsjoch eine der großen Herausforderungen. Seinen italienischen Namen „Passo del Rombo" – zu Deutsch etwa „Pass des Dröhnens" – könnte das Timmelsjoch jedoch eher den motorisierten Besuchern verdanken, die von Mitte Juni bis Mitte Oktober den steilen Berghang hinaufbrausen. Im Winter hingegen ist von den sportlich geschwungenen Serpentinen kaum mehr zu sehen als ein paar Spuren im Schnee.

— 2,509 metres
Usually closed in winter between October and June

The Timmelsjoch High Alpine Road between the Ötztal in Tyrol/Austria and Passeier valley in South Tyrol/Northern Italy, is the highest road in the Eastern Alps at an altitude of 2,509 metres. Particularly for ambitious cyclists, the route via 60 corners up to the Timmelsjoch at 1,800 metres is one of the greatest challenges. Its Italian name "Passo del Rombo" – meaning roughly "Rumbling Pass" – can be attributed to the motorists who roar up the steep flanks of the mountains from mid-June to mid-October. In winter, however, the only signs of the sweeping esses are tyre tracks left in the snow.

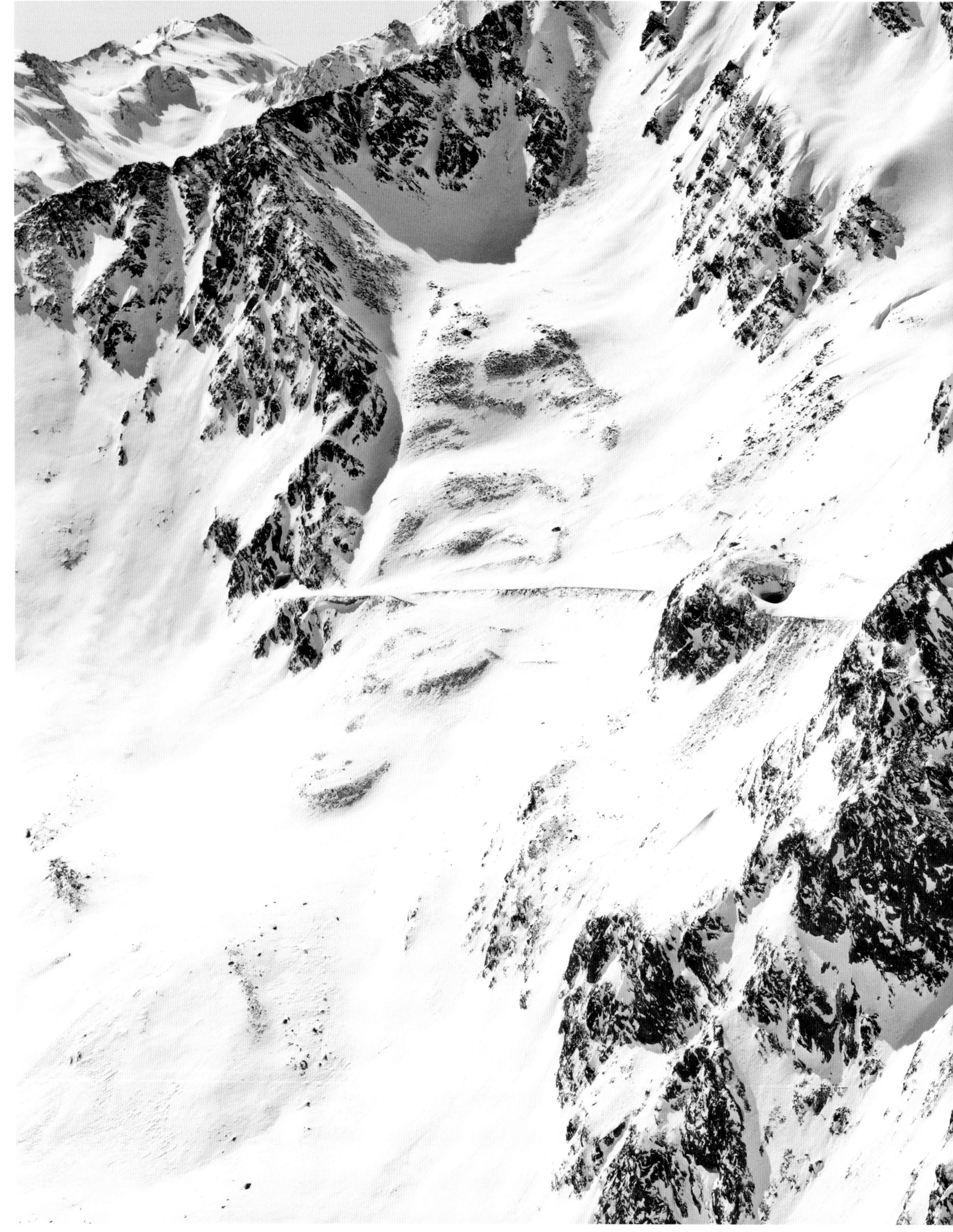

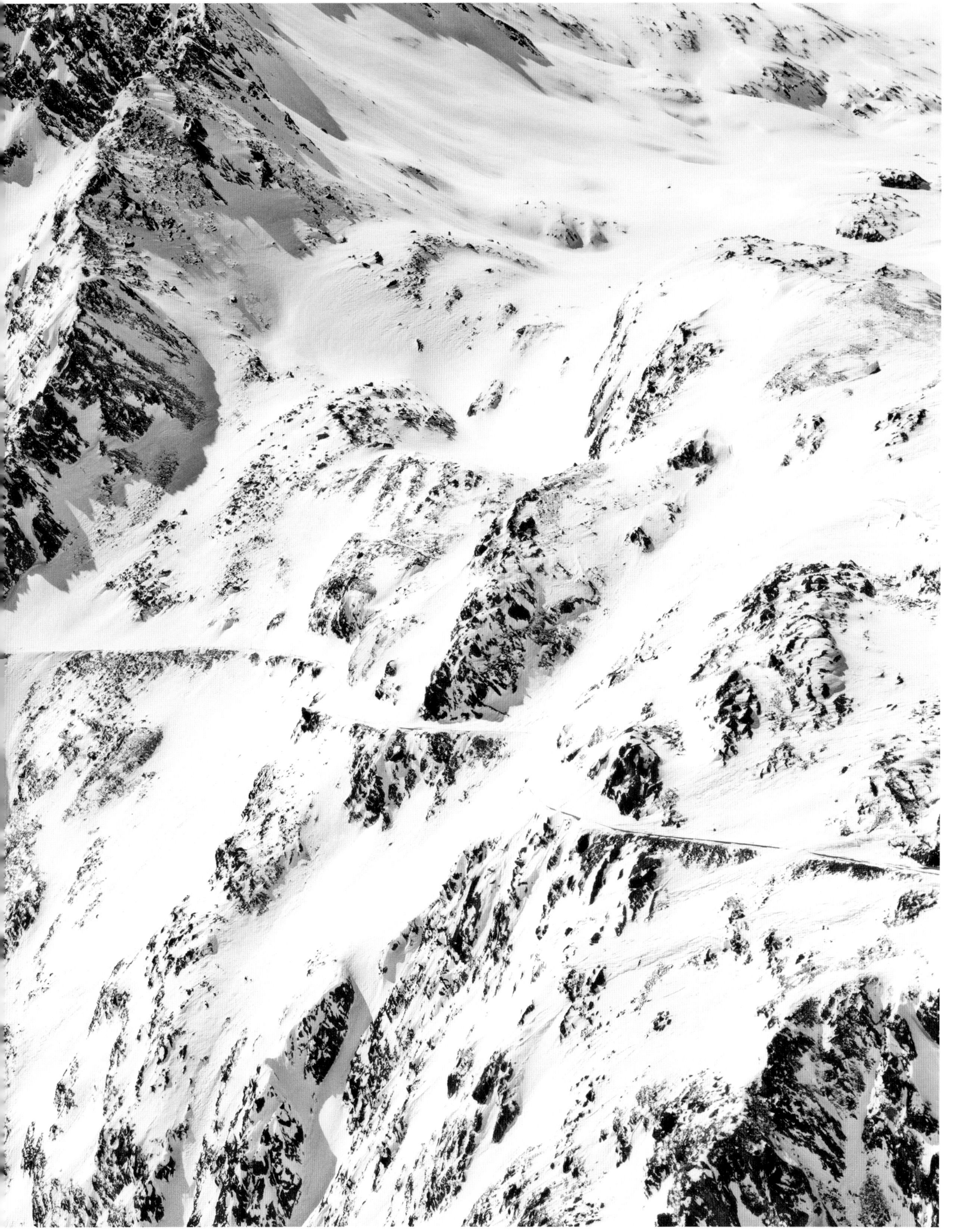

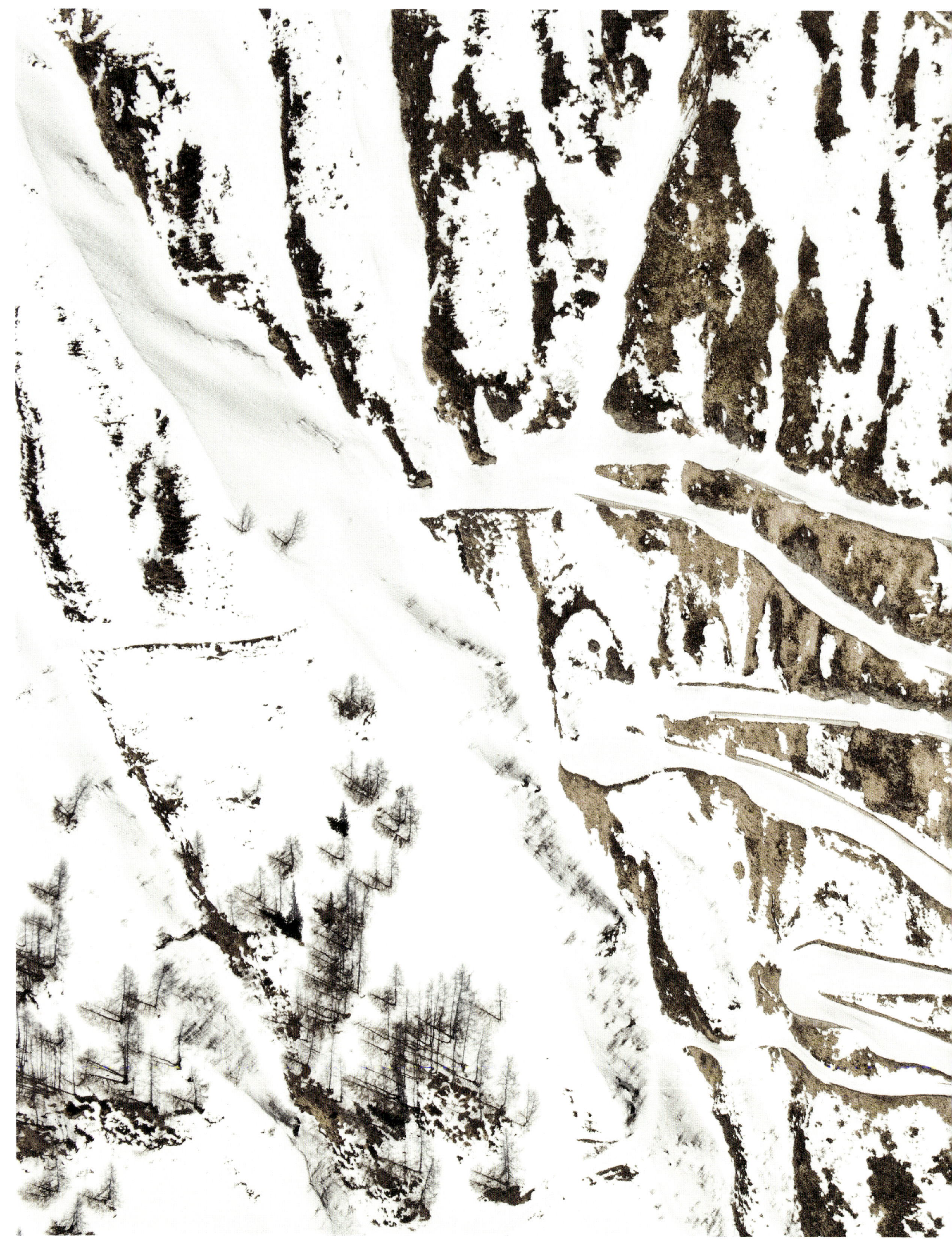

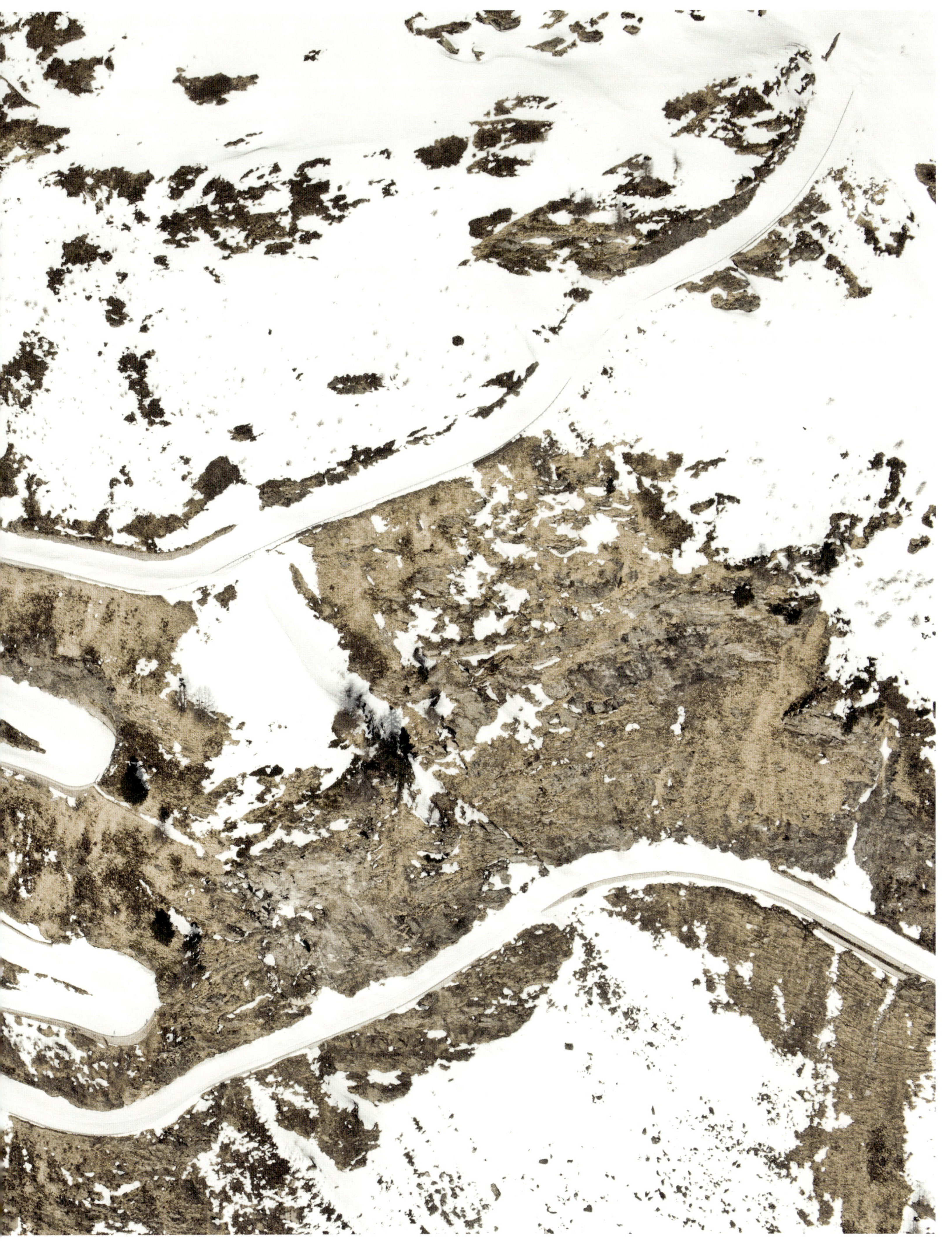

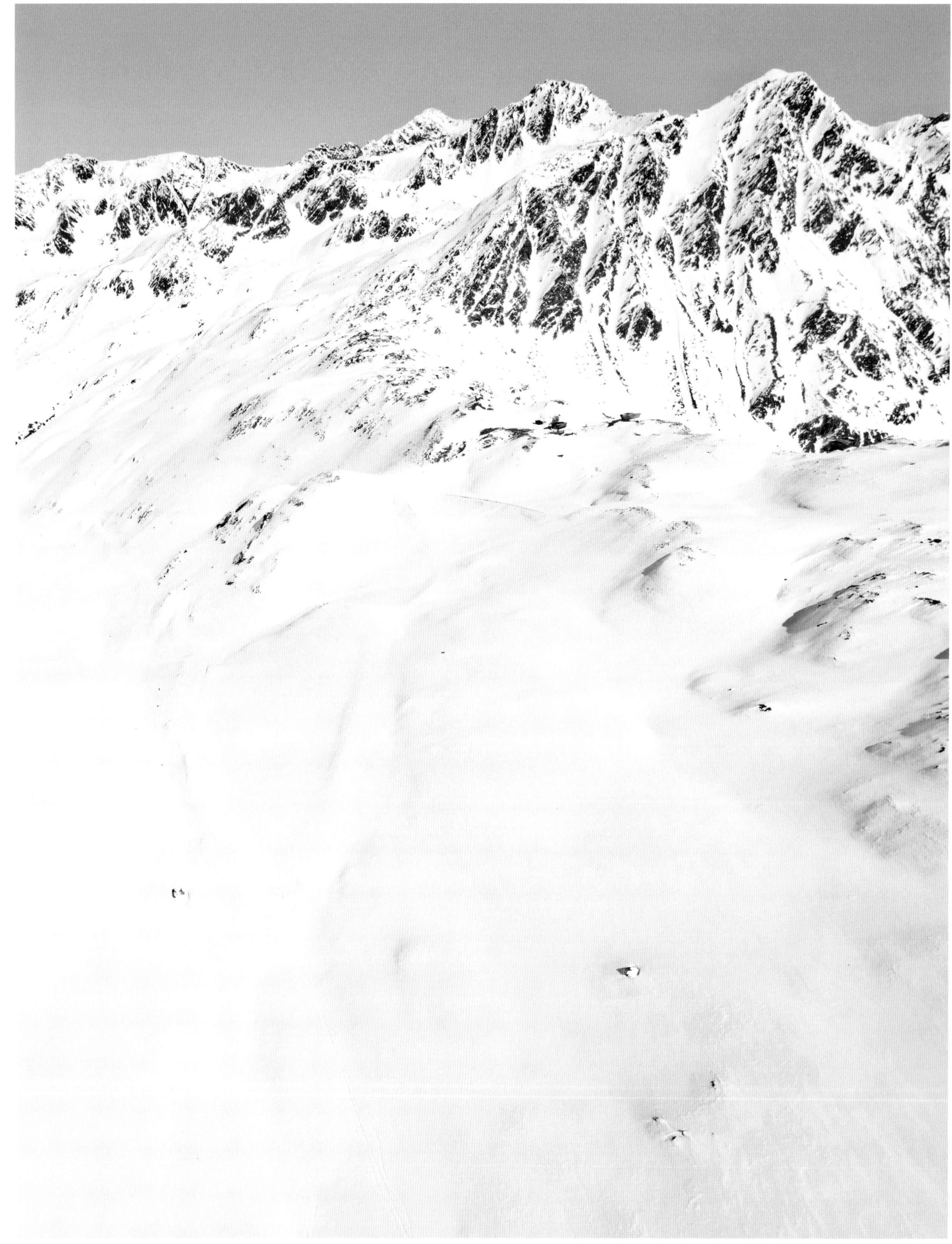

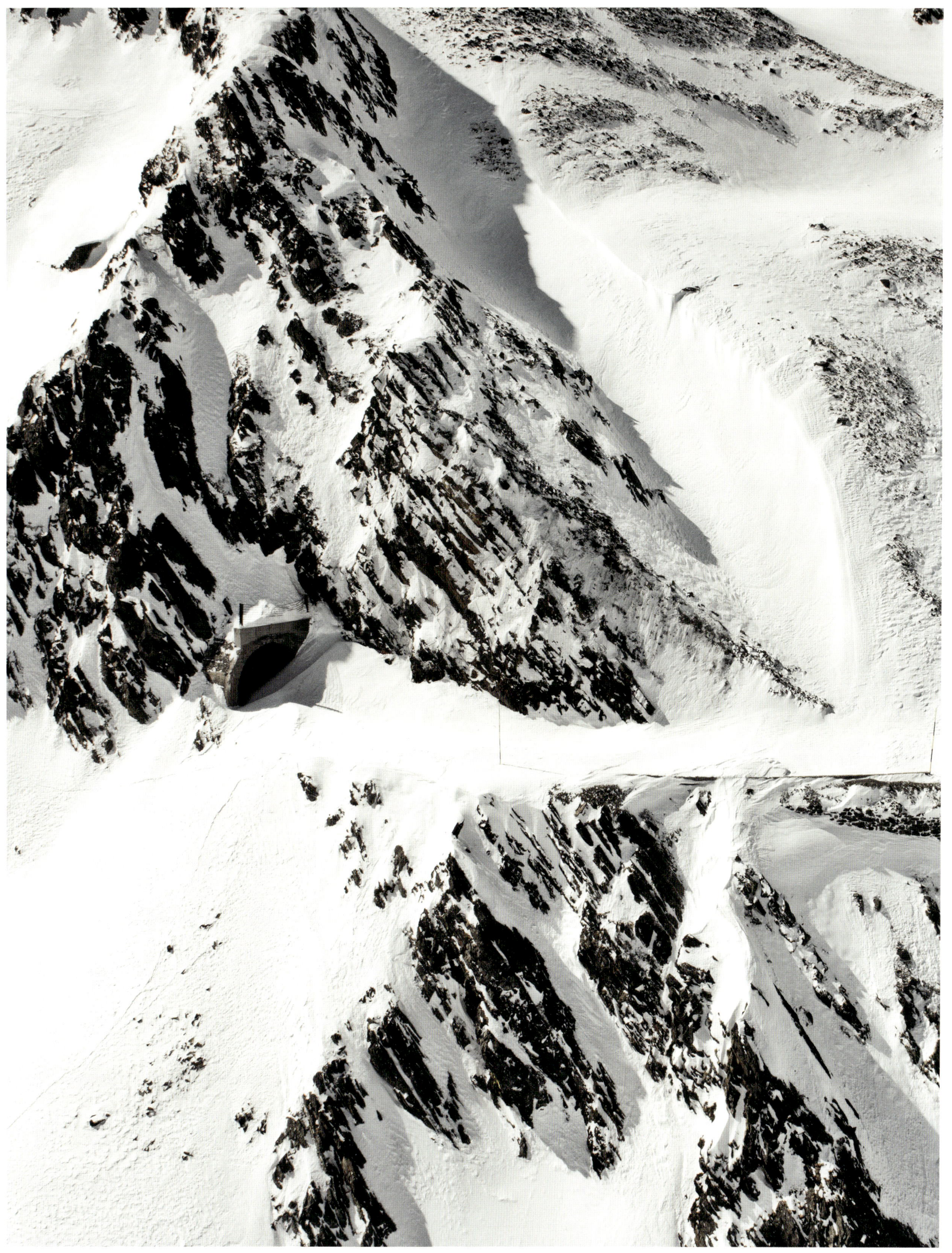

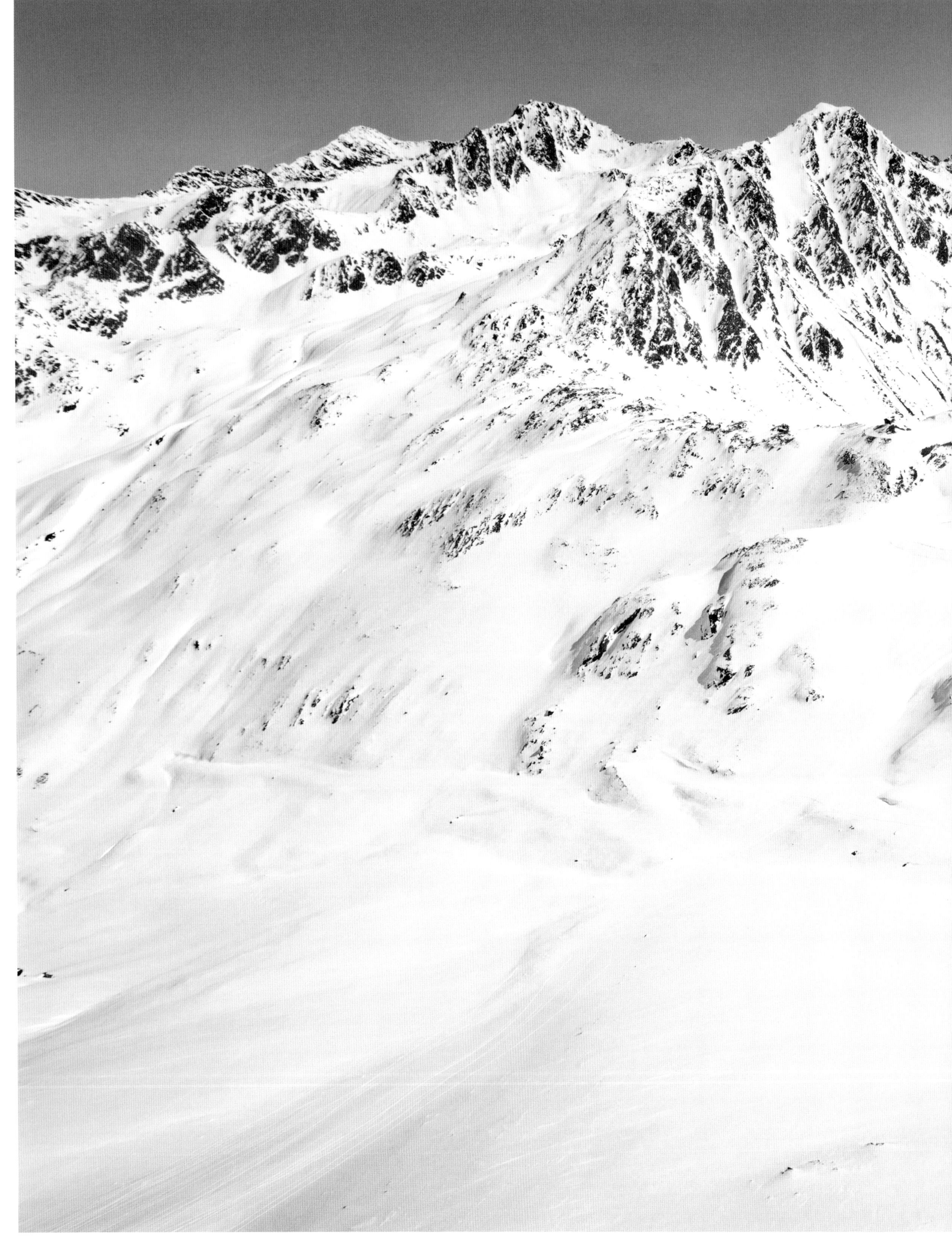

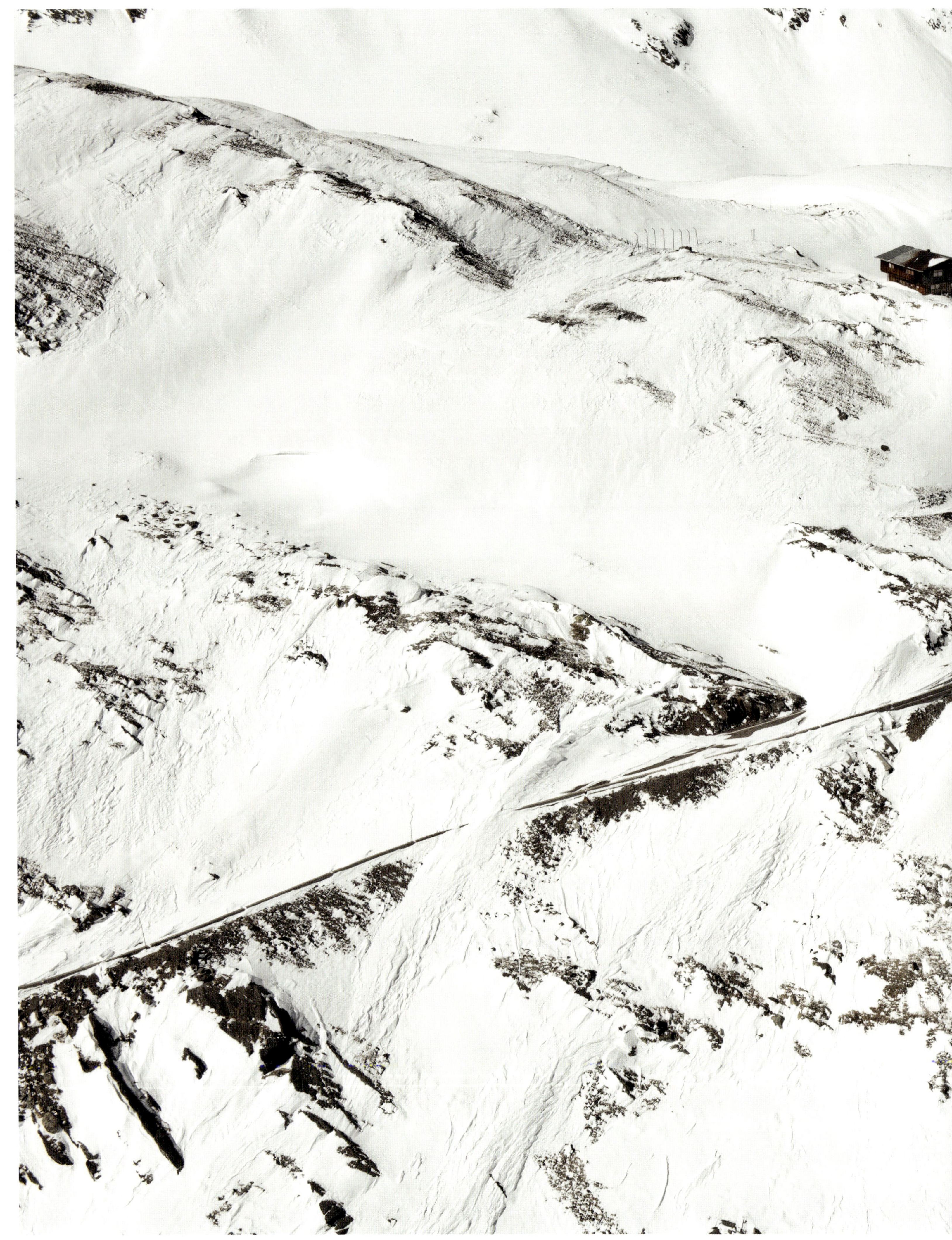

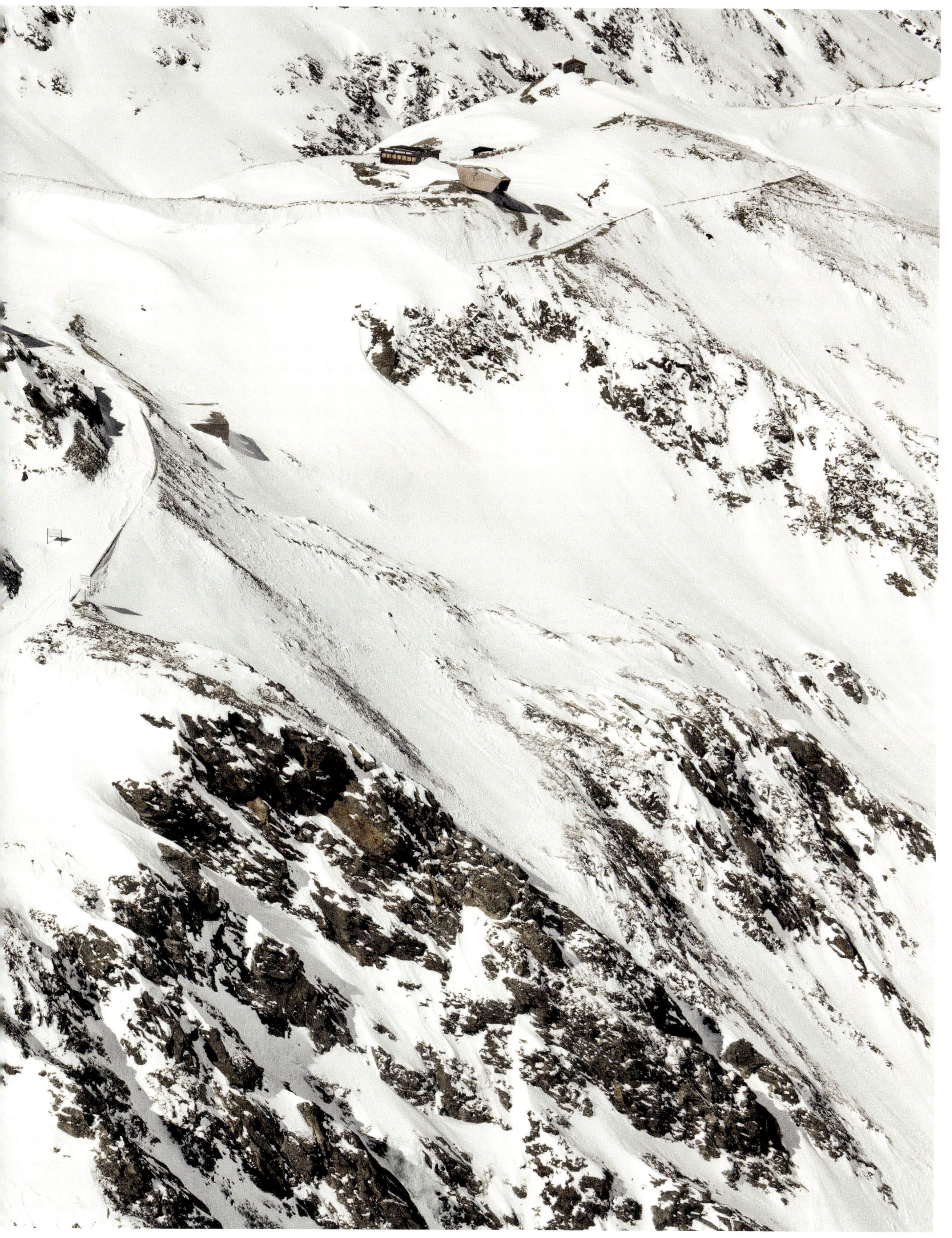

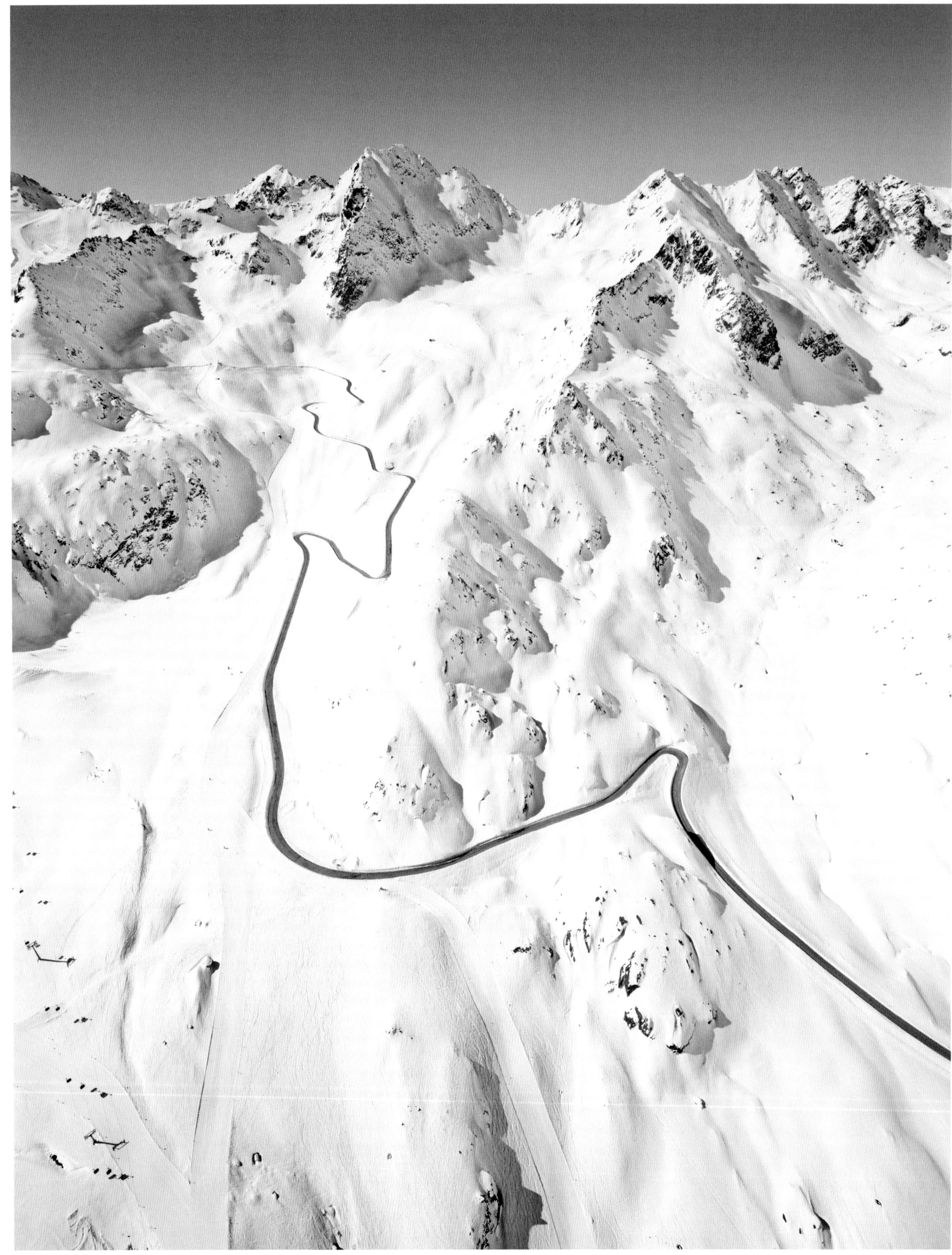

Kaunertaler Gletscherstraße

— 2750 Meter
Meist ganzjährig geöffnet

Die Kaunertaler Gletscherstraße gilt als eine der schönsten Hochalpenstraßen Österreichs. Von Feichten im Kaunertal führt die Fahrt in 29 elegant geschwungenen Kehren durch sechs Klimastufen und über 1500 Höhenmeter hinauf zum Gepatschstausee und weiter zum 1980 eröffneten Gletscher-Skigebiet. Die Straße wird im Winter zwar geräumt, seine Schneeketten sollte man aber dennoch griffbereit halten.

— 2,750 metres
Usually open all year round

The Kaunertal Glacier Road is considered one of the most beautiful high Alpine roads in Austria. From Feichten in the Kaunertal, the drive leads in 29 elegantly snaking corners through six climatic zones and up 1,500 metres to the Gepatsch reservoir, and on to the glacier ski area, which opened in 1980. Although the road in winter is cleared, motorists should still have their snow chains handy.

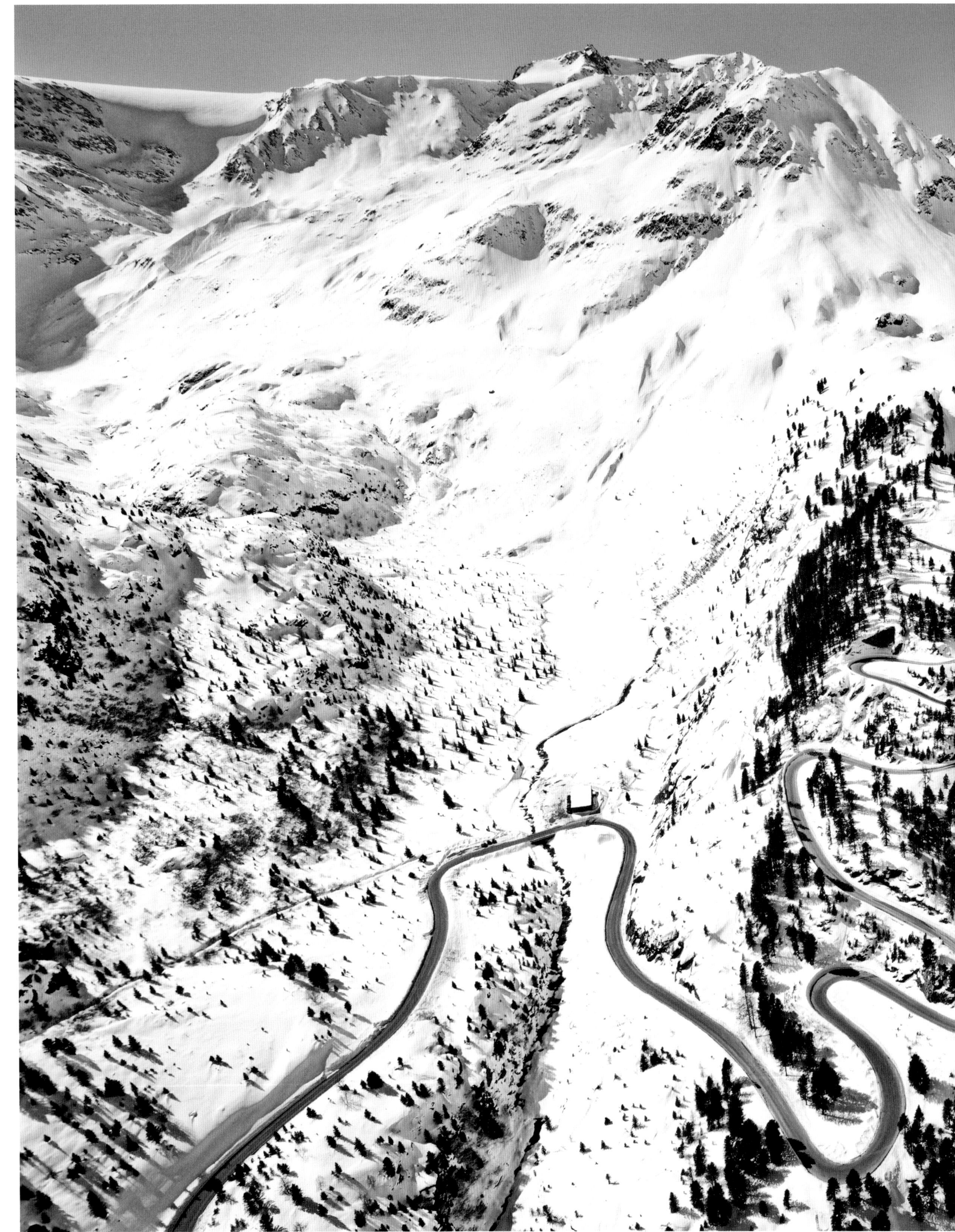

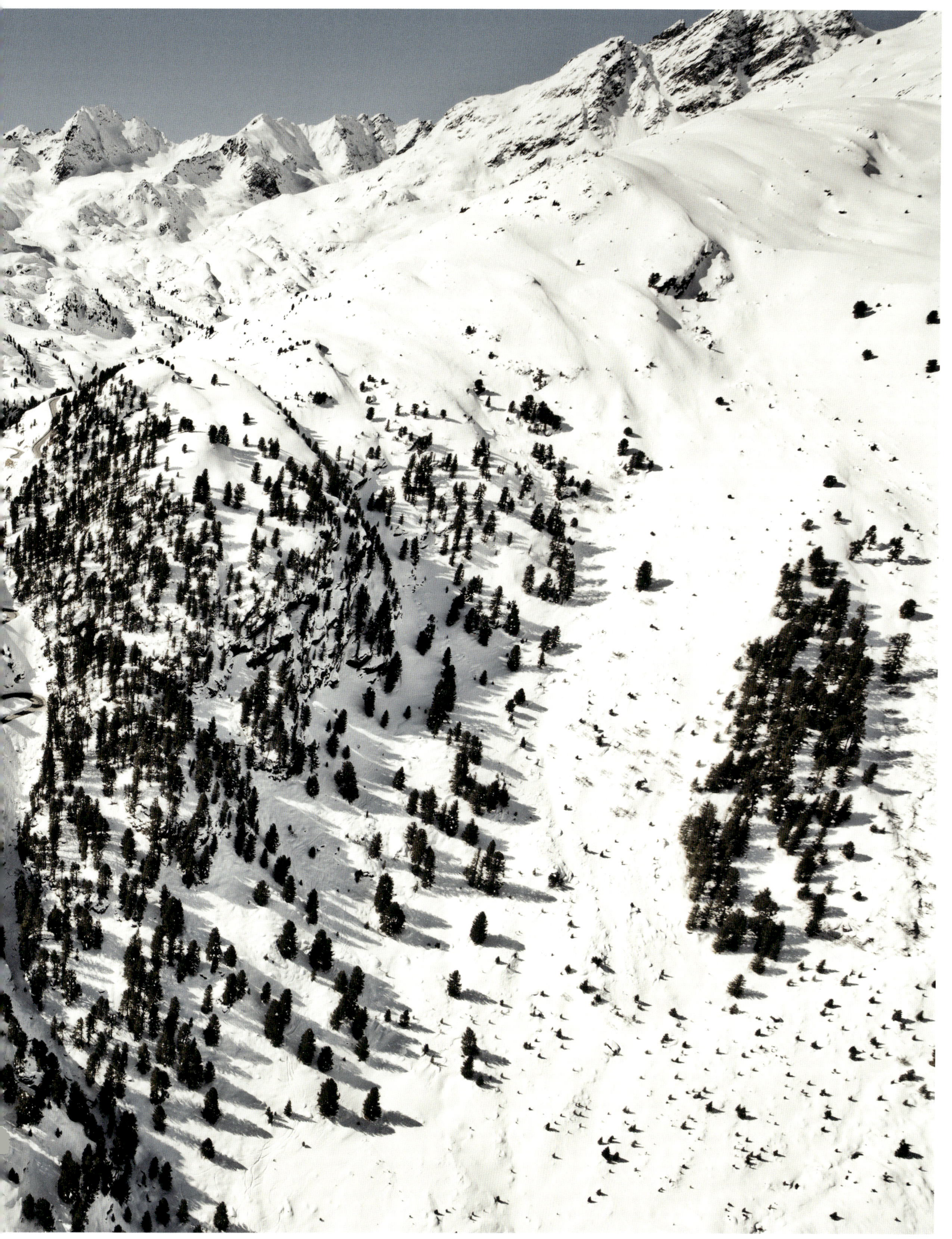

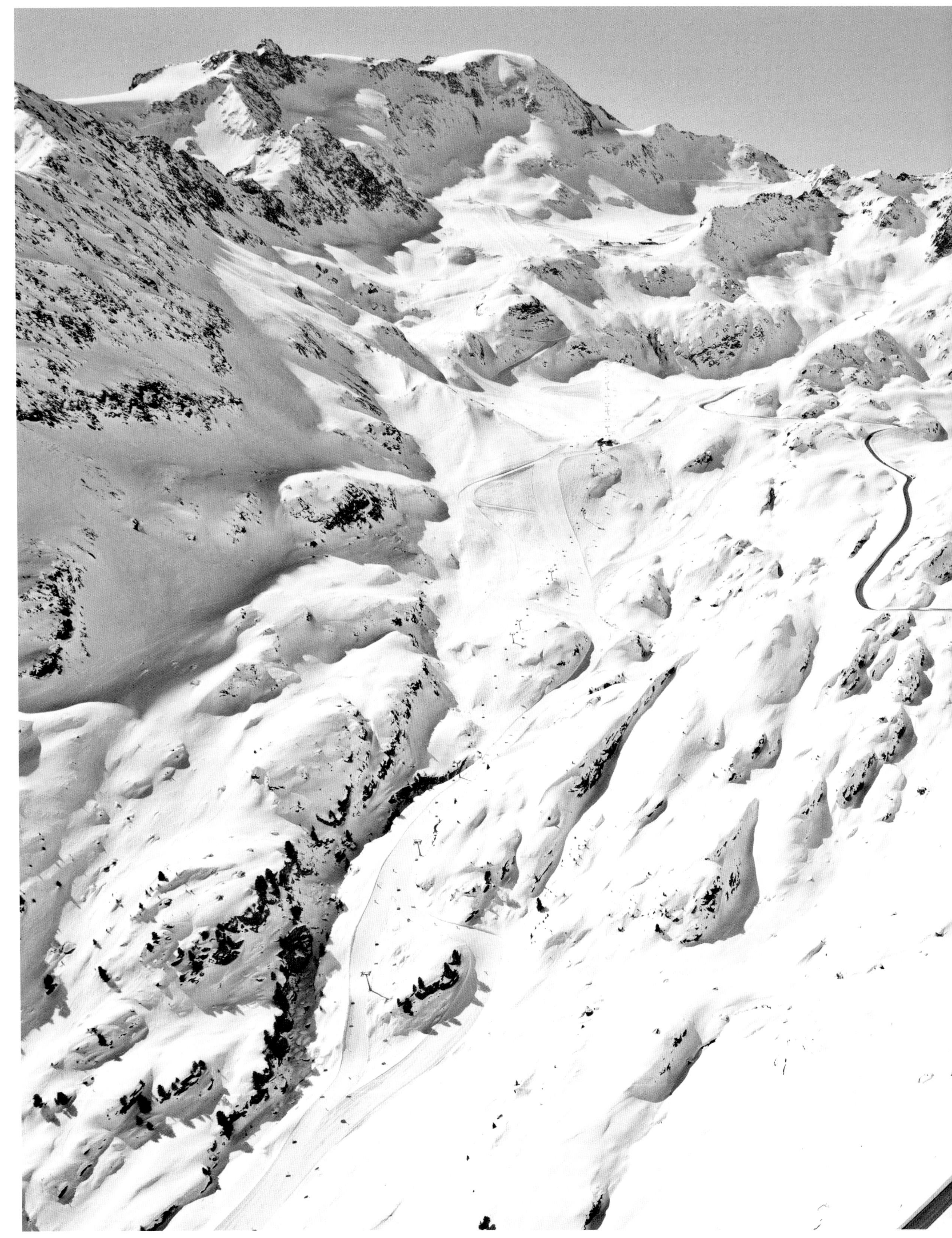

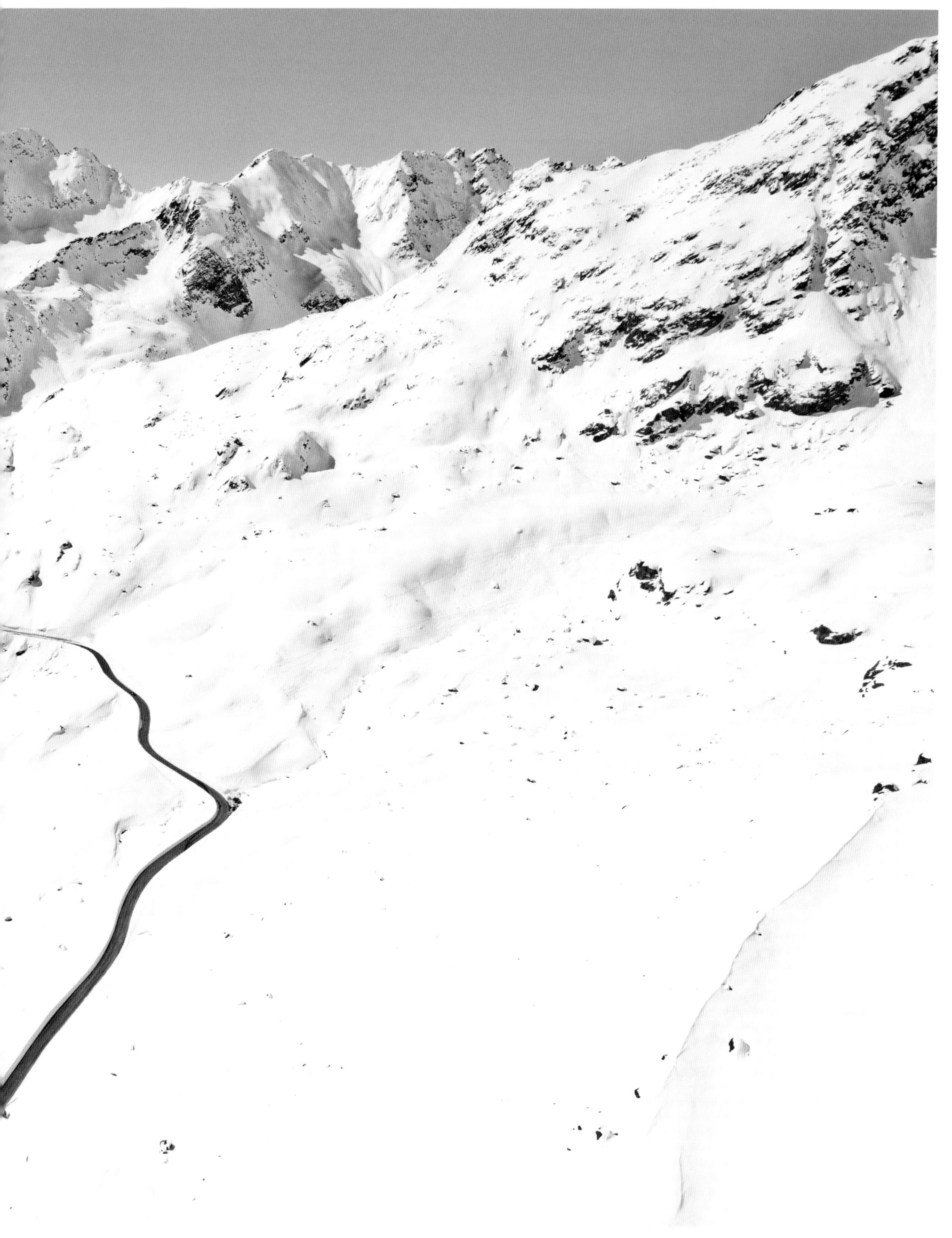

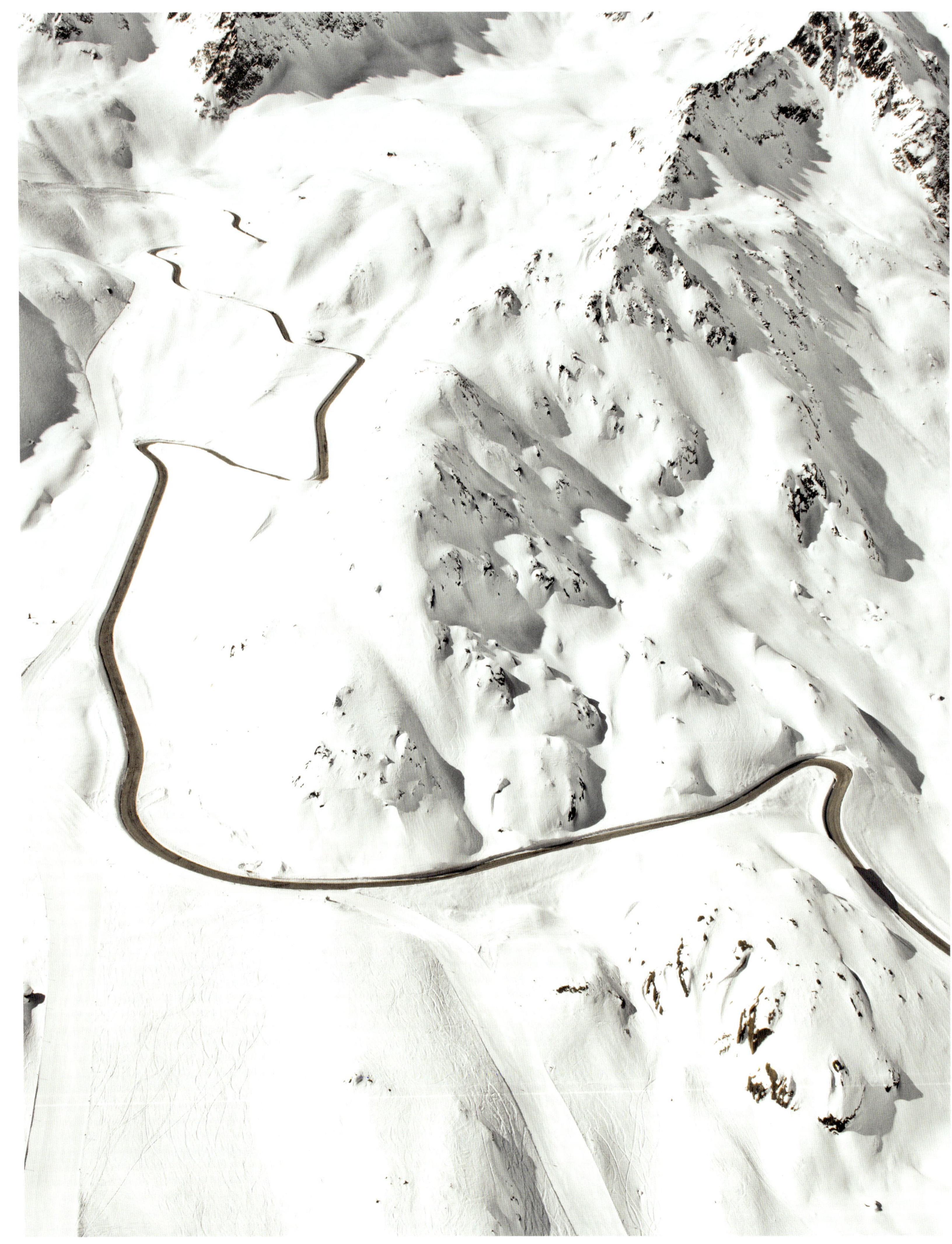

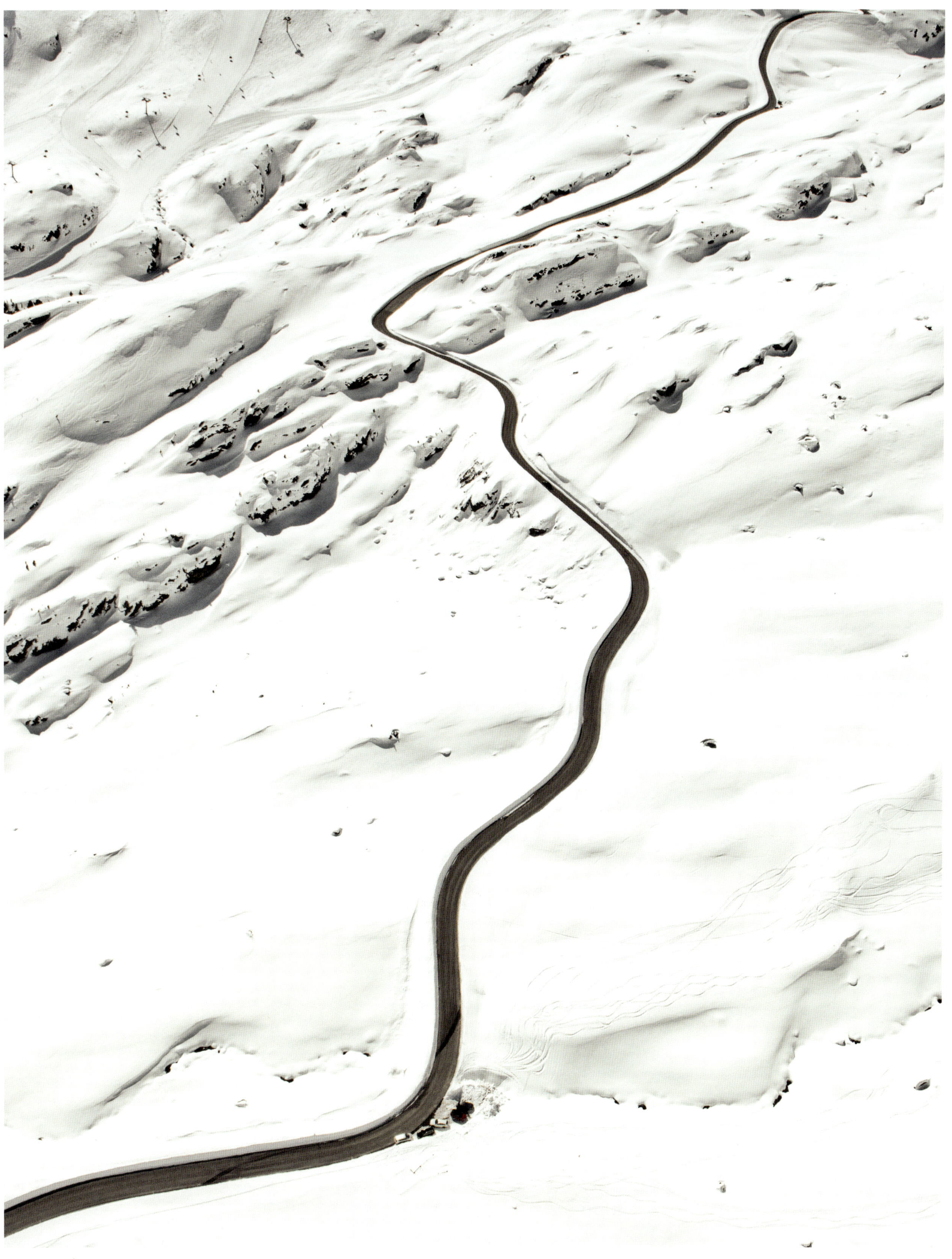

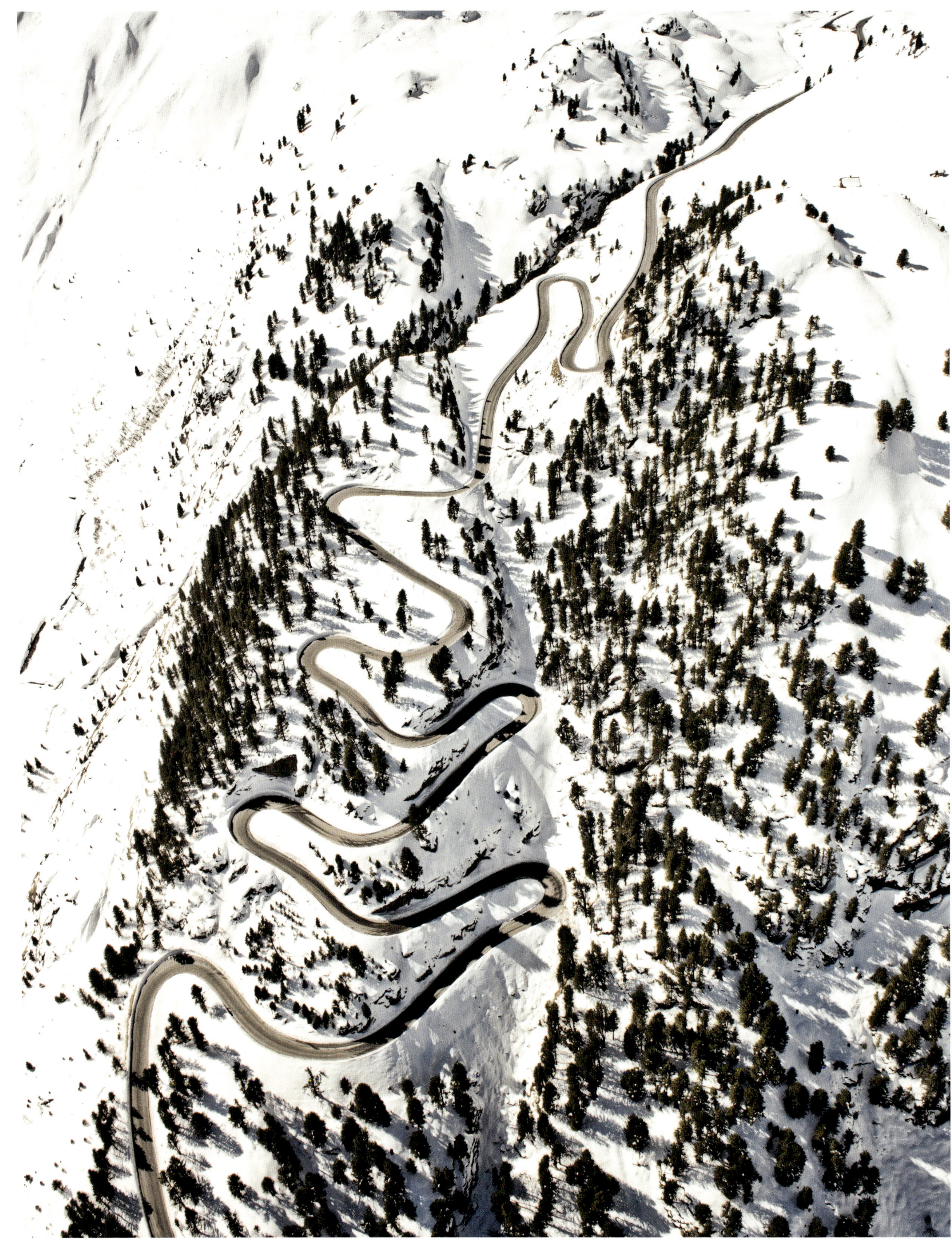

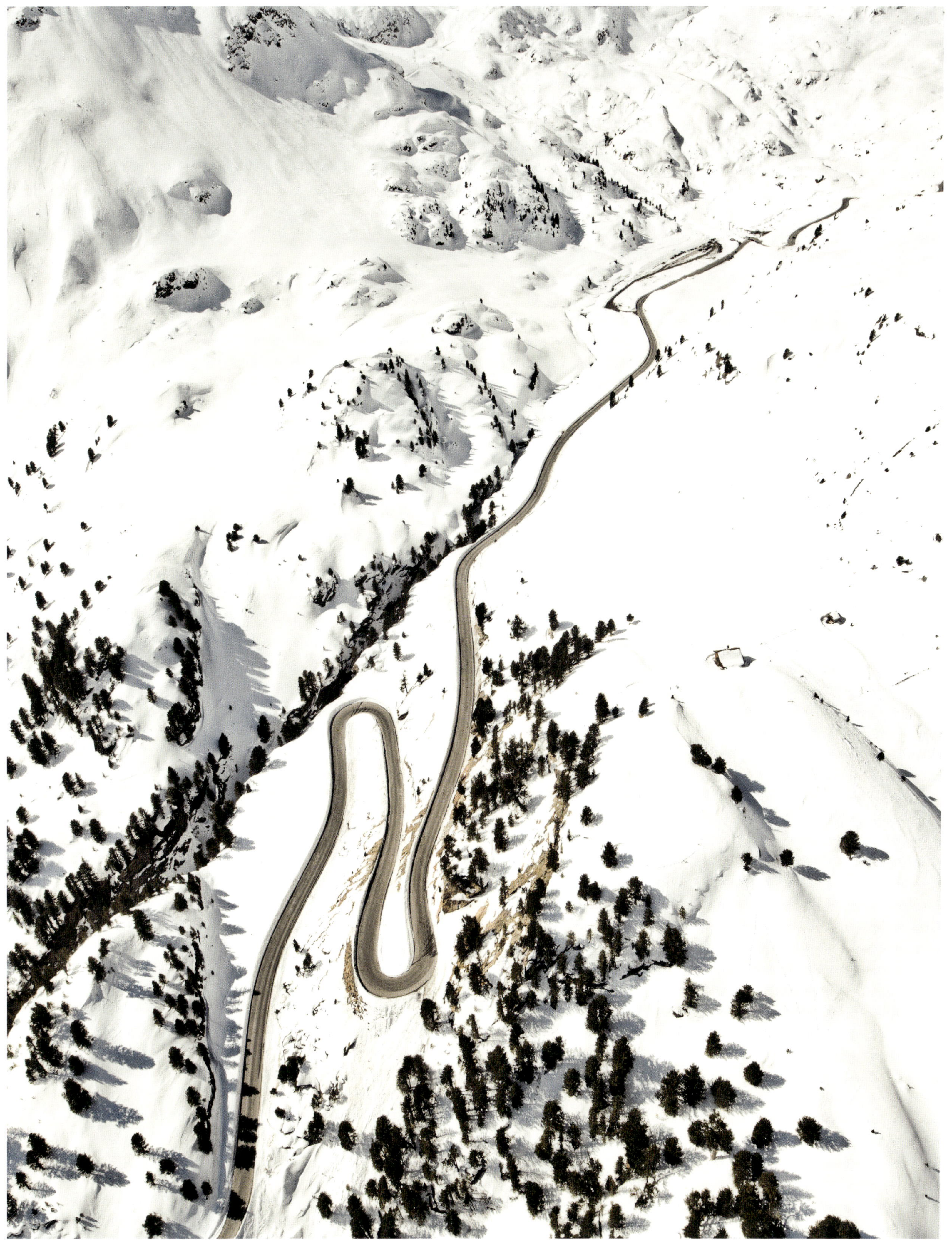

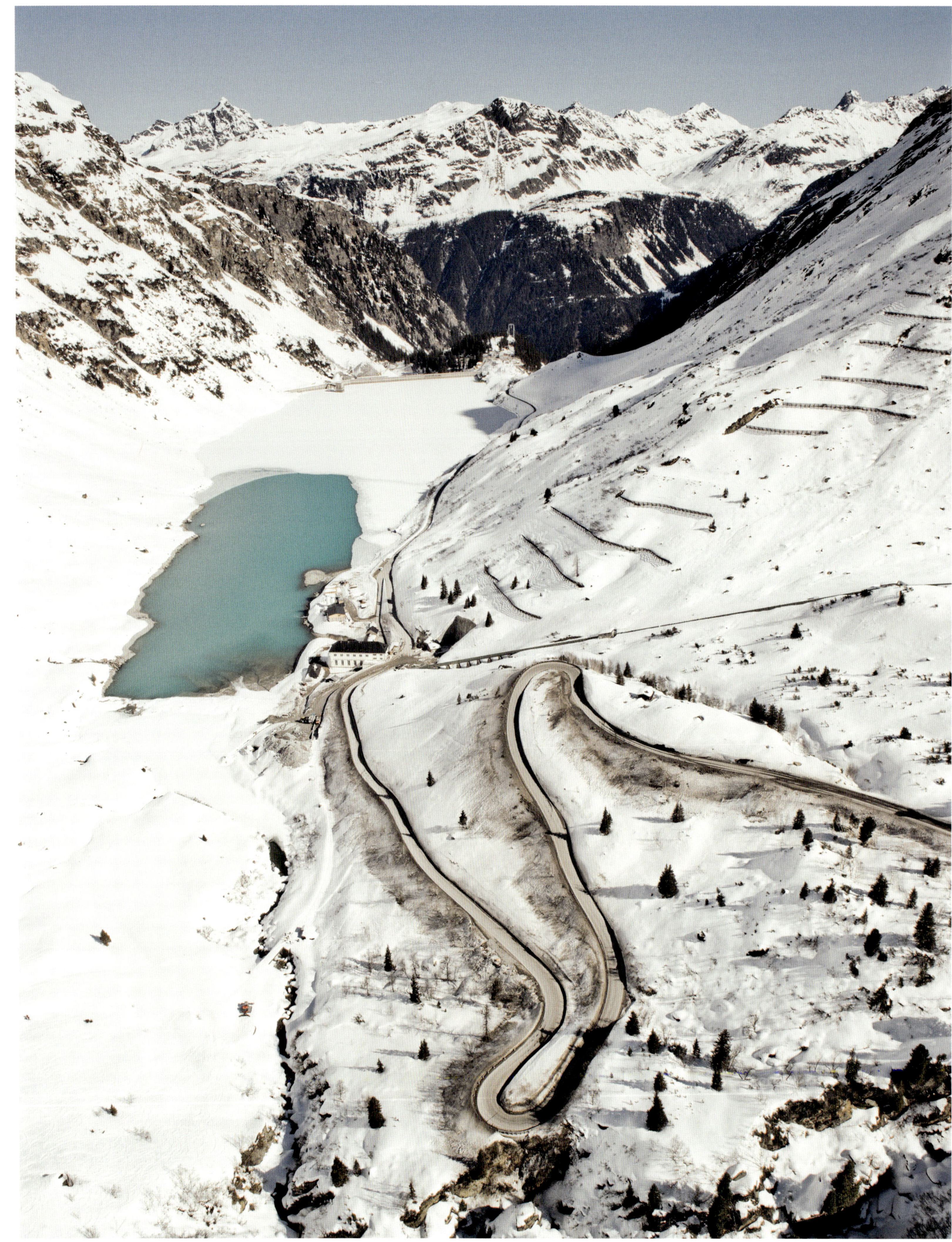

Silvretta-Hochalpenstraße

— 2032 Meter
Wintersperre meist von November bis Mai

Die Silvretta-Hochalpenstraße führt vom Vorarlberger Montafon über rund 25 Kilometer ins Tiroler Paznauen und gilt als eine der schönsten und beliebtesten Bergrouten Österreichs. Vor allem die 32 Haarnadelkurven der Westrampe ziehen Motor- und Radsportler magisch an. Im Winter bleibt die Straße geschlossen, das Skigebiet Silvretta-Bielerhöhe erreicht man dieser Zeit ab Partenen mit einer Seilbahn.

— 2,032 metres
Usually closed in winter from November to May

The Silvretta High Alpine Road leads over almost 25 kilometres, from the Montafon valley in Vorarlberg into the Paznaun valley in Tyrol, and is considered one of Austria's most picturesque and popular mountain roads. Above all, the 32 hairpins on the western slope hold a magical fascination for car and motorbike enthusiasts. The road is closed in winter, but during these months the Silvretta-Bielerhöhe can be reached from Partenen via cable car.

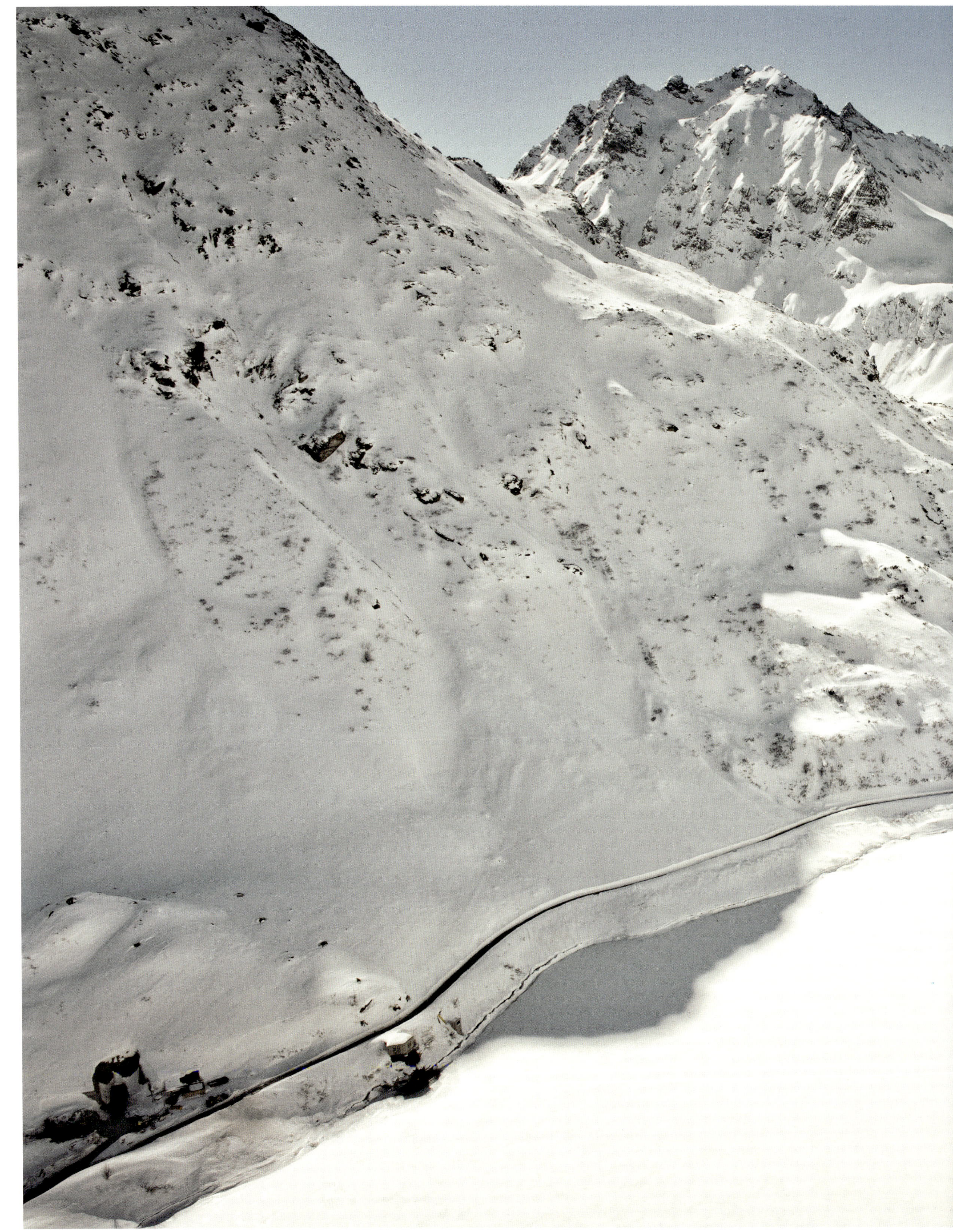

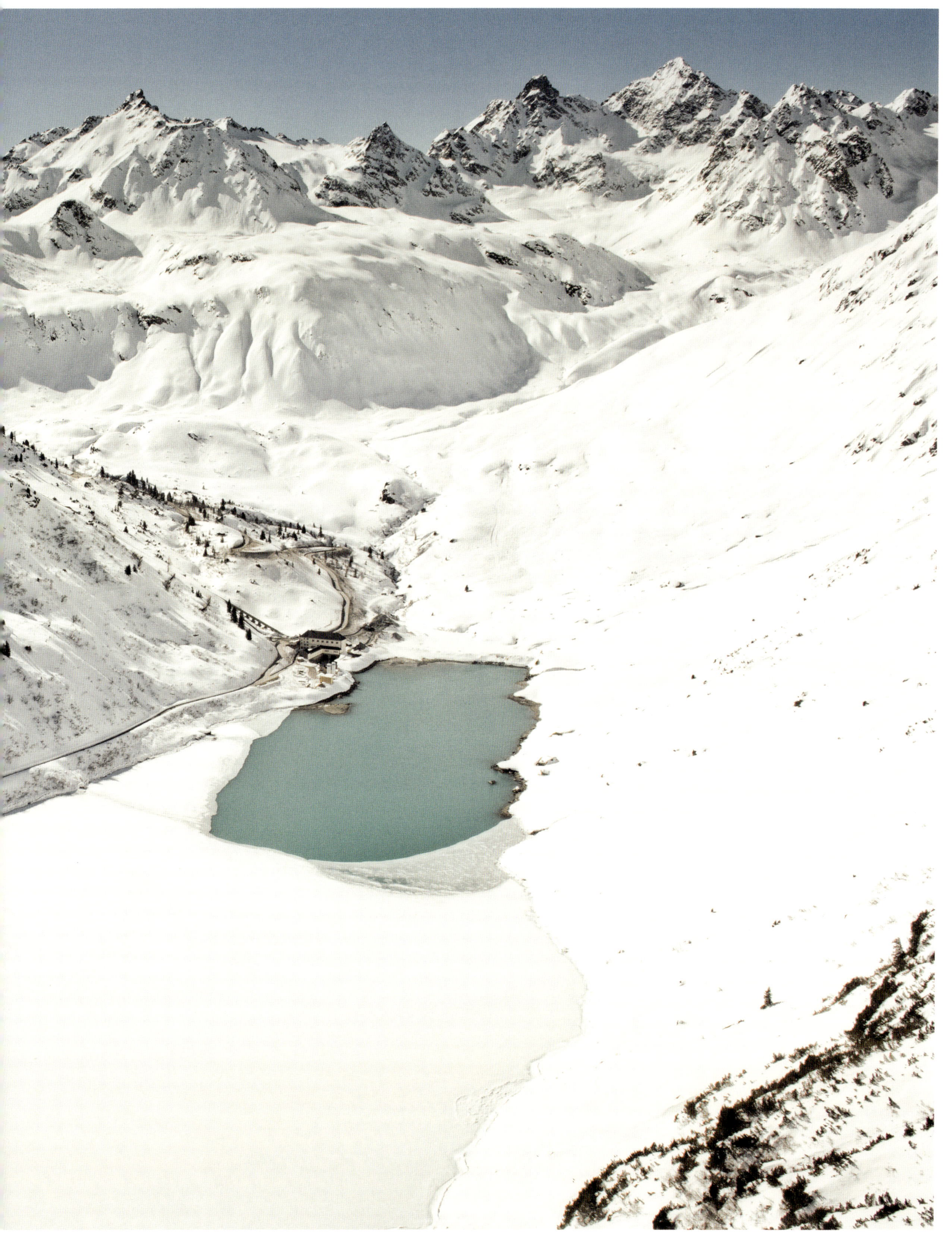

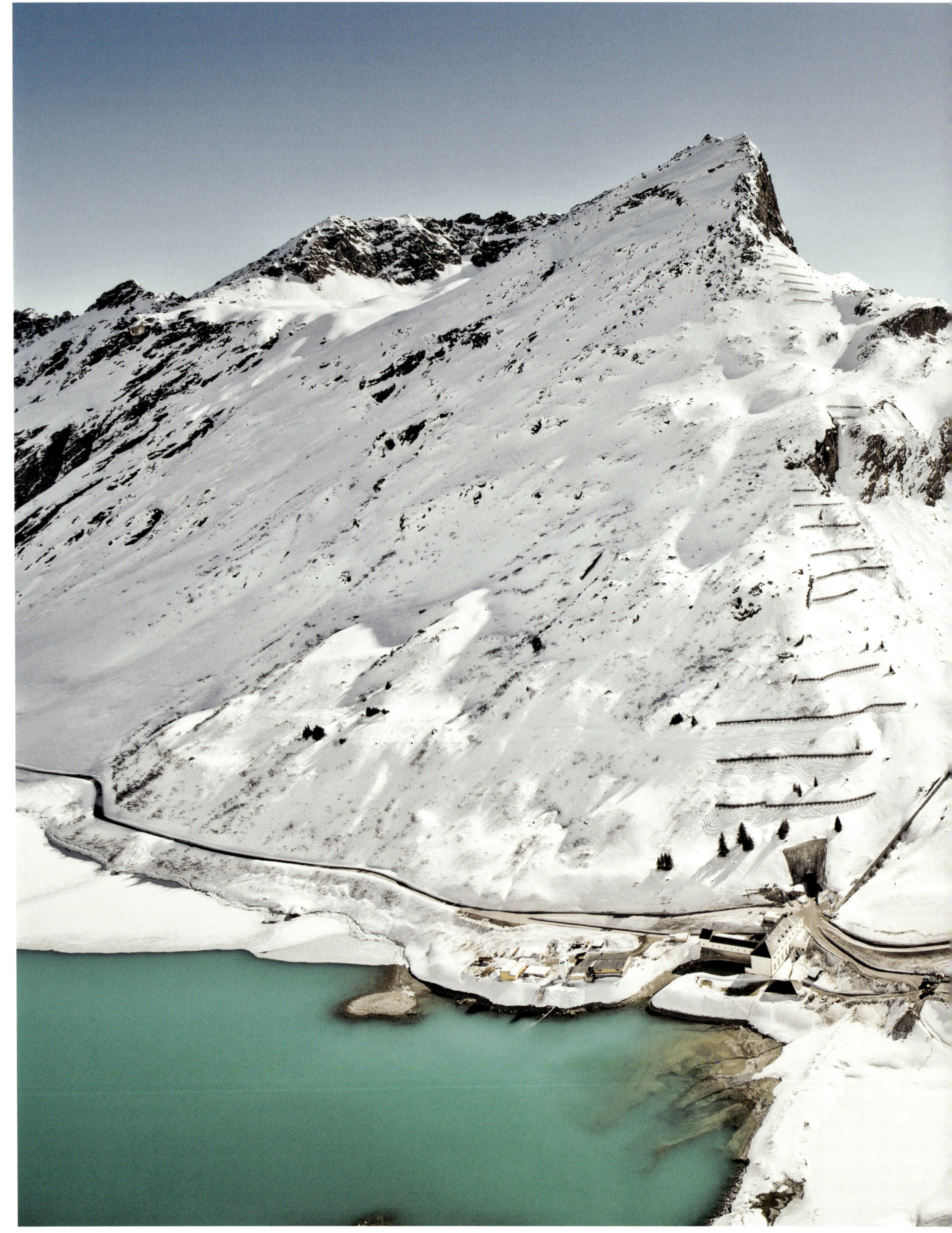

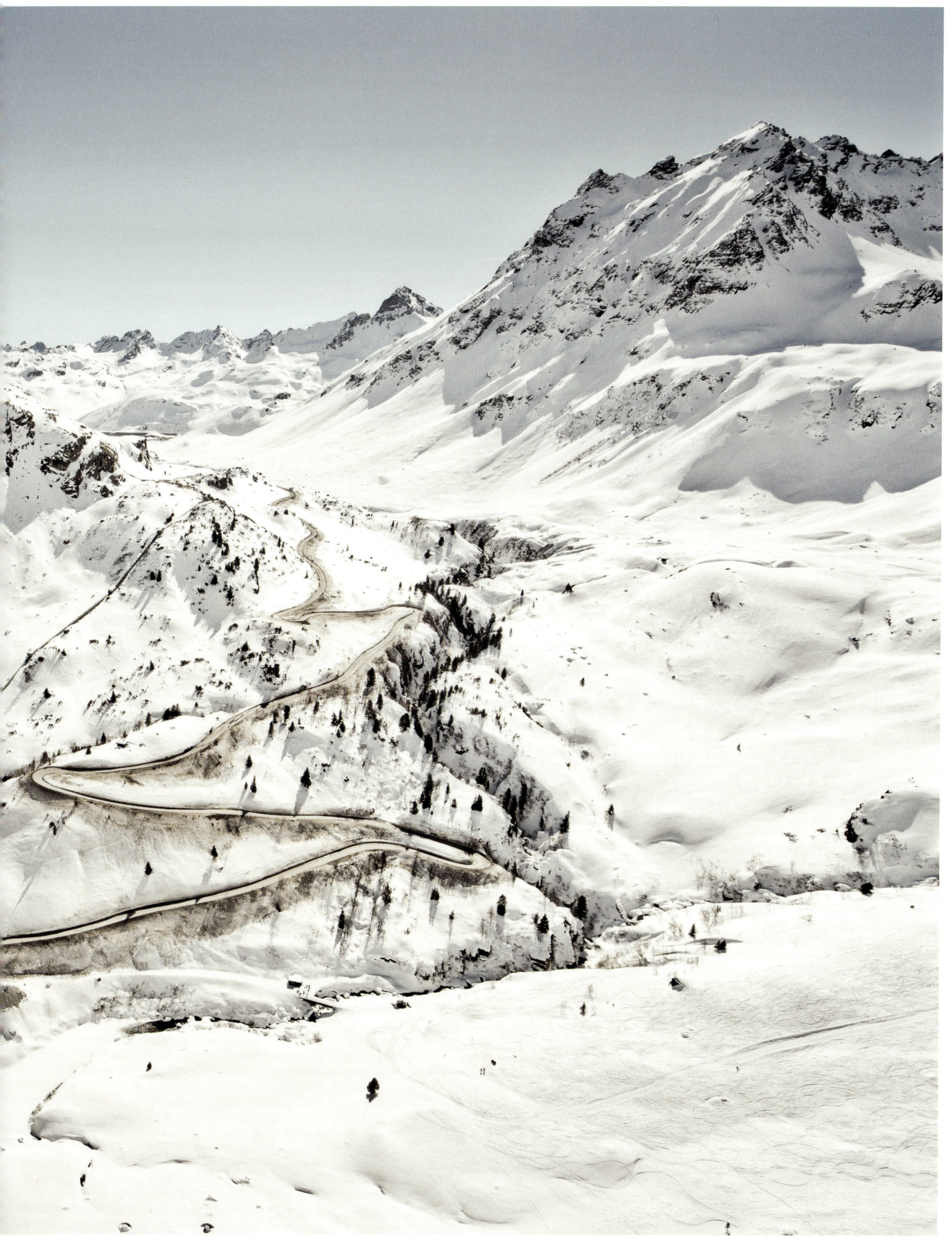

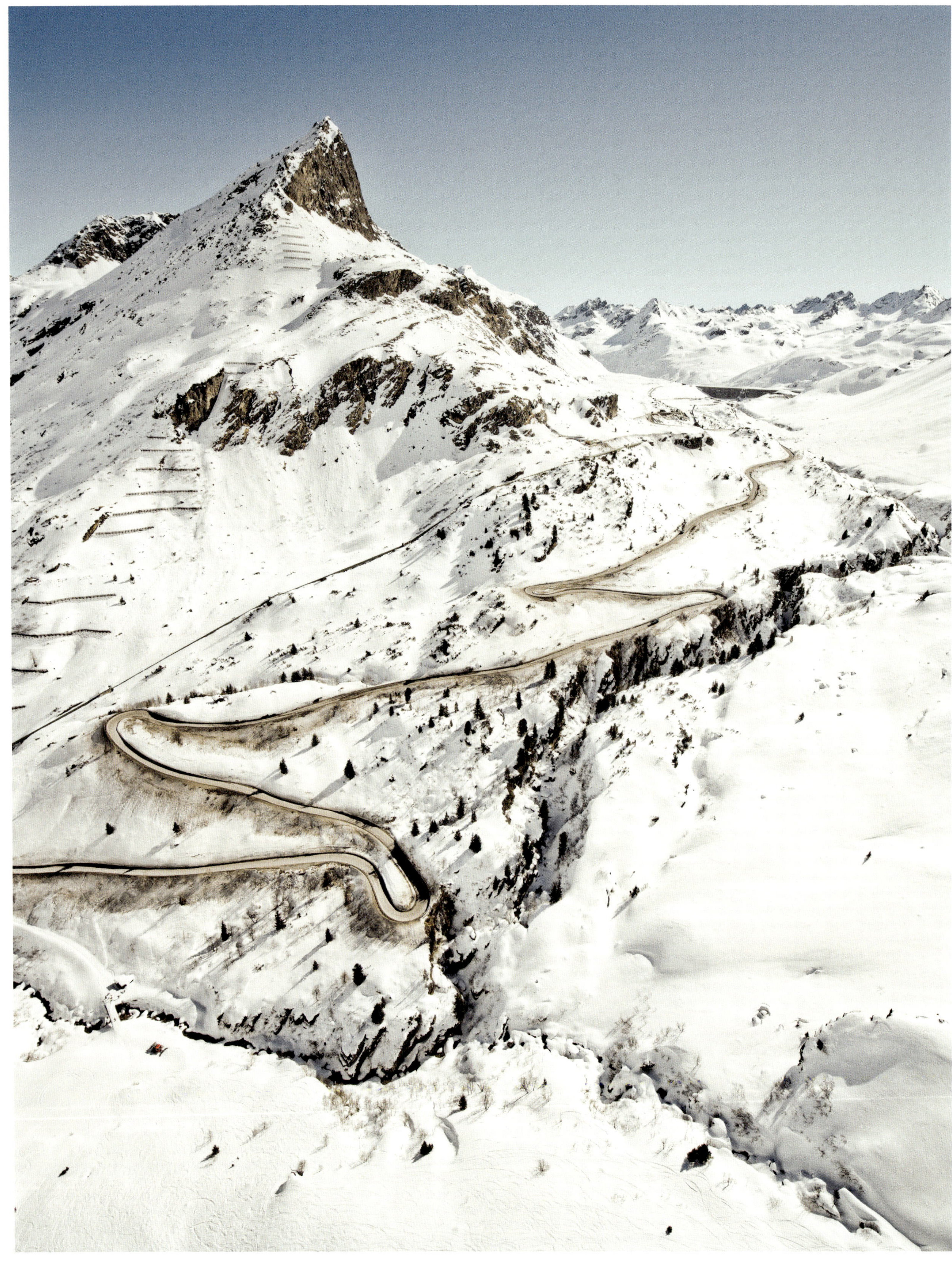

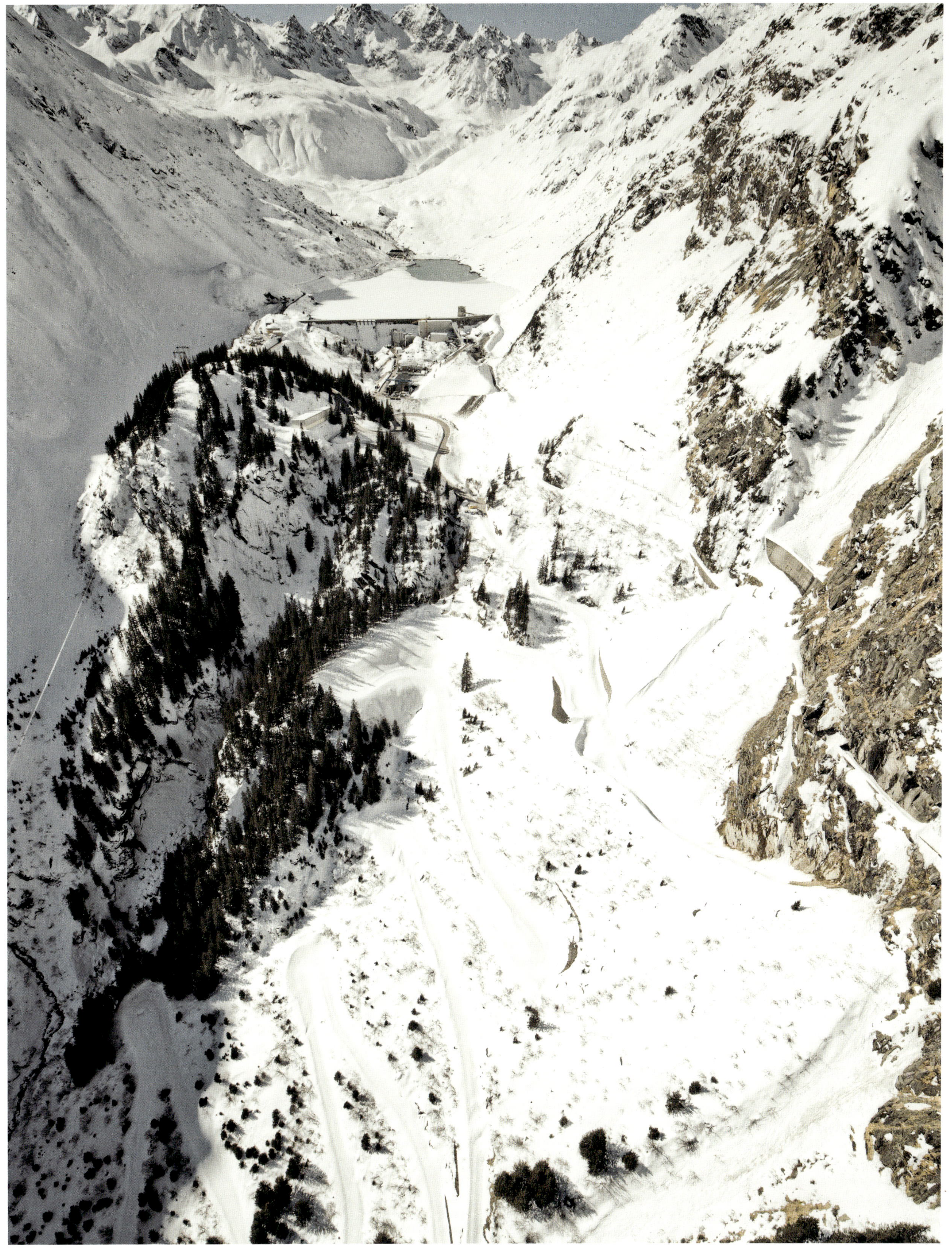

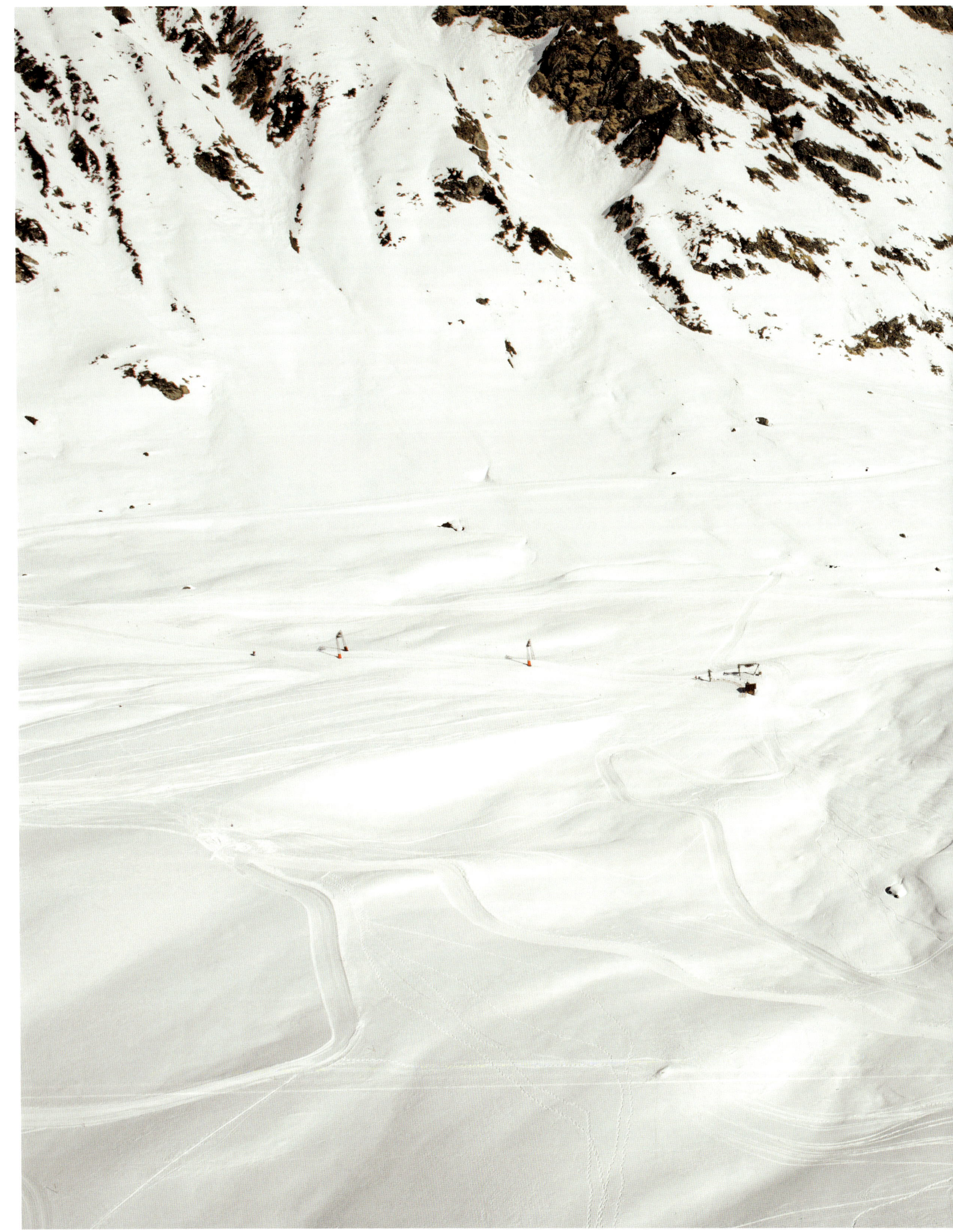

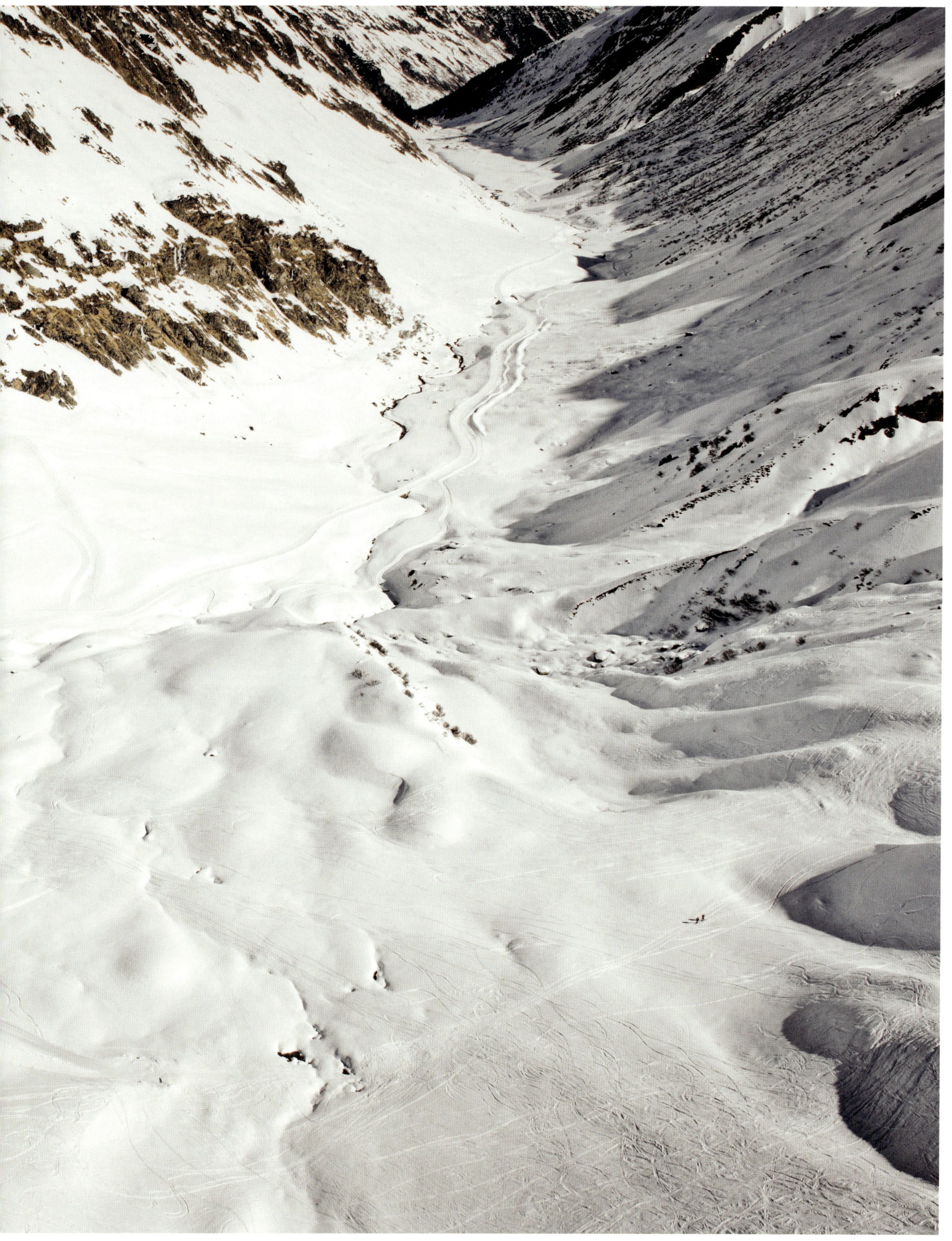

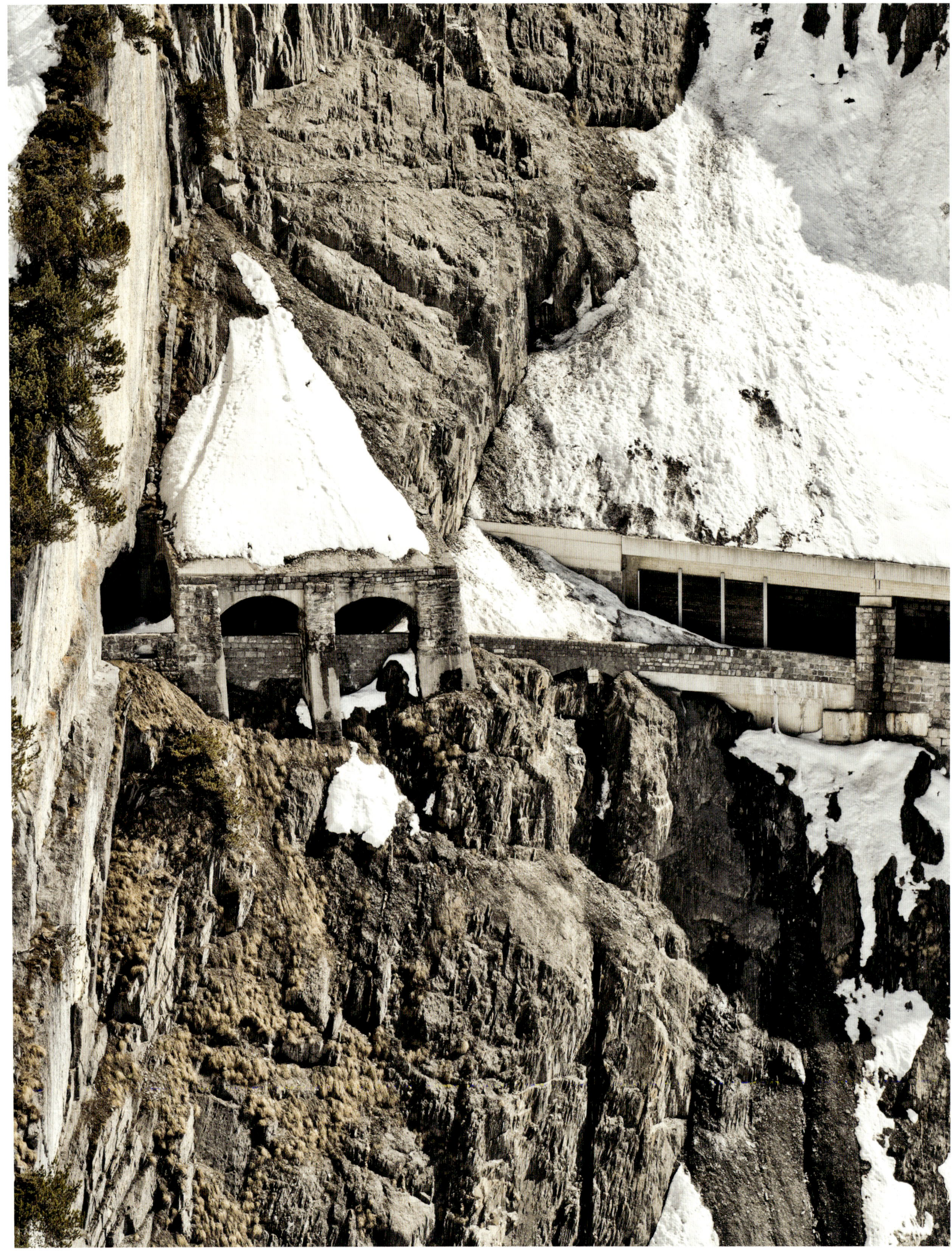

Flexenpass

— 1773 Meter
Ganzjährig geöffnet

Der Flexenpass verdankt seinen Namen dem lateinischen Wort flexio für Biegung – und mit schwungvollen Kurven und Kehren ist die Strecke wahrlich gesegnet. Die 1909 vollendete Passstraße führt von Lech und Zürs hinüber nach Stuben und zur Arlbergstraße und ist vor allem für ihre Panorama-Galerien in schwindelerregenden Höhen bekannt. Als Zugang zu einigen der beliebtesten Wintersportgebiete Österreichs wird die Straße auch im Winter geräumt . So kann man sich schon vor der Piste in Wedelstimmung fahren – als automobile Skigymnastik zum Aufwärmen sozusagen.

— 1,773 metres
Open all year round

Flexen Pass gets its name from the Latin word flexio, or bending – and with its sweeping corners and curves, this stretch of road is truly blessed. Completed in 1909, the pass leads from Lech and Zürs across to Stuben and on to the Arlberg road. It is especially renowned for its panoramic galleries at dizzying heights. To allow access to several of Austria's most popular winter sport areas, the road is cleared in winter, which also provides a great way to warm up in preparation for carving a track down the ski slopes – automotive ski gymnastics, so to speak.

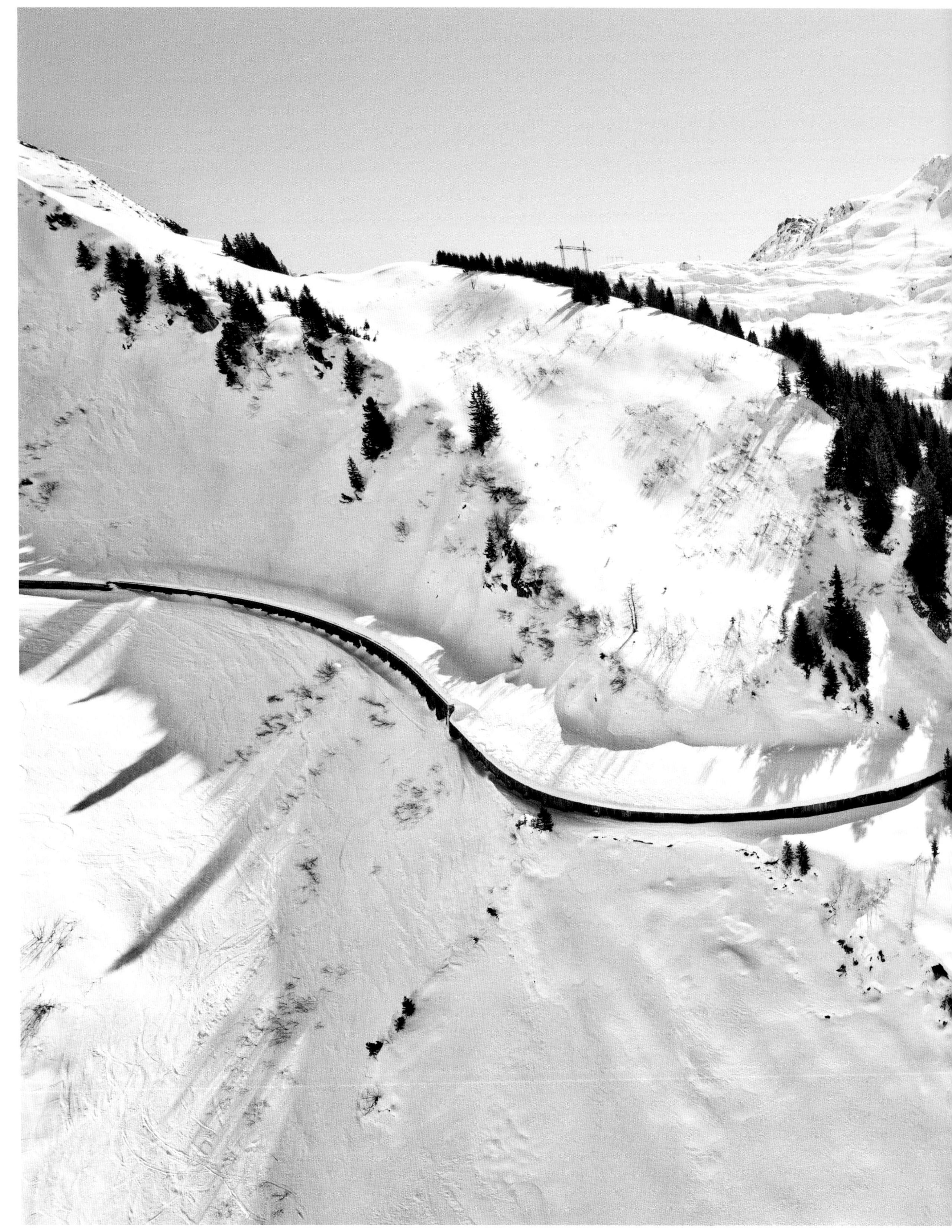

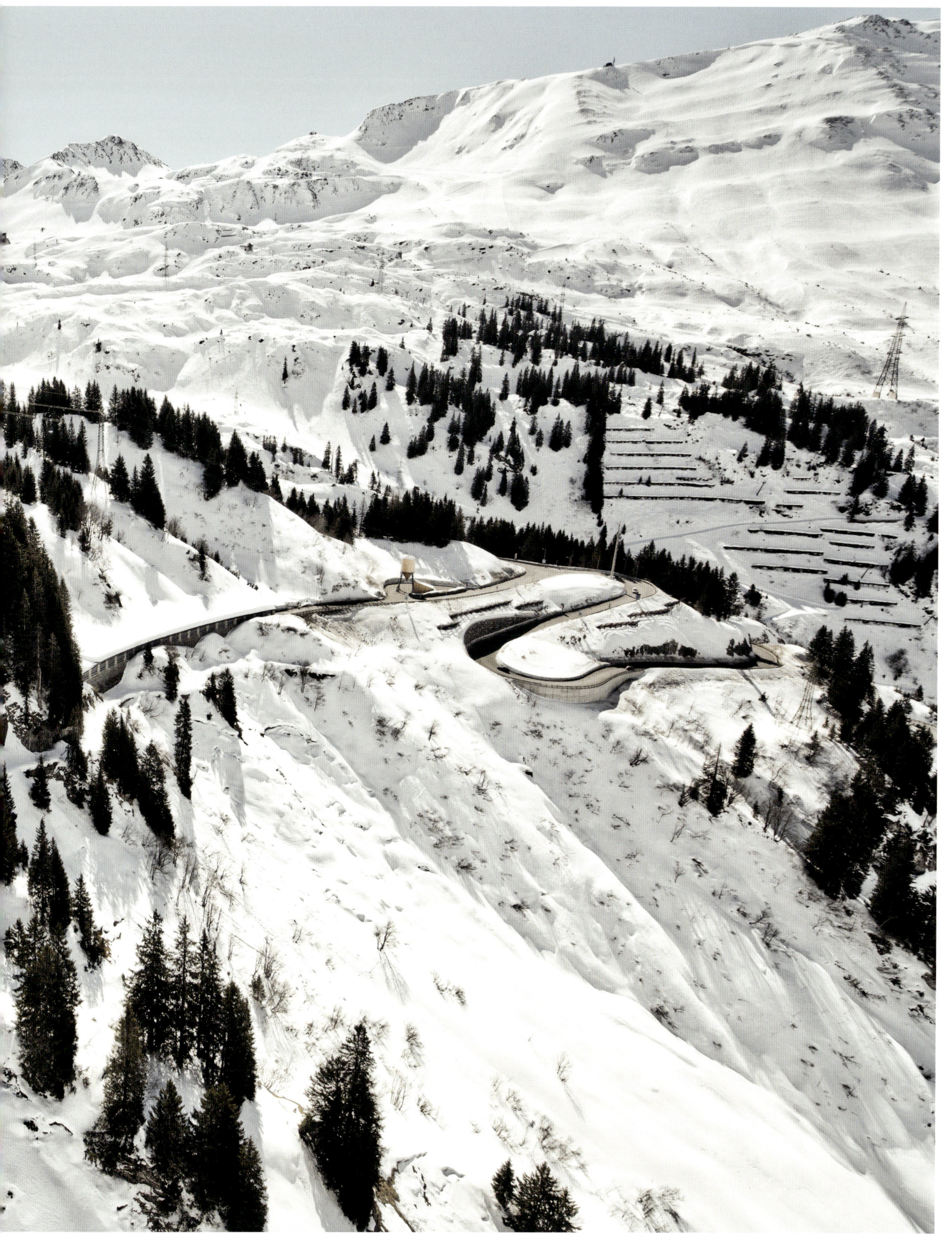

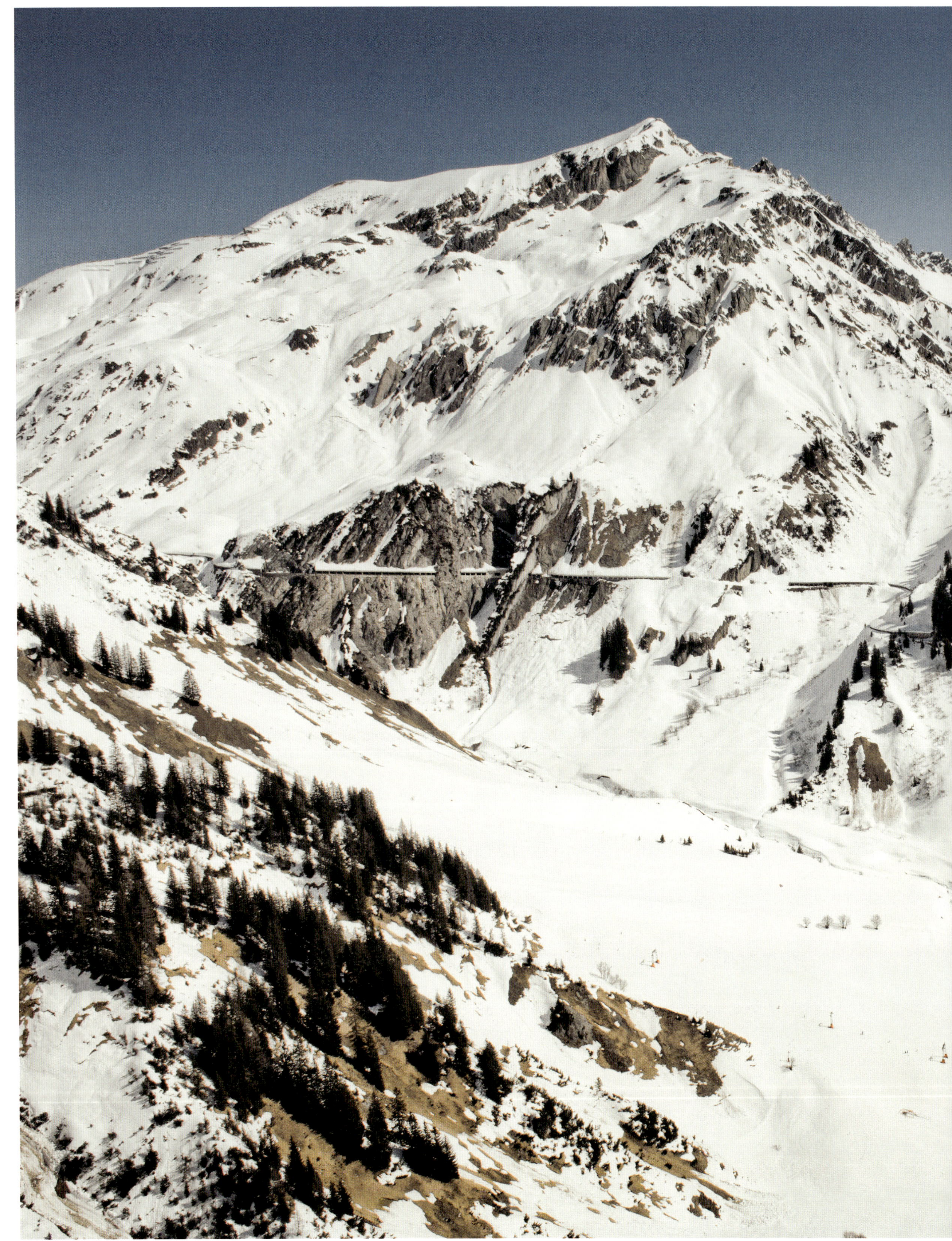

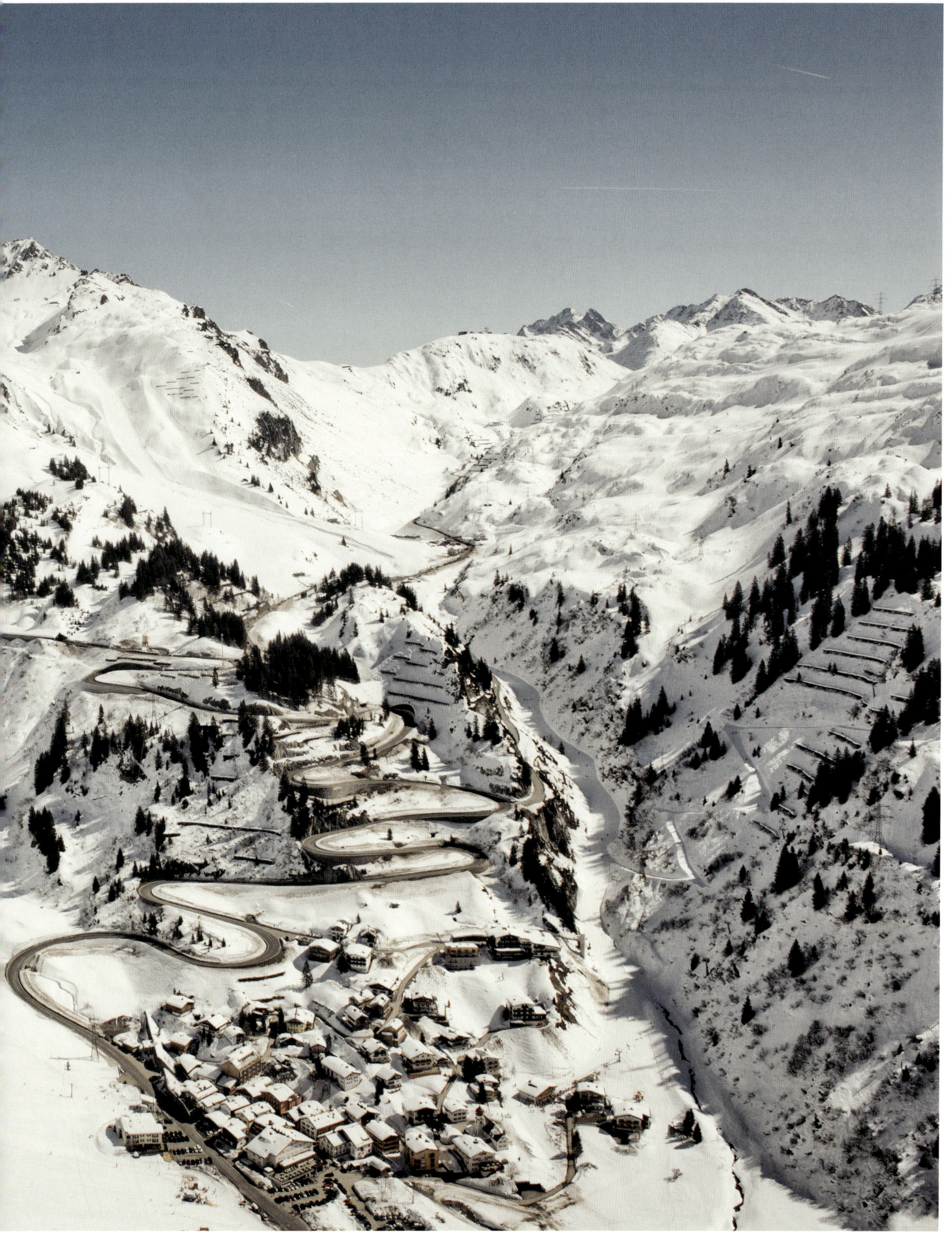

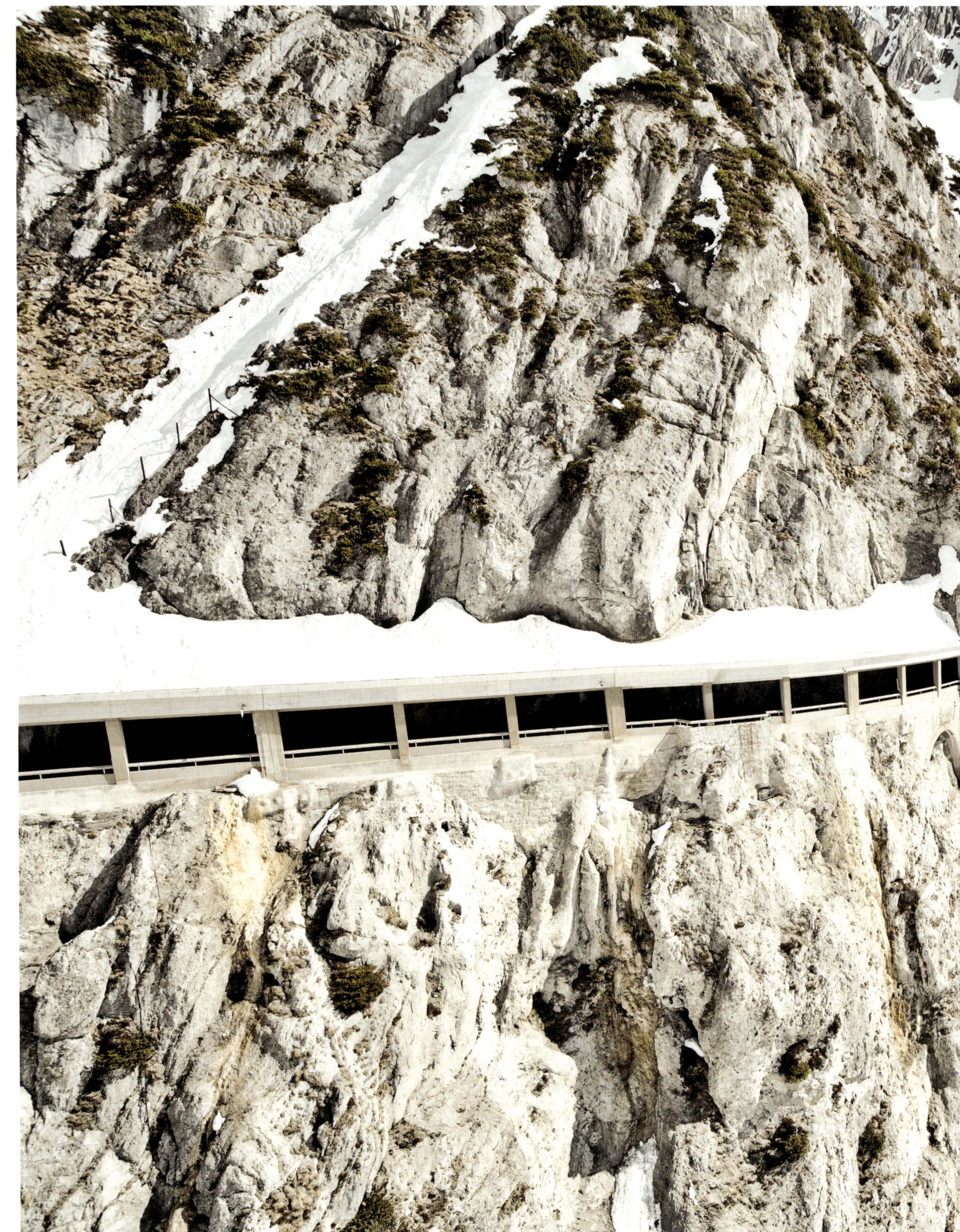

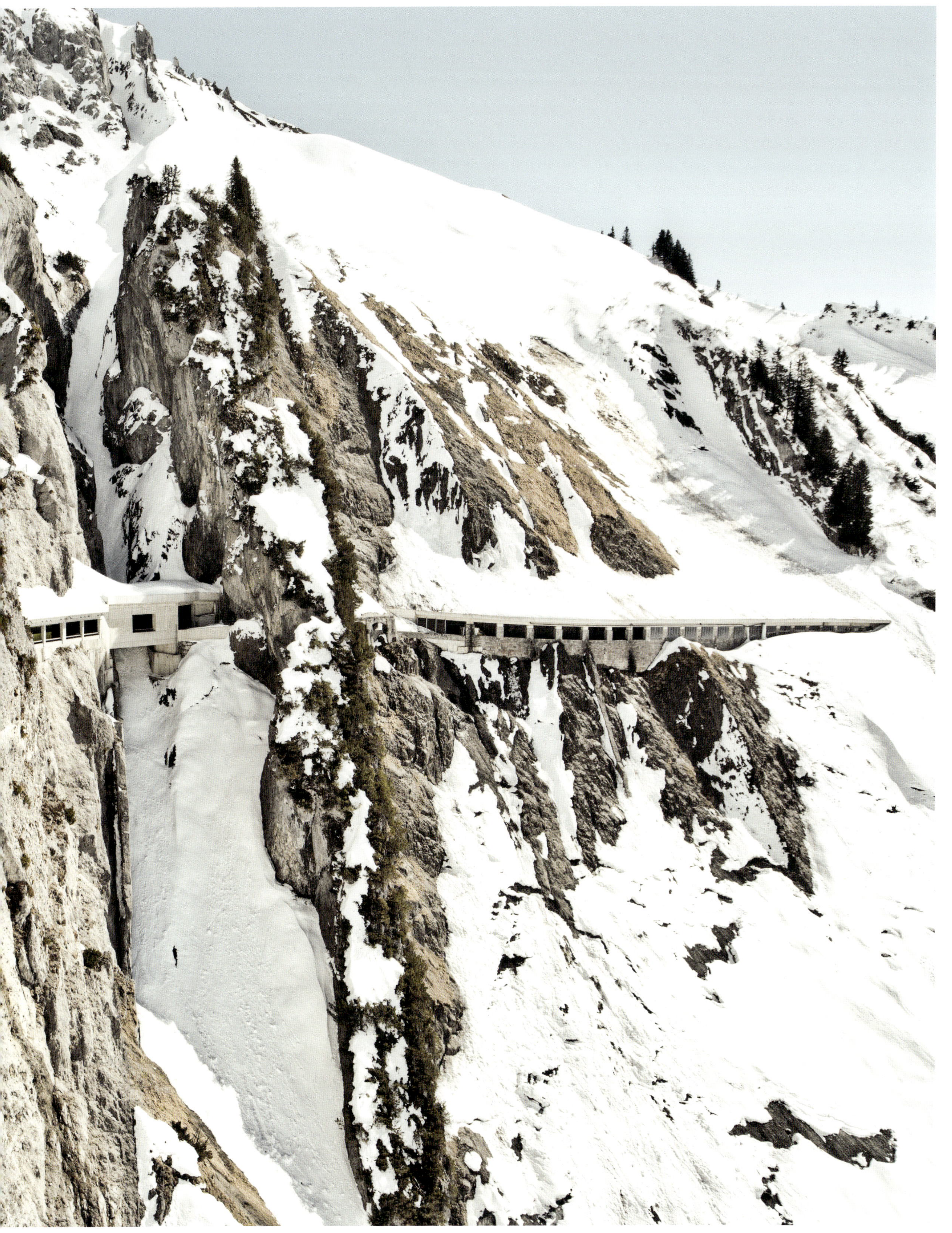

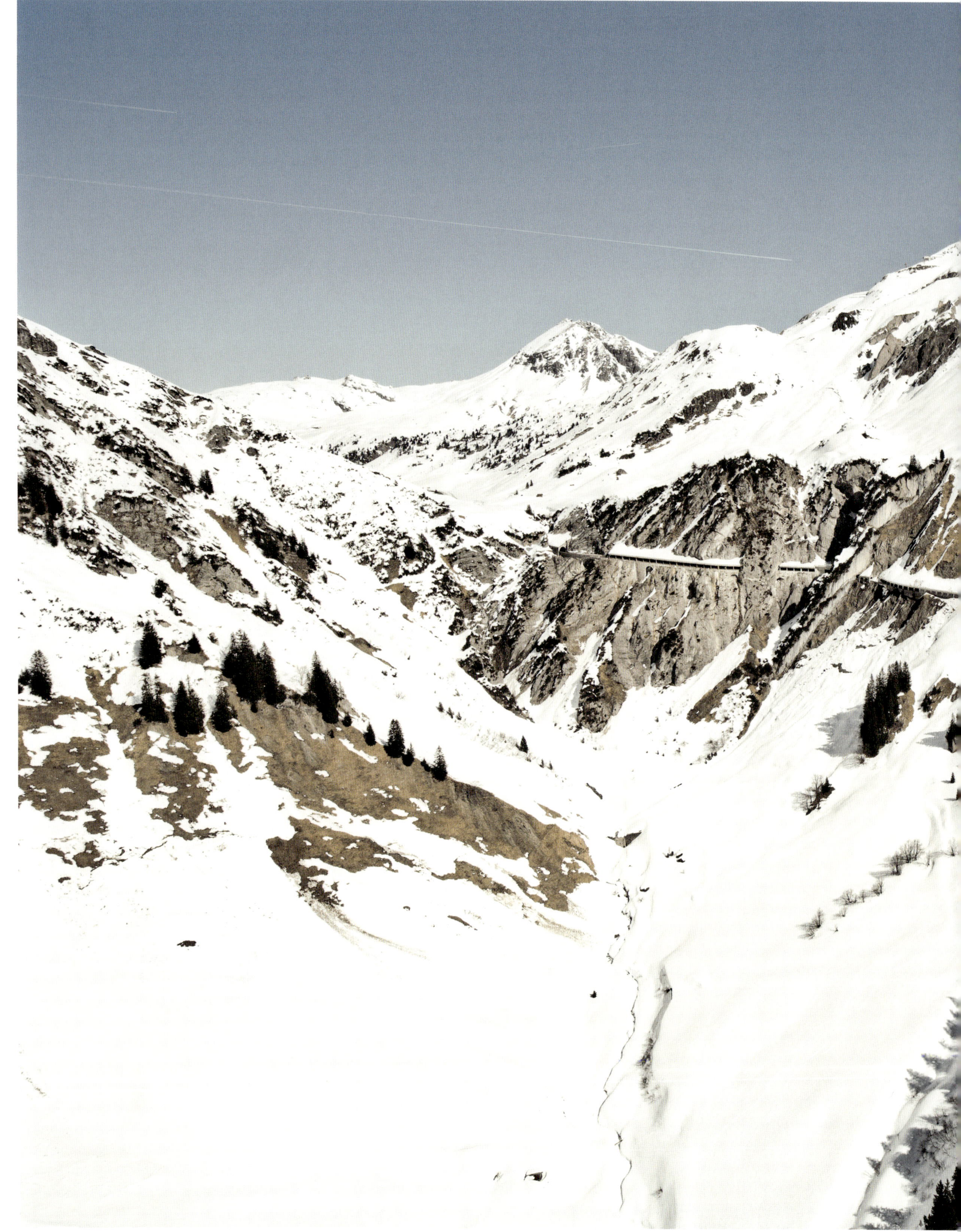

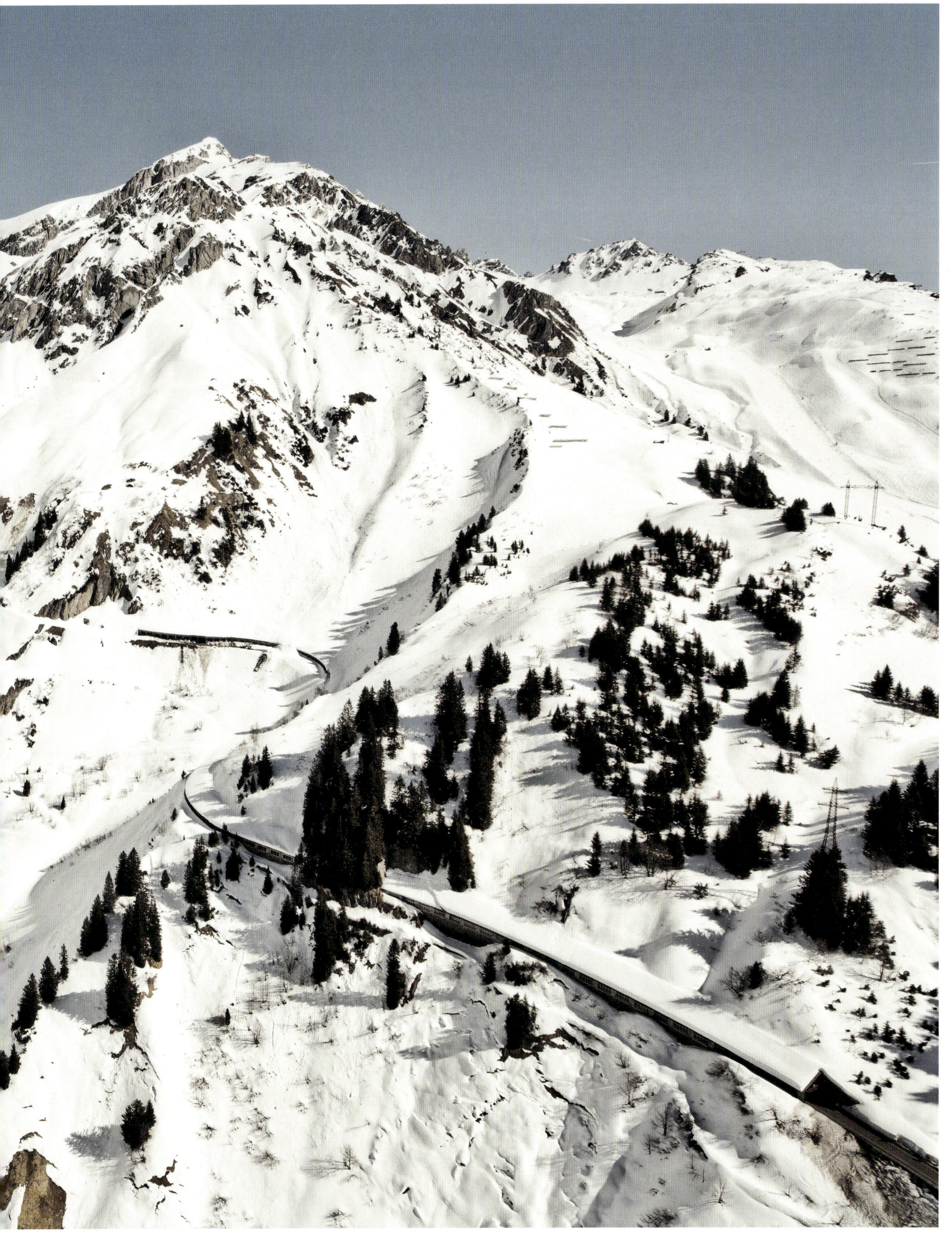

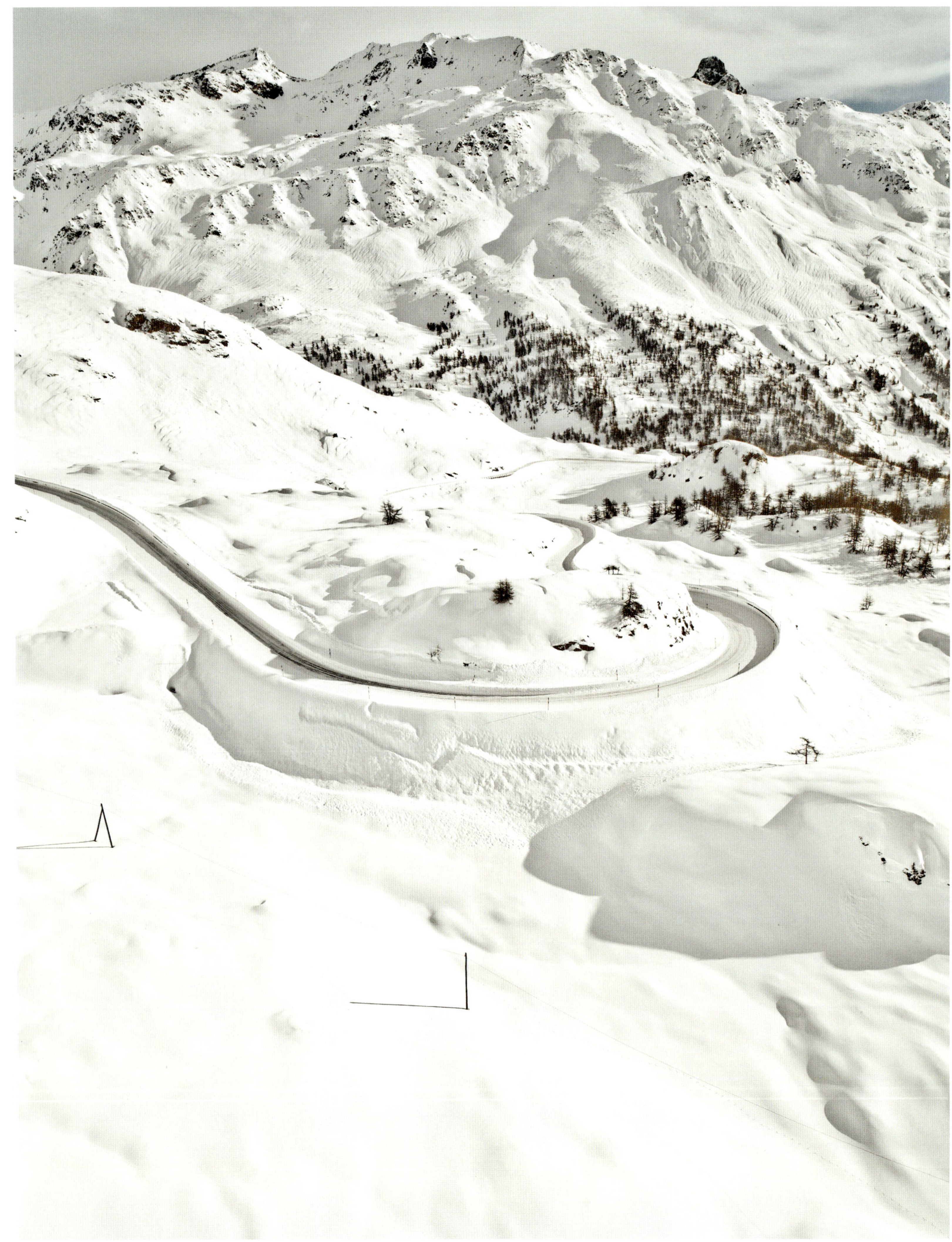

Berninapass

— 2328 Meter
Meist ganzjährig geöffnet

Der Berninapass verbindet das Oberengadin im Schweizerischen Kanton Graubünden mit dem Puschlav und dem italienischen Veltlin. Die Passstraße, die von Pontresina in eleganten Schwüngen hinauf zum Ospizio Bernina und zur Scheitelhöhe, dann weiter hinab nach San Carlo di Poschiavo führt, gehört zu den höchsten ganzjährig geöffneten Straßen der Alpen. Mit etwas Glück überholt man hier im Winter auch den berühmten roten Bernina-Express, der sich vor dem Gipfel des Piz Palü und dem zugefrorenen Lago Bianco durch die weiße Wunderlandschaft windet.

— 2,328 metres
Usually open all year round

The Bernina Pass connects the Upper Engadine in the Swiss canton of Grisons with Val Poschiavo and the Italian Valtellina. The pass road up from Pontresina, sweeping elegantly up to the Ospizio Bernina and on to the summit before twisting down to Sand Carlo di Poschiavo, is one of the highest all-season roads in the Alps. With a little luck, one can also overtake the famous red Bernina Express in winter as it winds its way through the white wonderland towards the summit of Piz Palü and the frozen Lago Bianco.

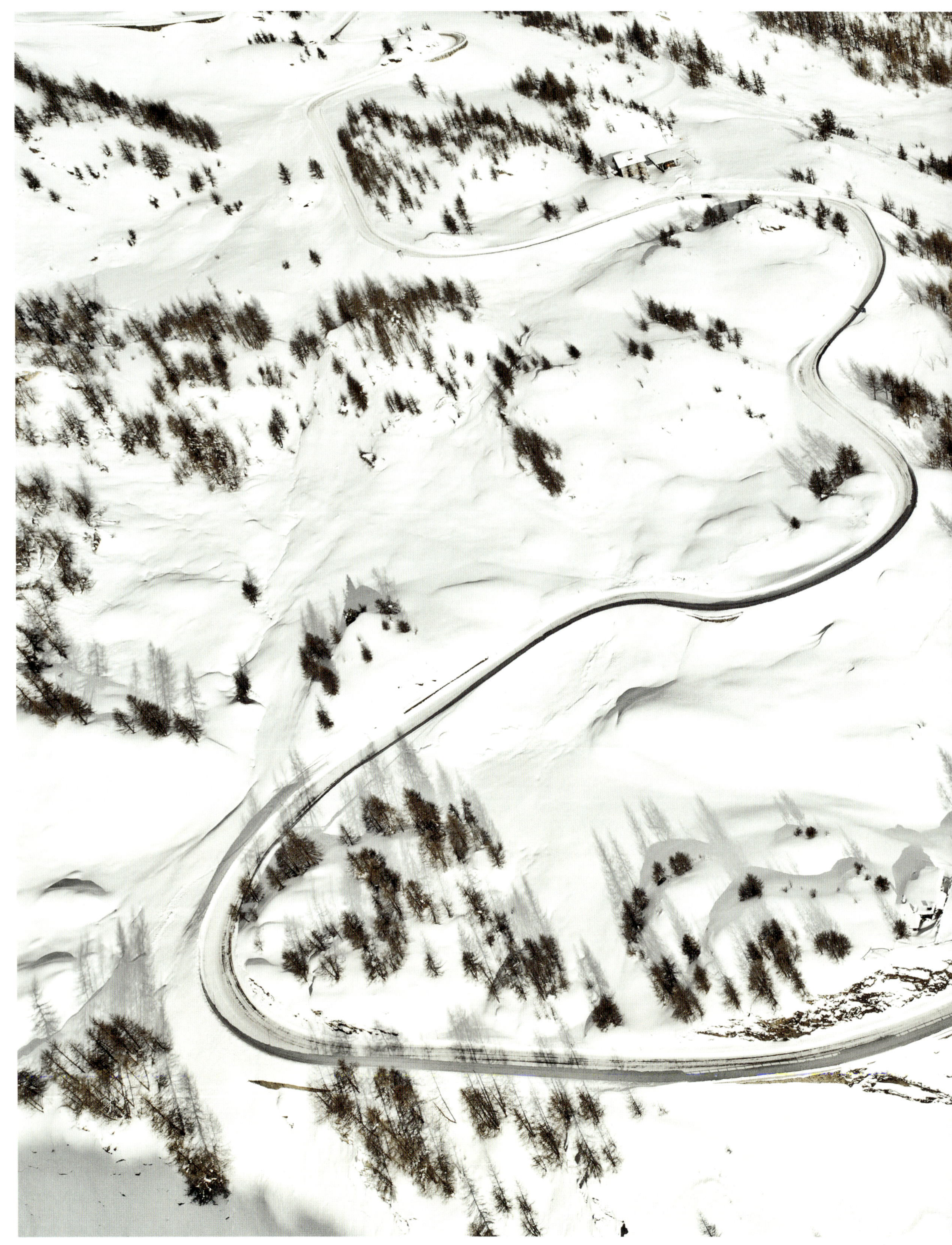

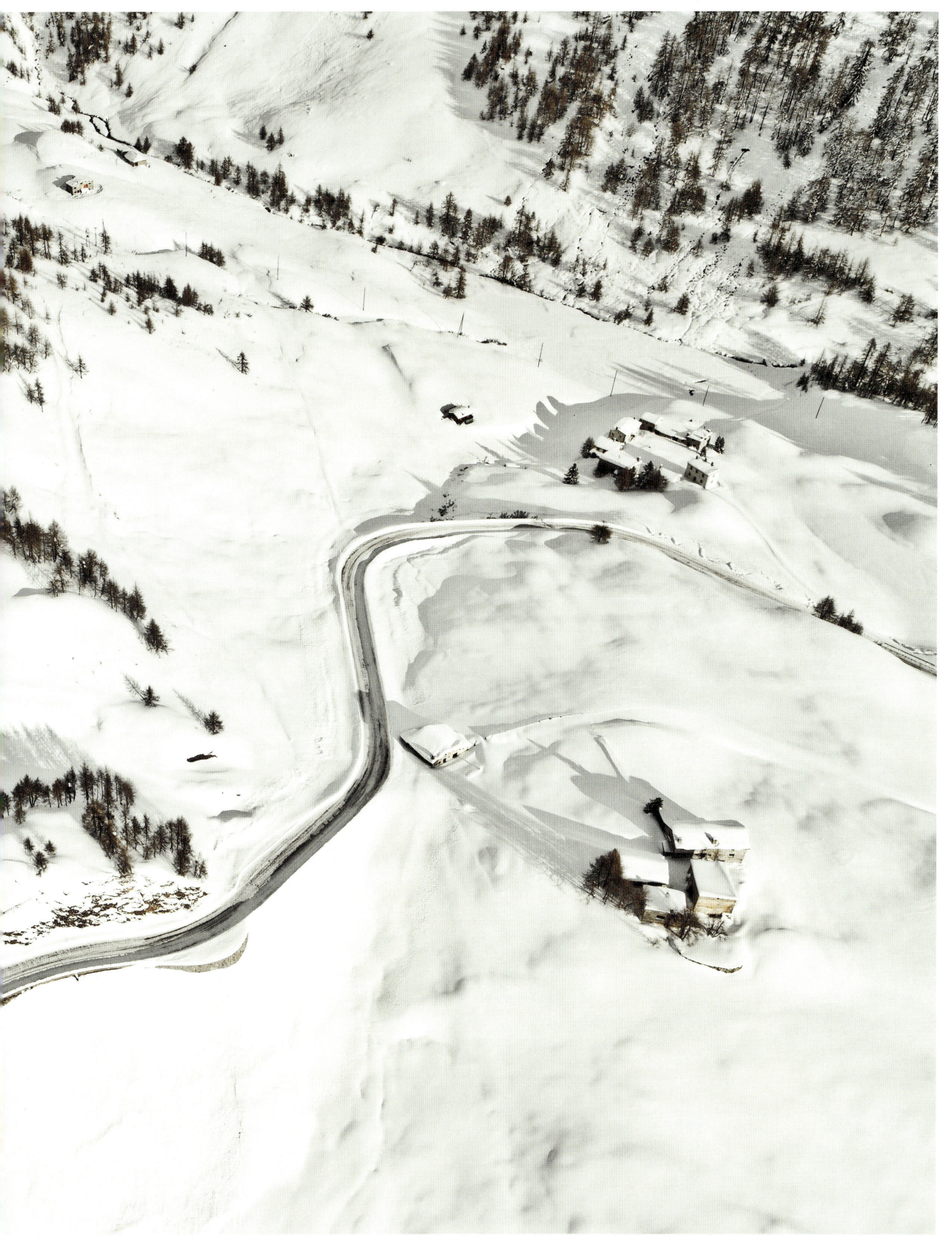

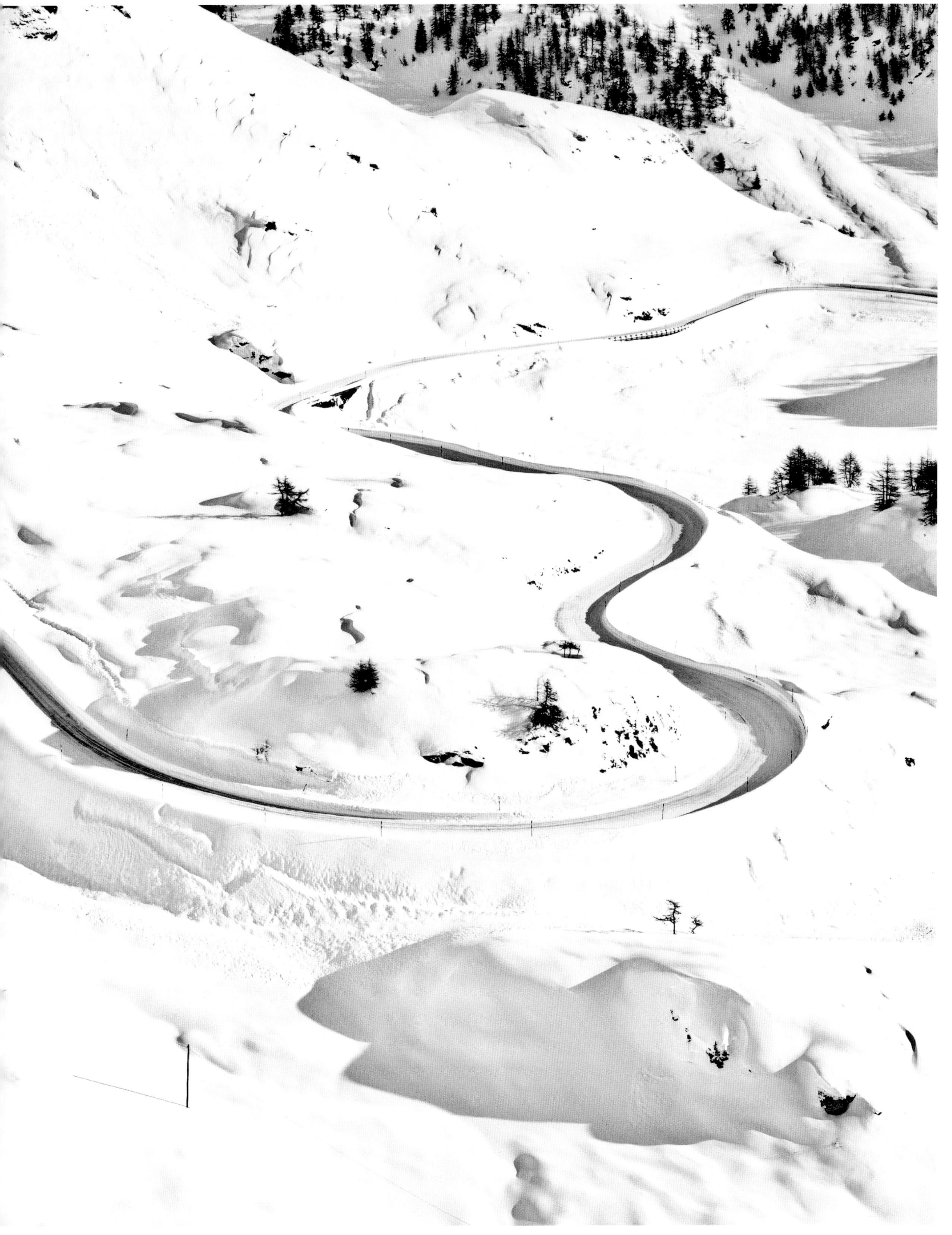

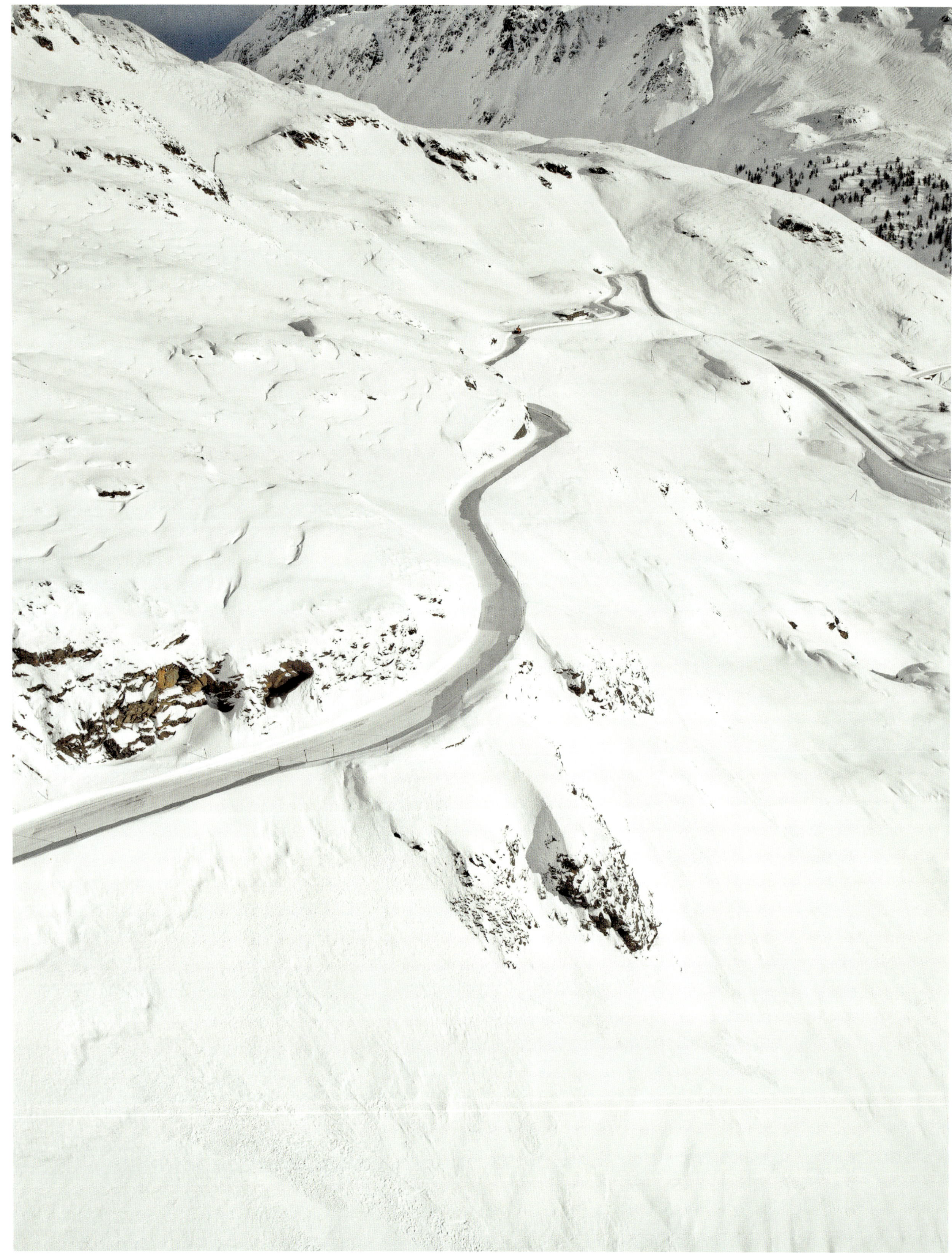

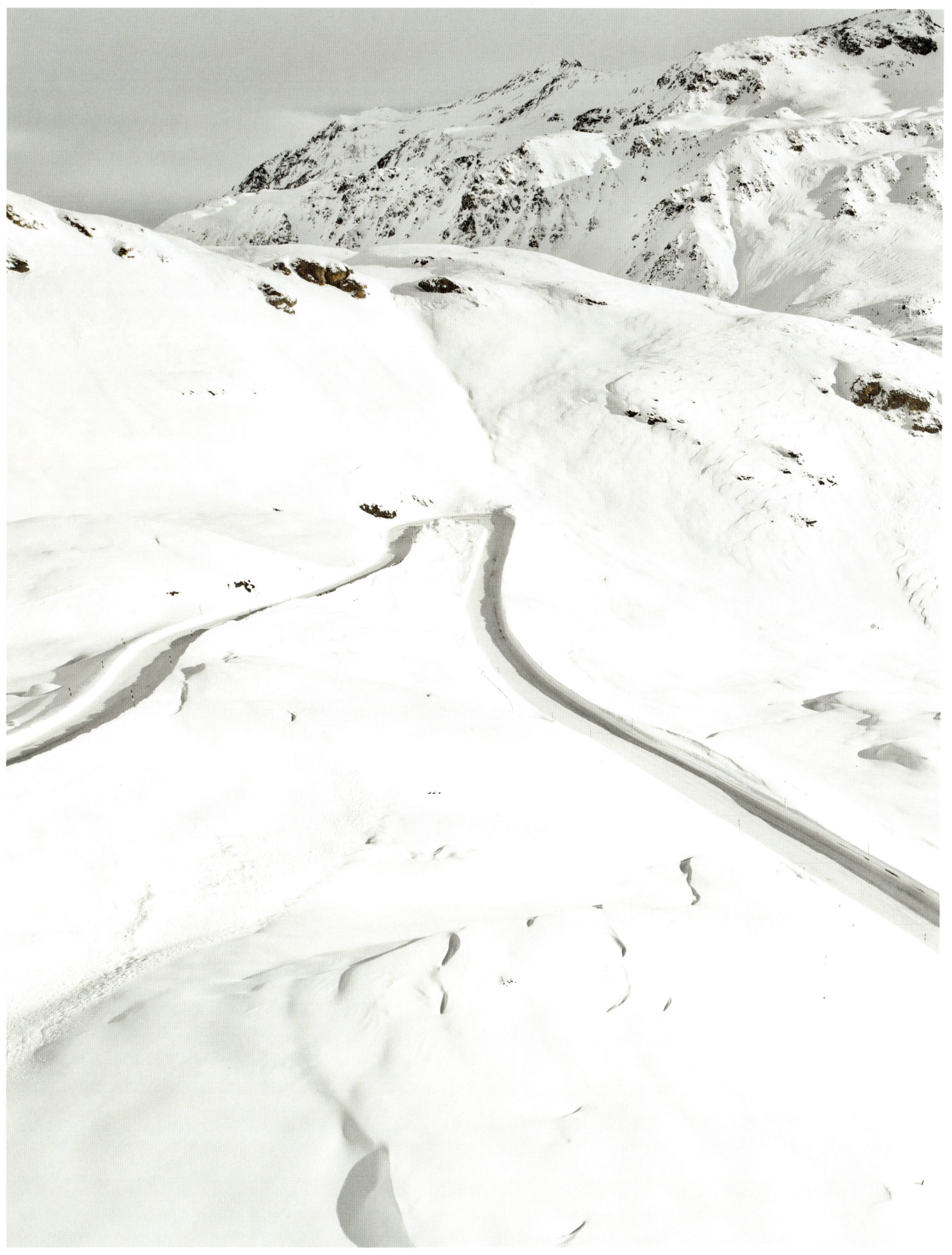

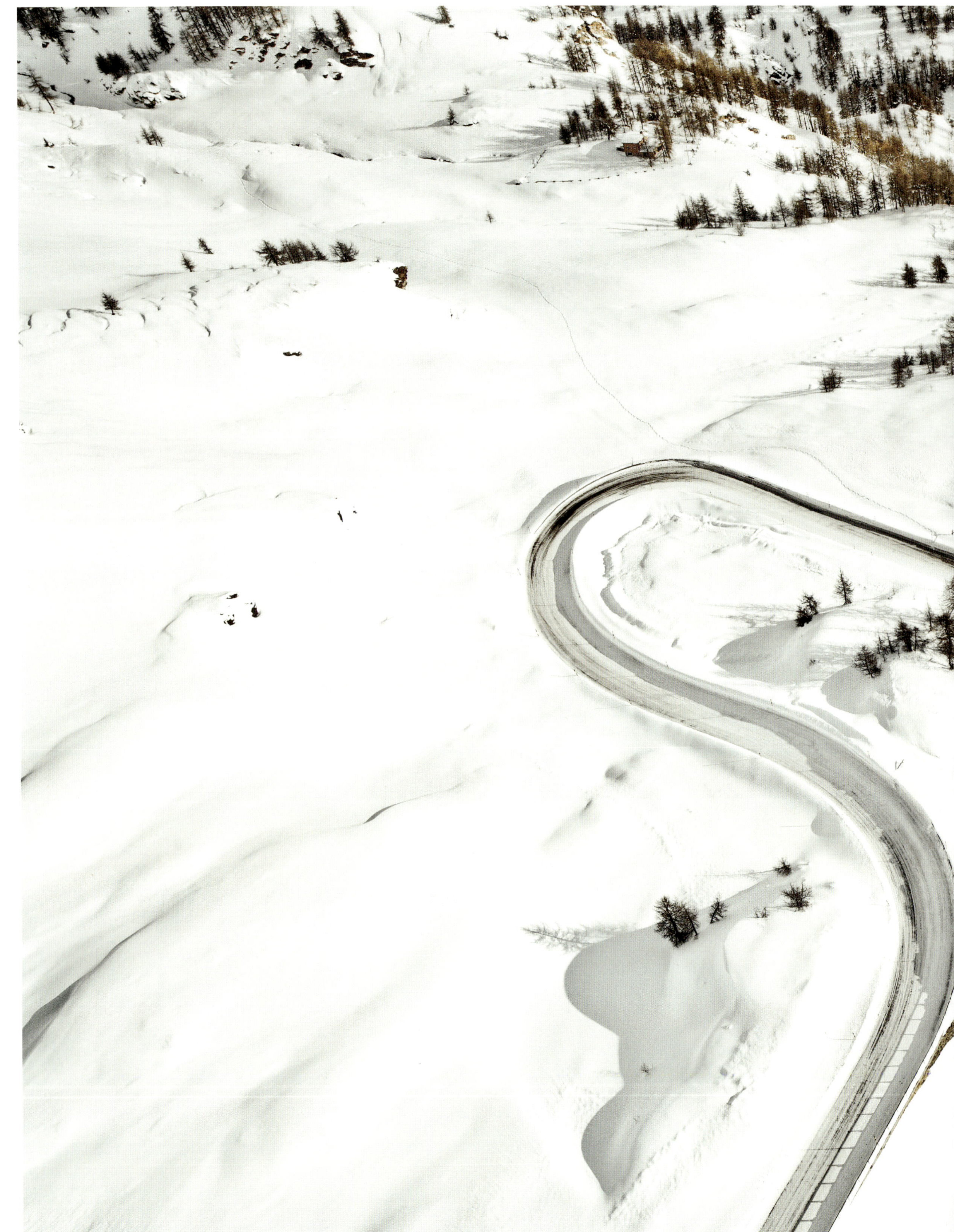

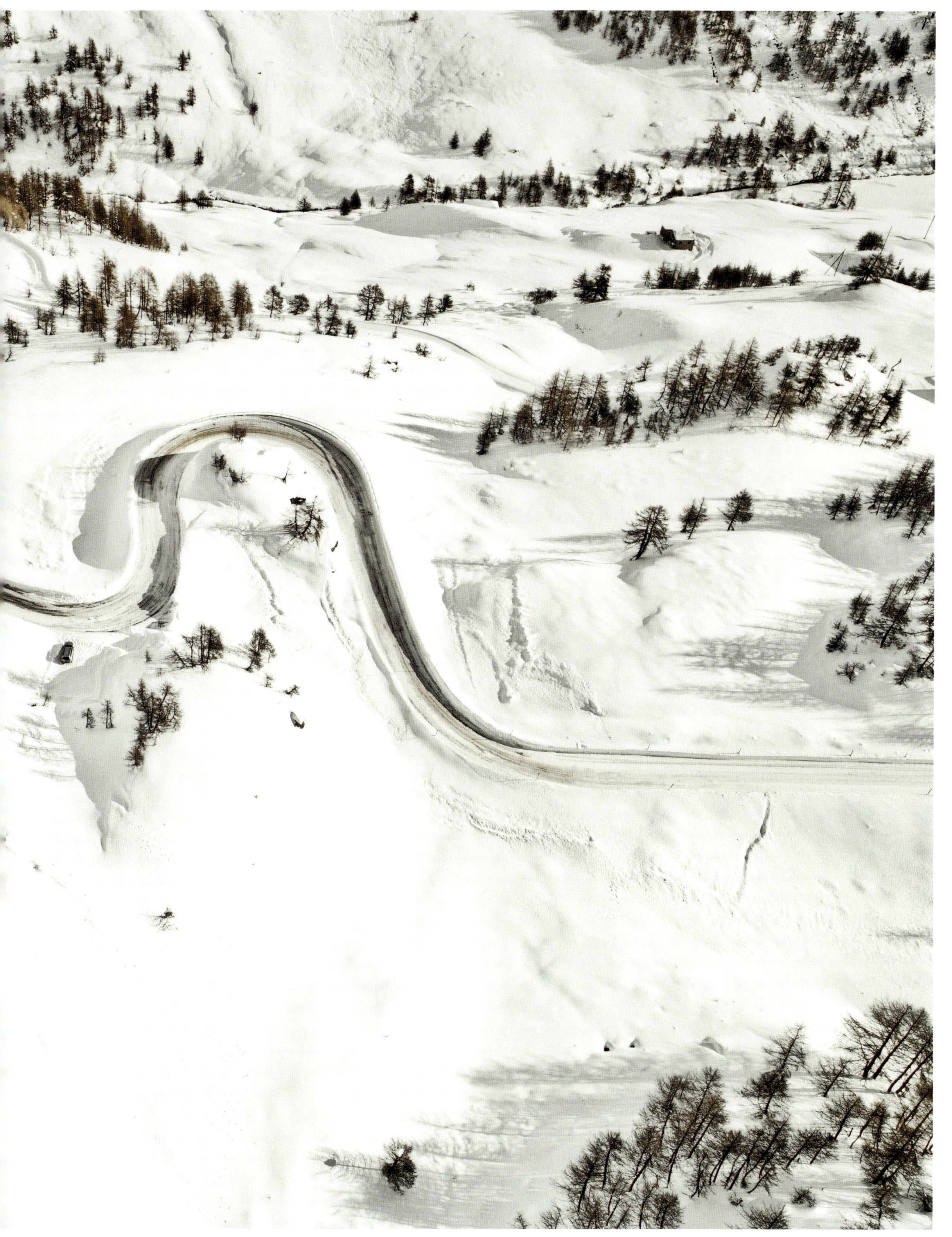

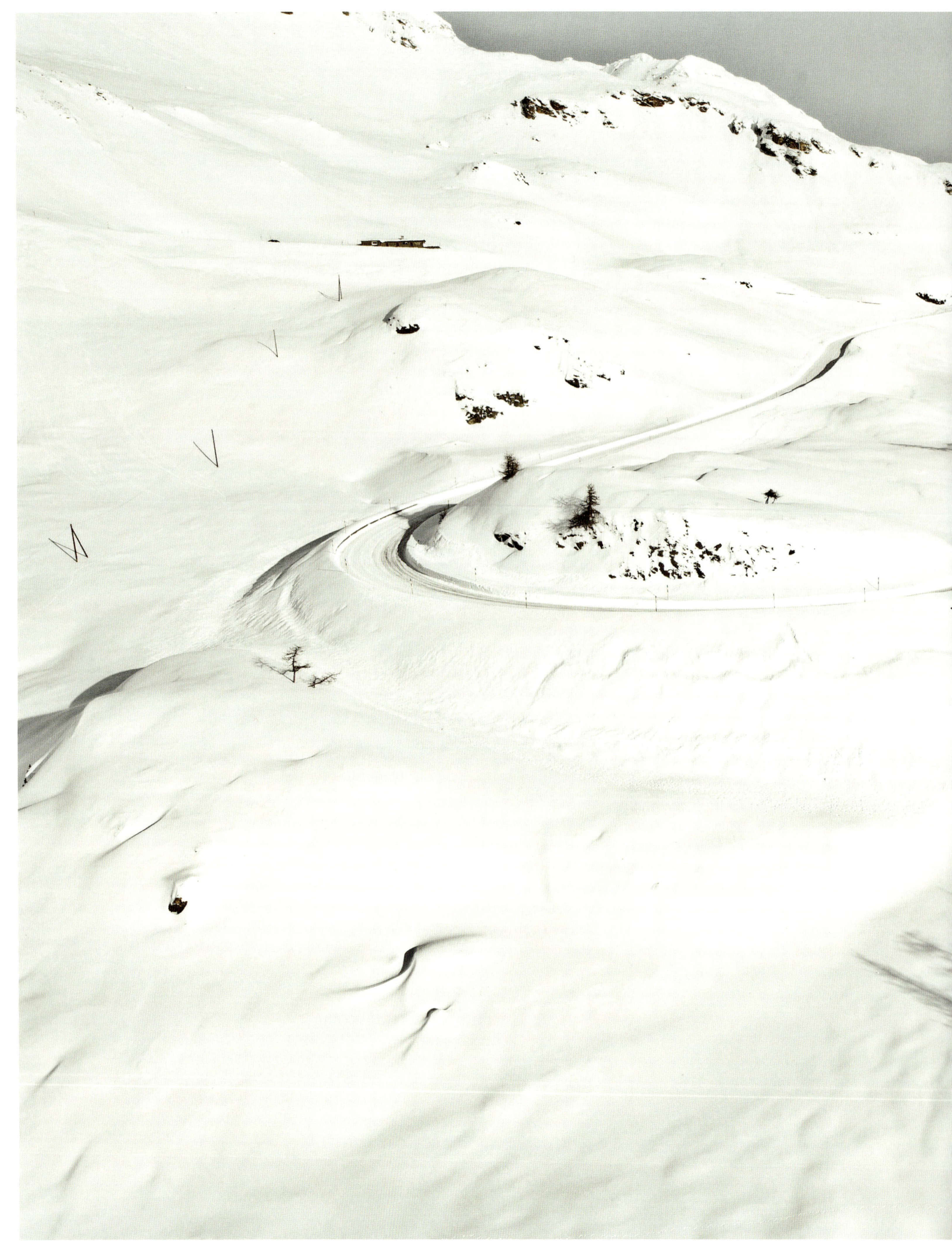

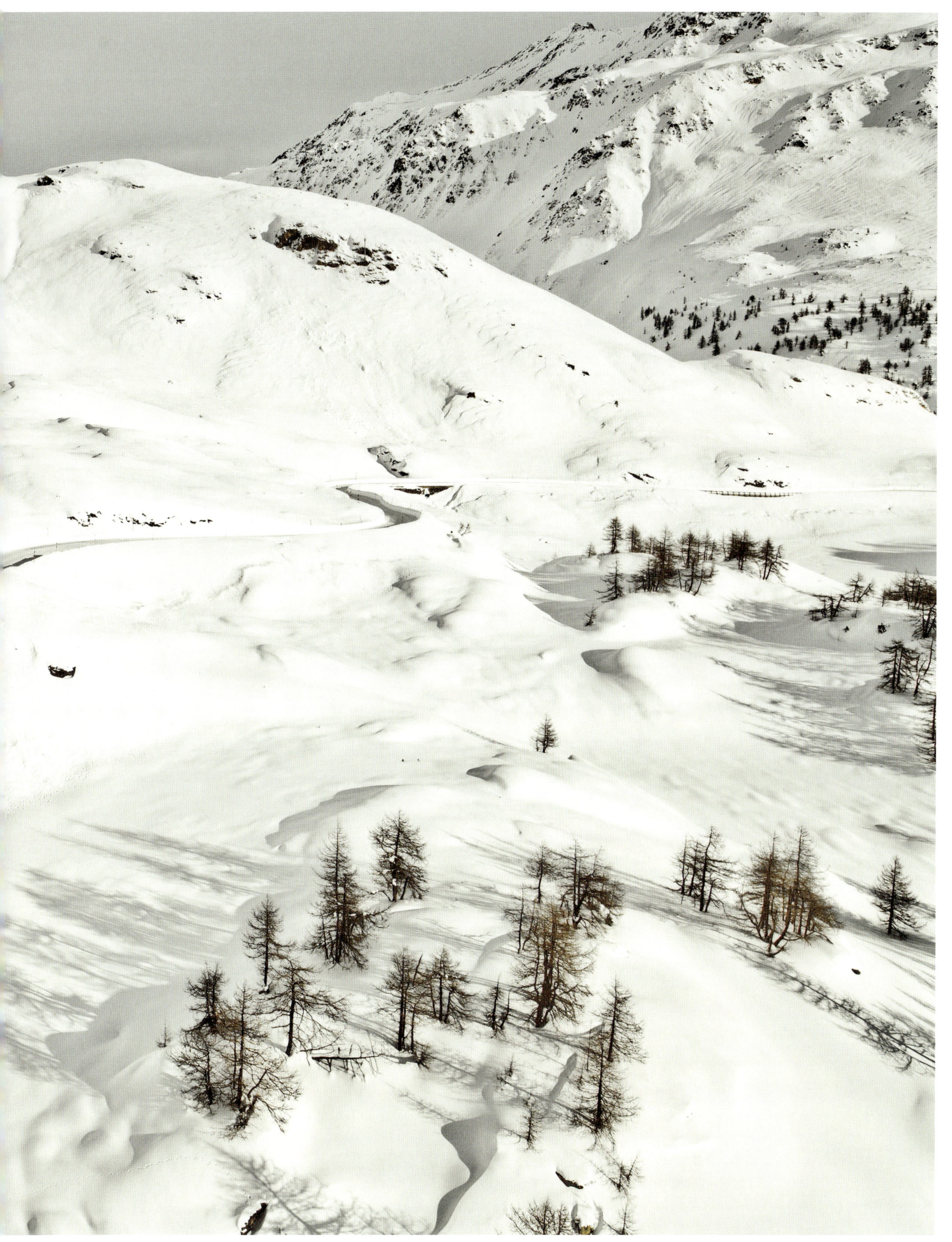

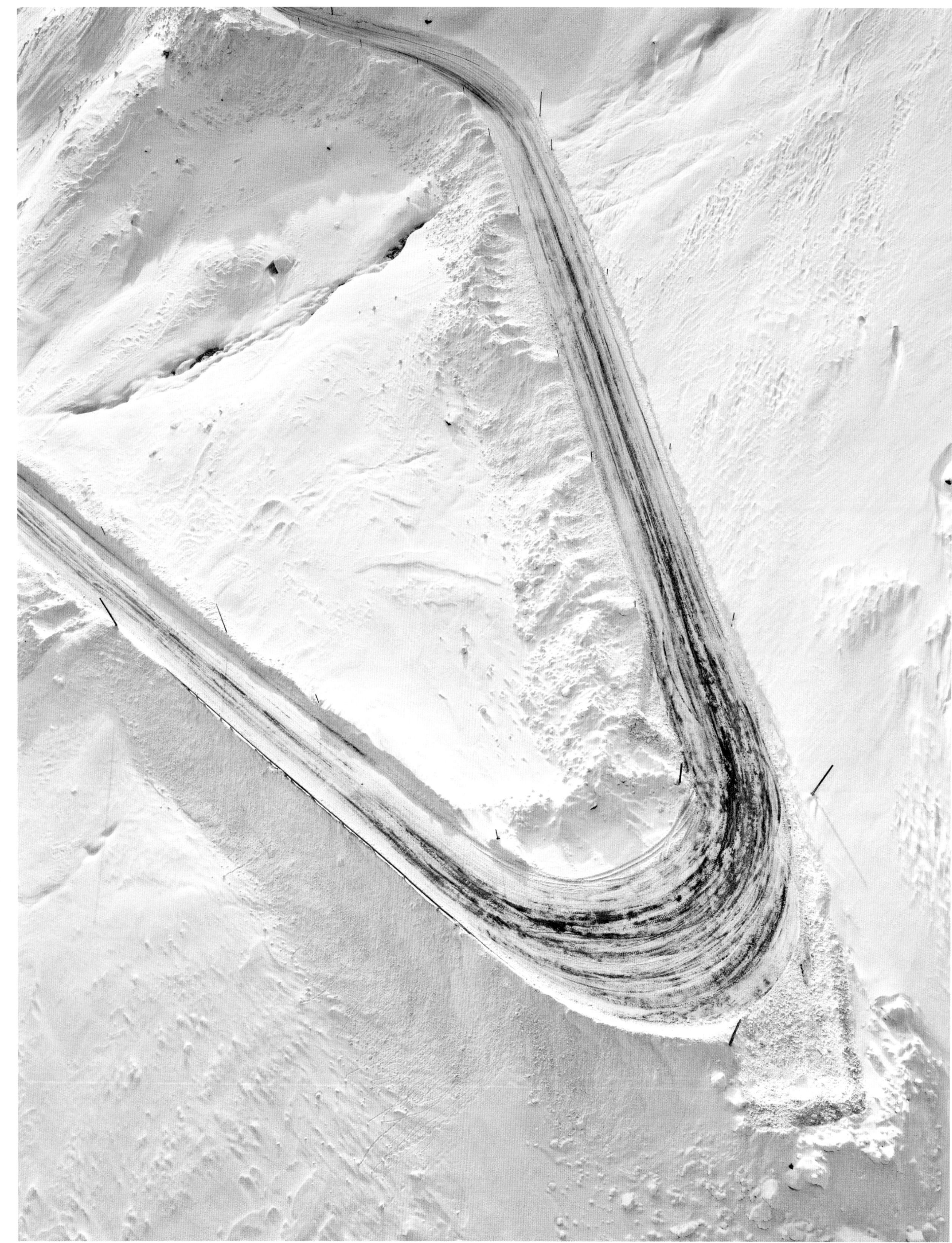

Flüelapass

— 2383 Meter
Wintersperre meist von November bis Mai

Der Flüelapass führt von Davos im Landwassertal nach Susch im Unterengadin. Im Winter wird die Passstraße zwar nicht mehr geräumt, sie ist von Januar bis Mai aber dennoch befahrbar – vorausgesetzt, man besitzt einen Wagen mit Allradantrieb, der auf den verschneiten und steilen Serpentinen nicht hoffnungslos ins Rutschen gerät. Manch ein Automobilhersteller testet auf dem Flüelapass die Wintertauglichkeit seiner neuesten Modelle.

— 2,383 metres
Usually closed in winter from November to May

The Flüela Pass leads from Davos in the Landwasser valley to Susch in Lower Engadine. In winter the pass is no longer cleared, but it is still navigable from January to May – provided one has an all-wheel-drive vehicle that doesn't begin to hopelessly slip and slide on the snowy, steep switchbacks. Some car manufacturers use the Flüela Pass to test the winter performance of their latest models.

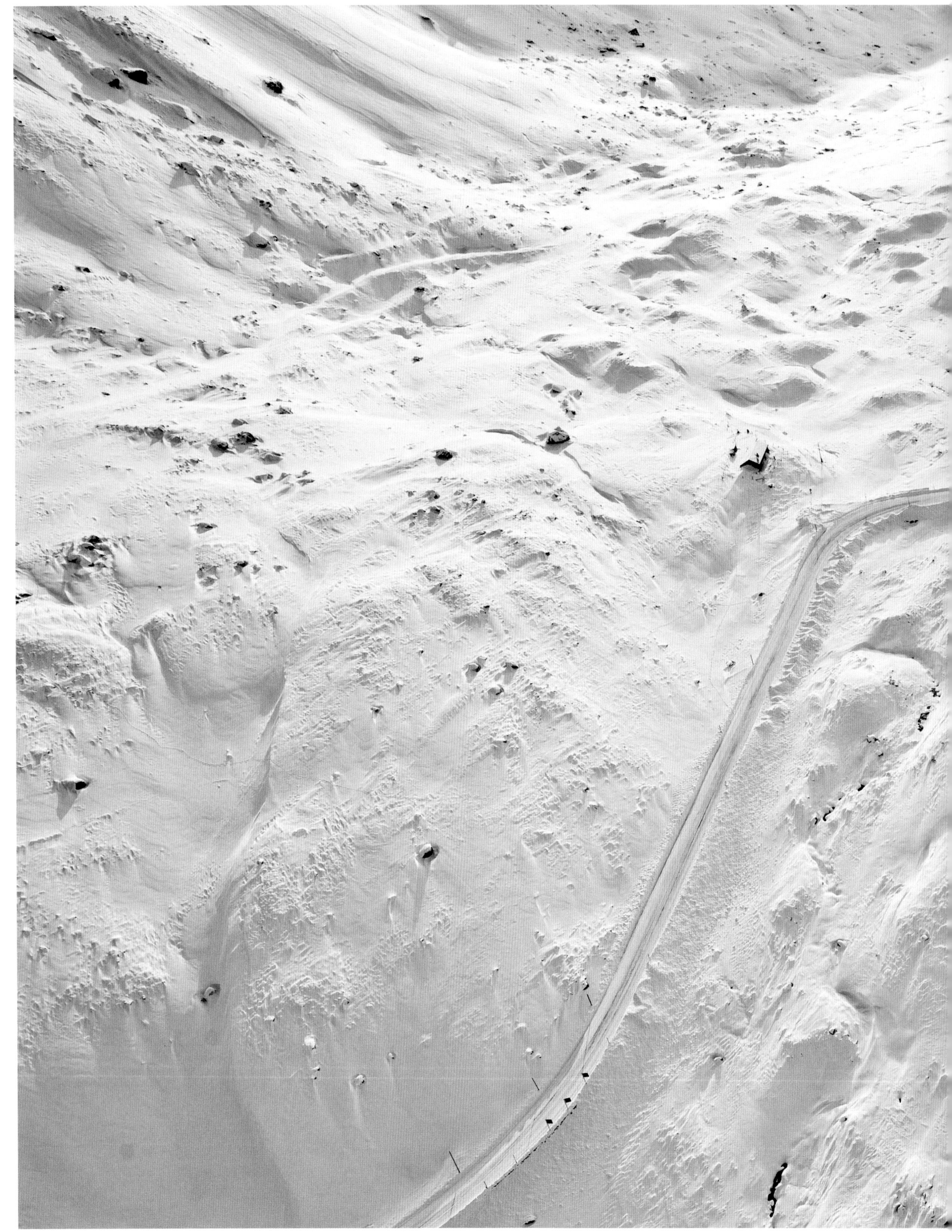

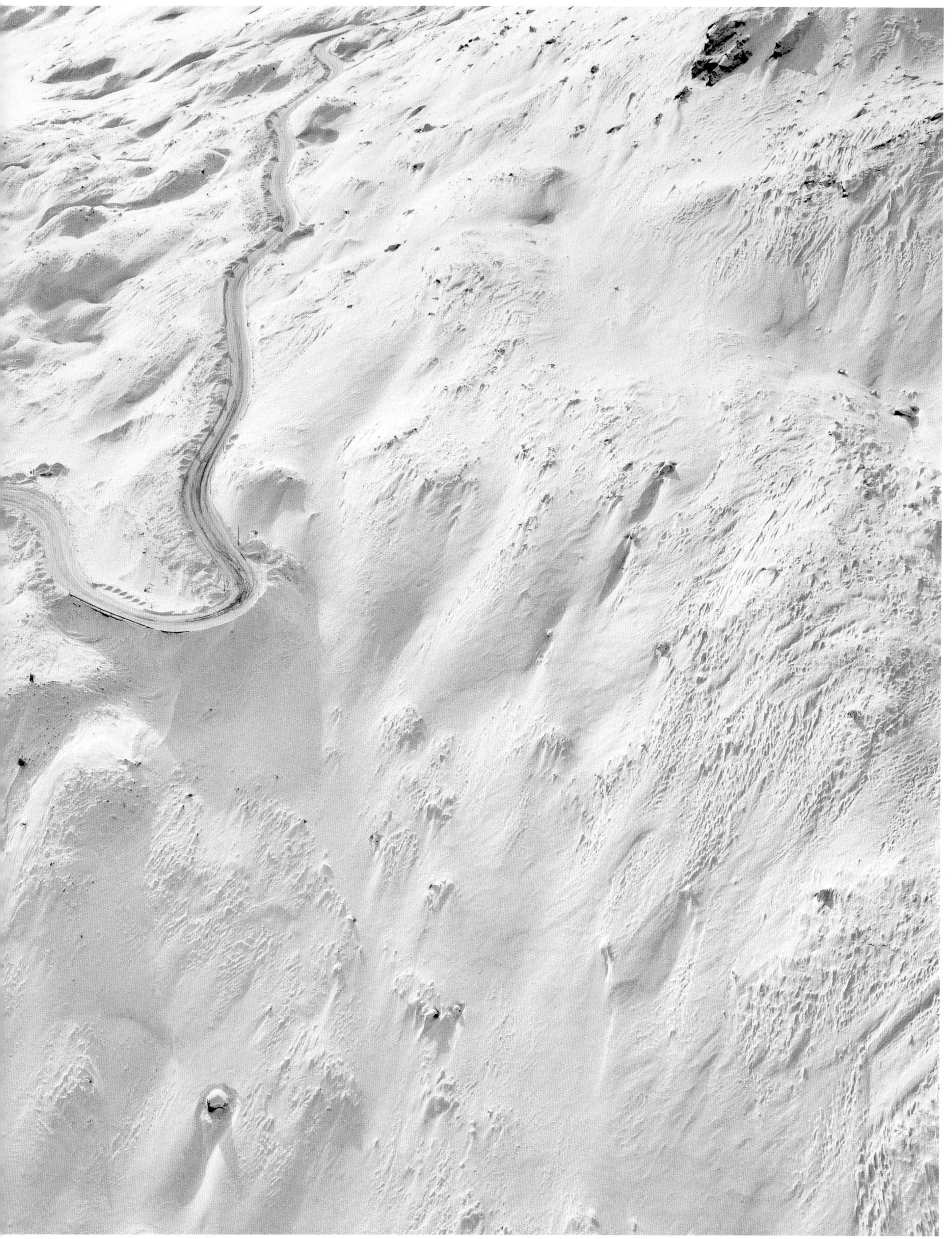

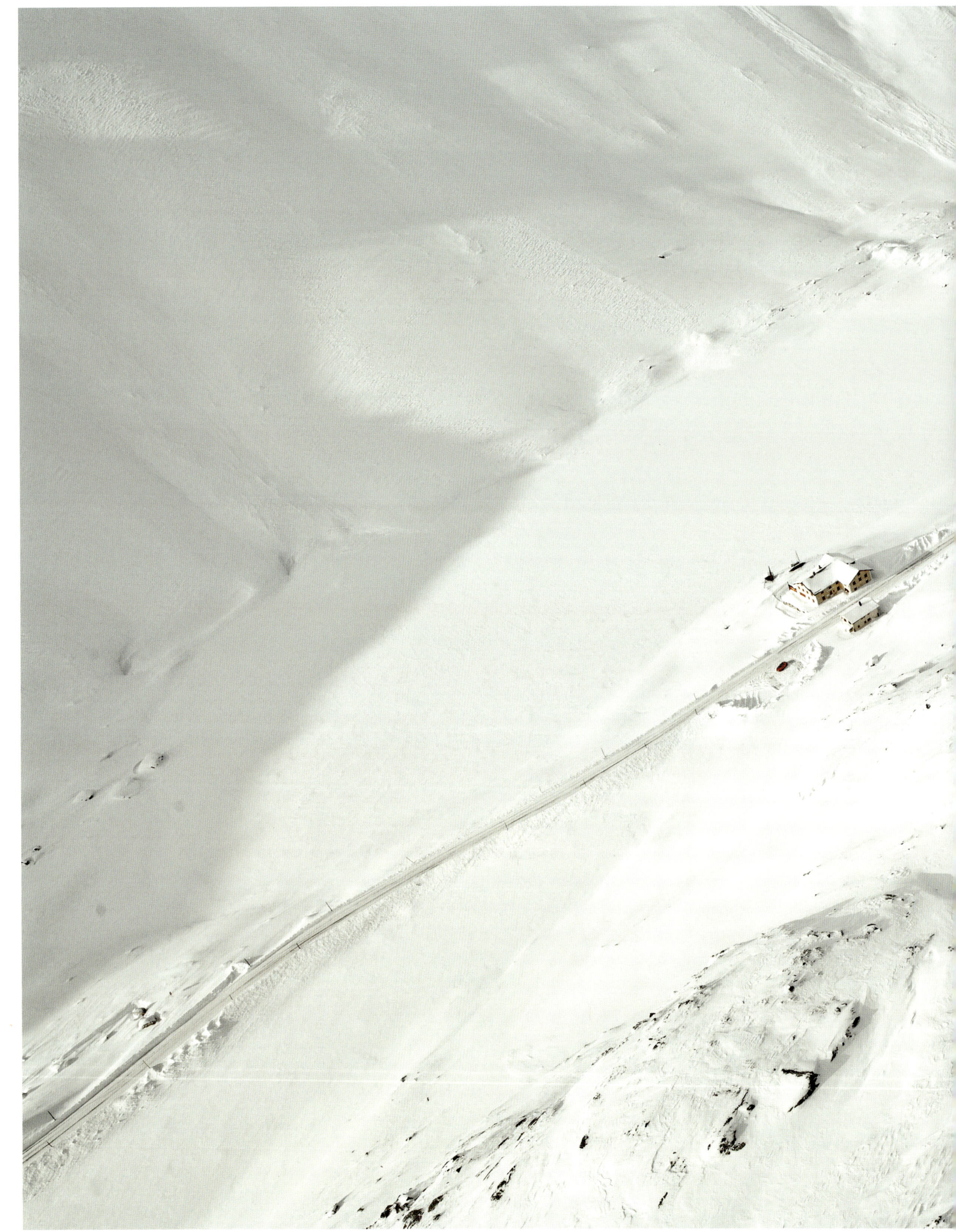

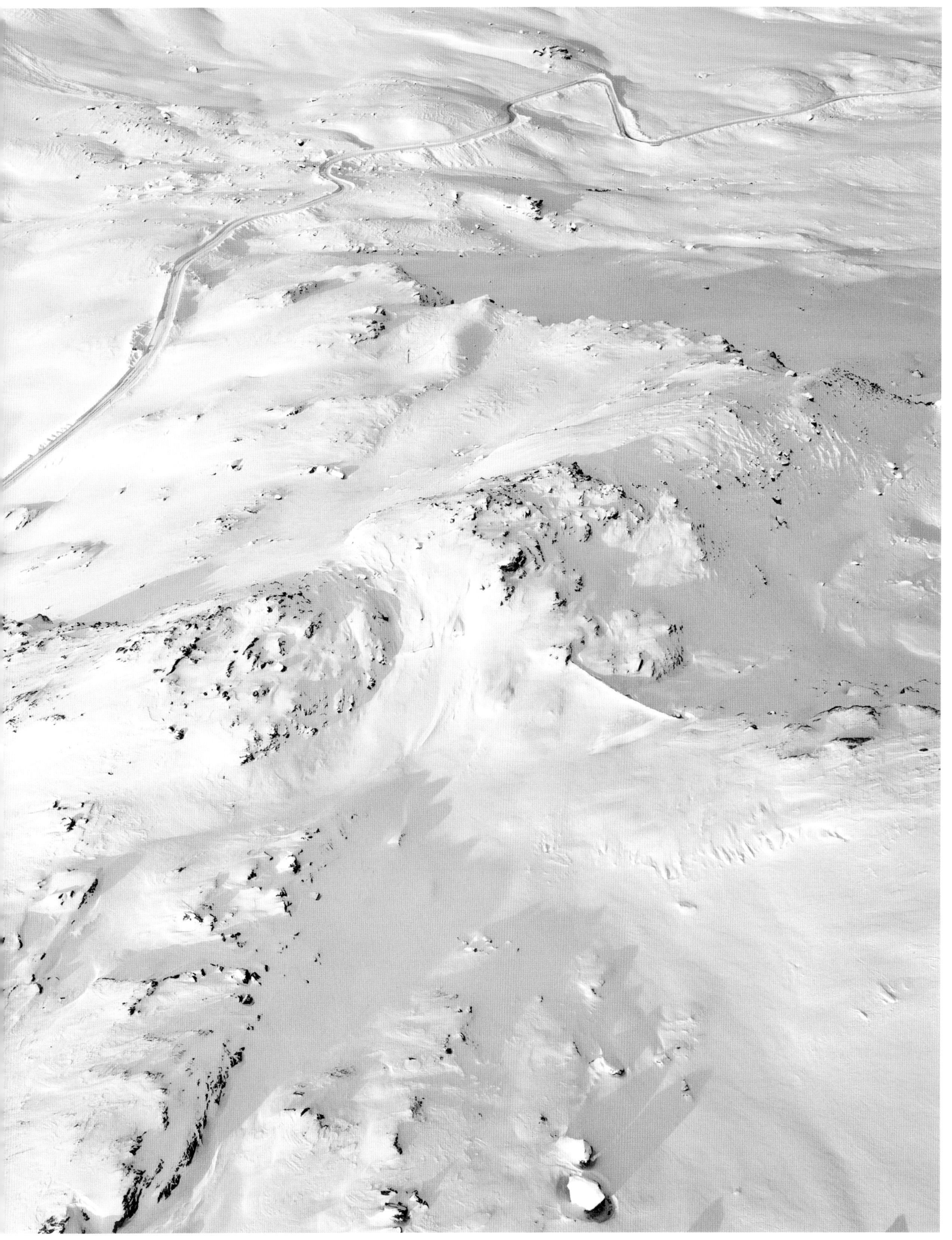

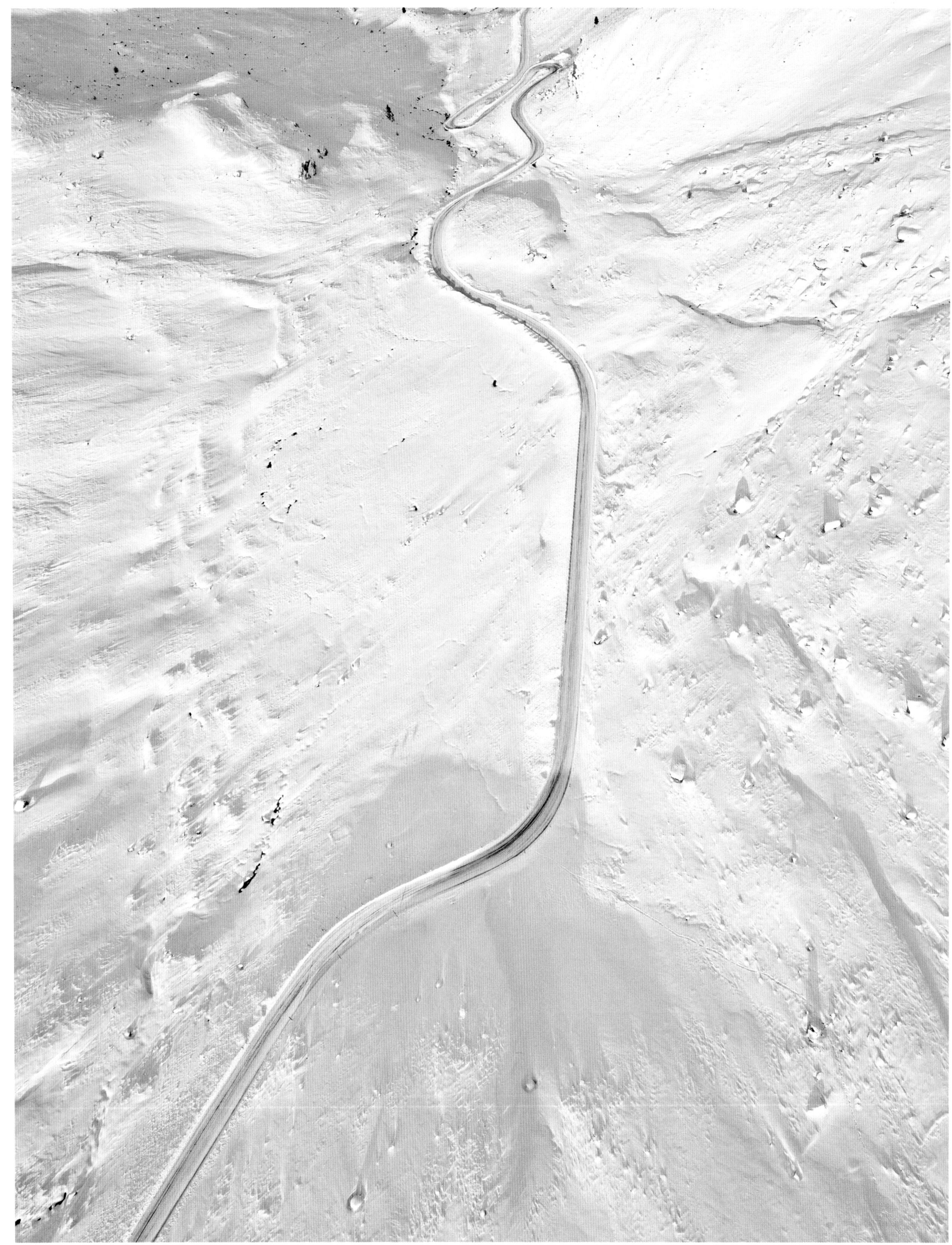

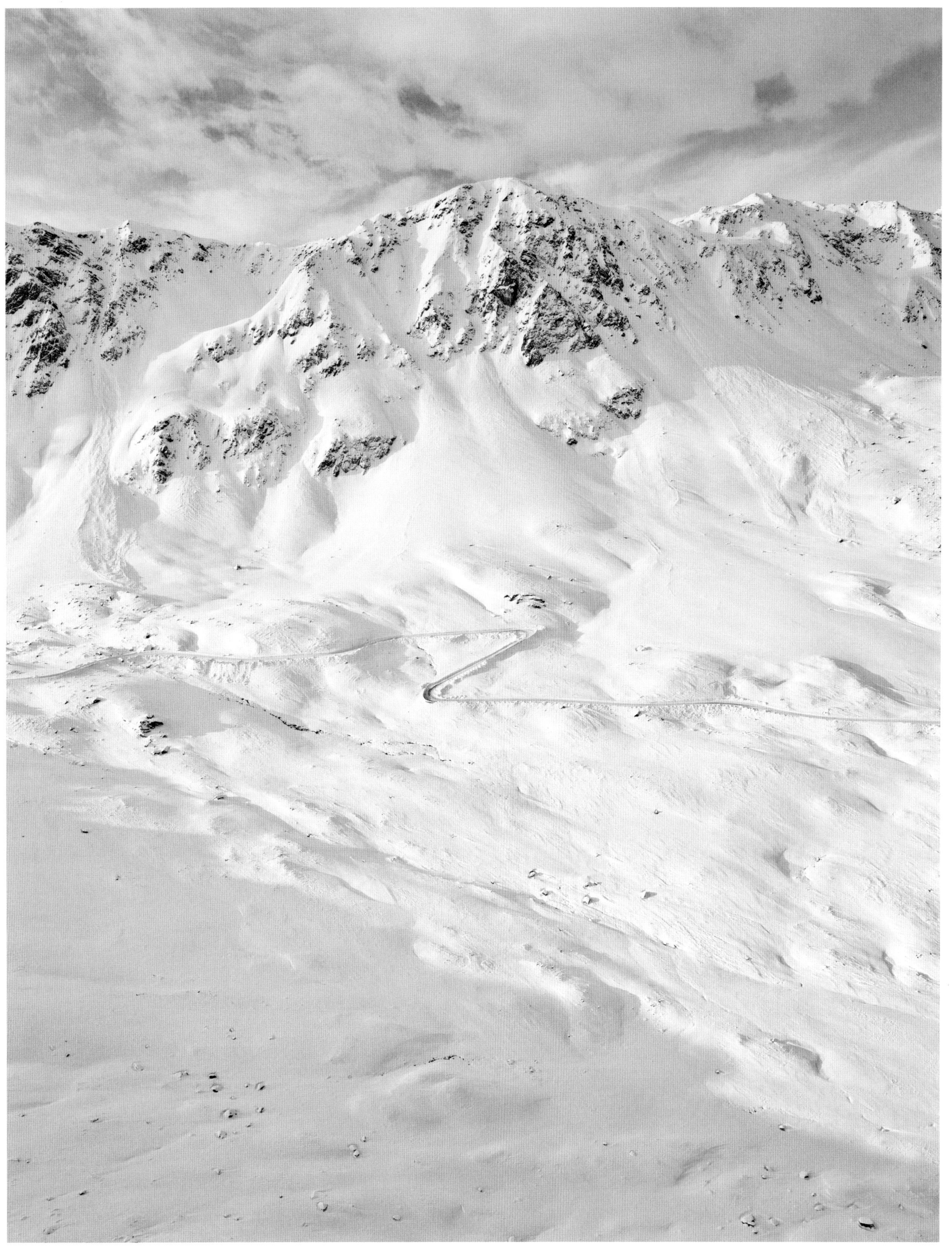

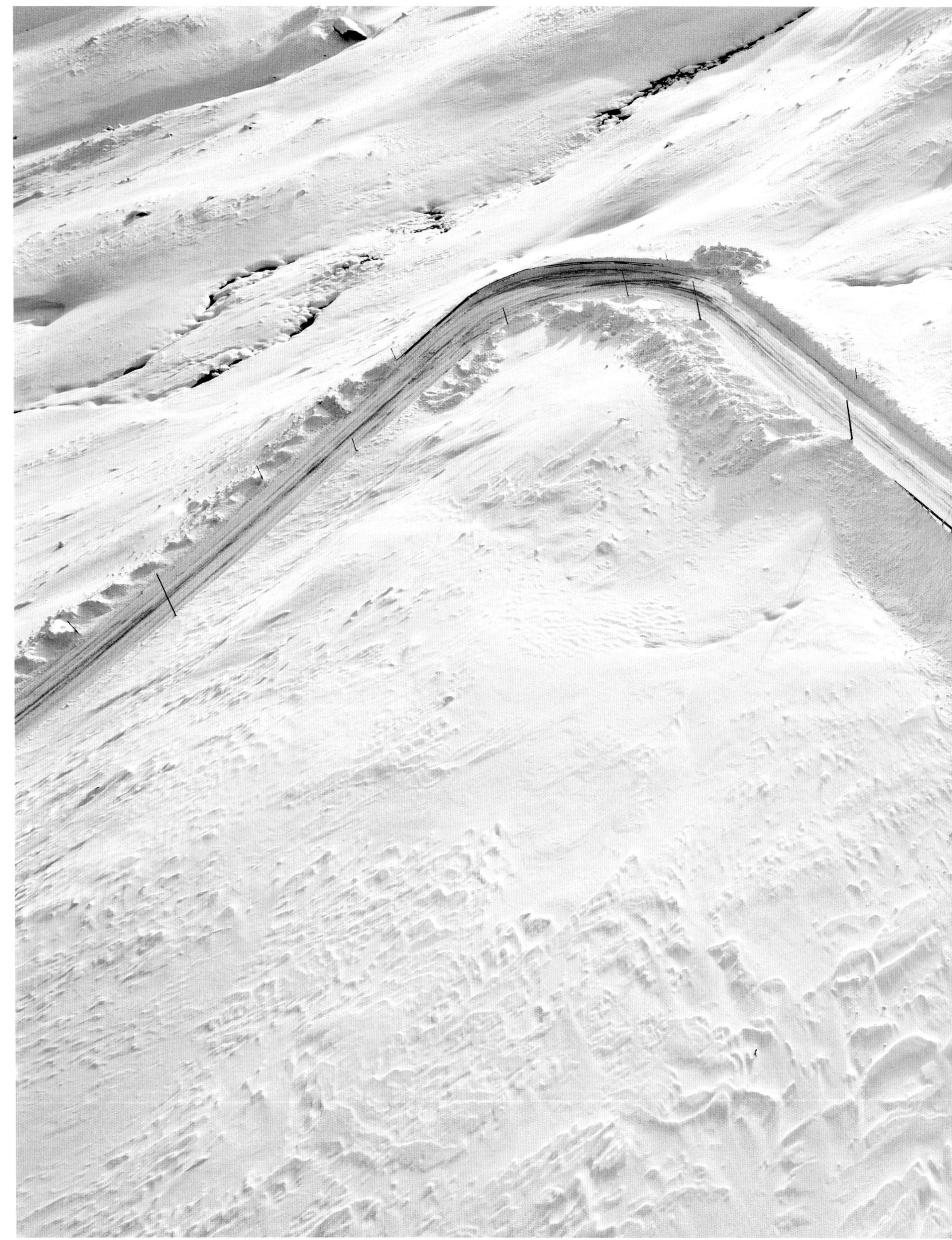

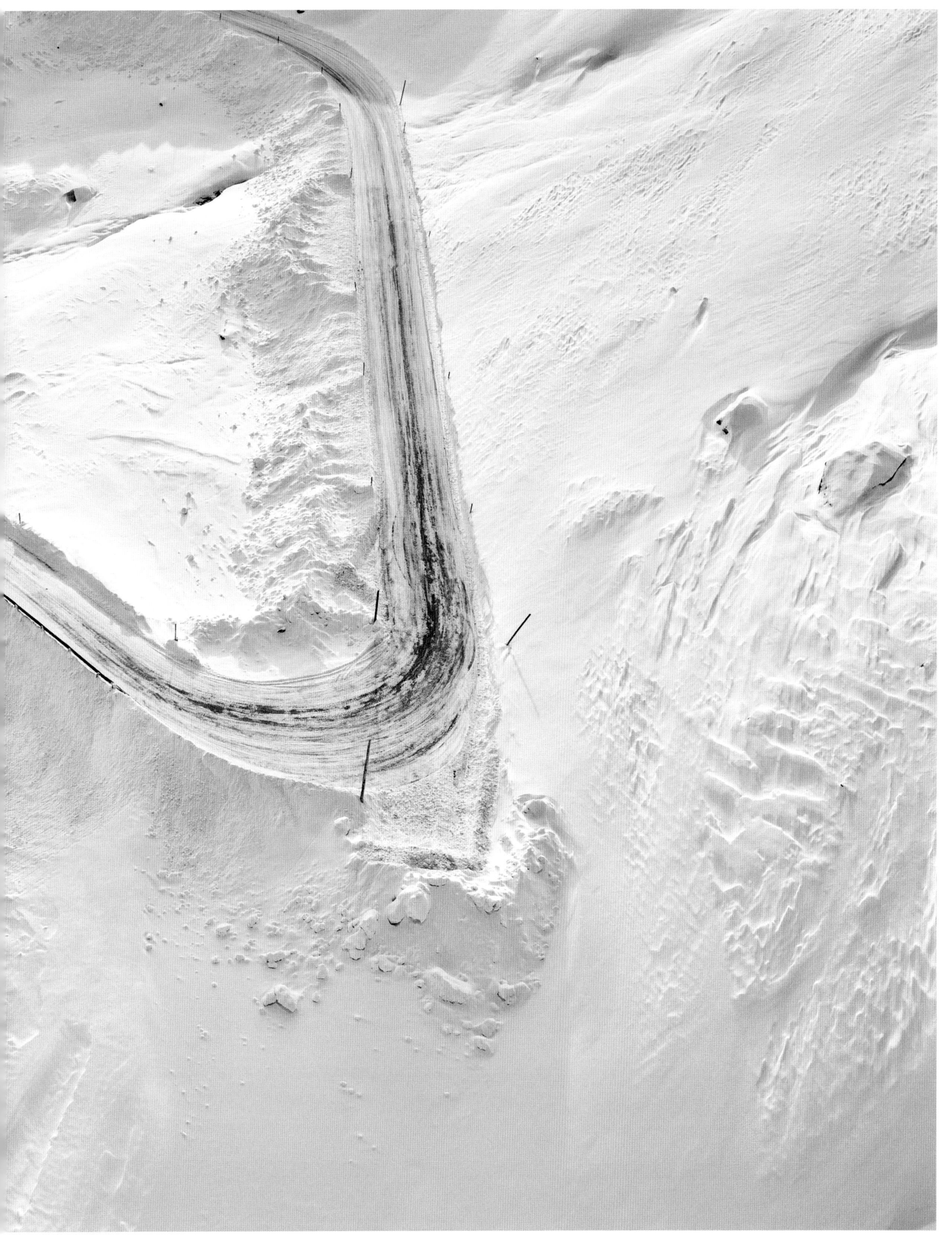

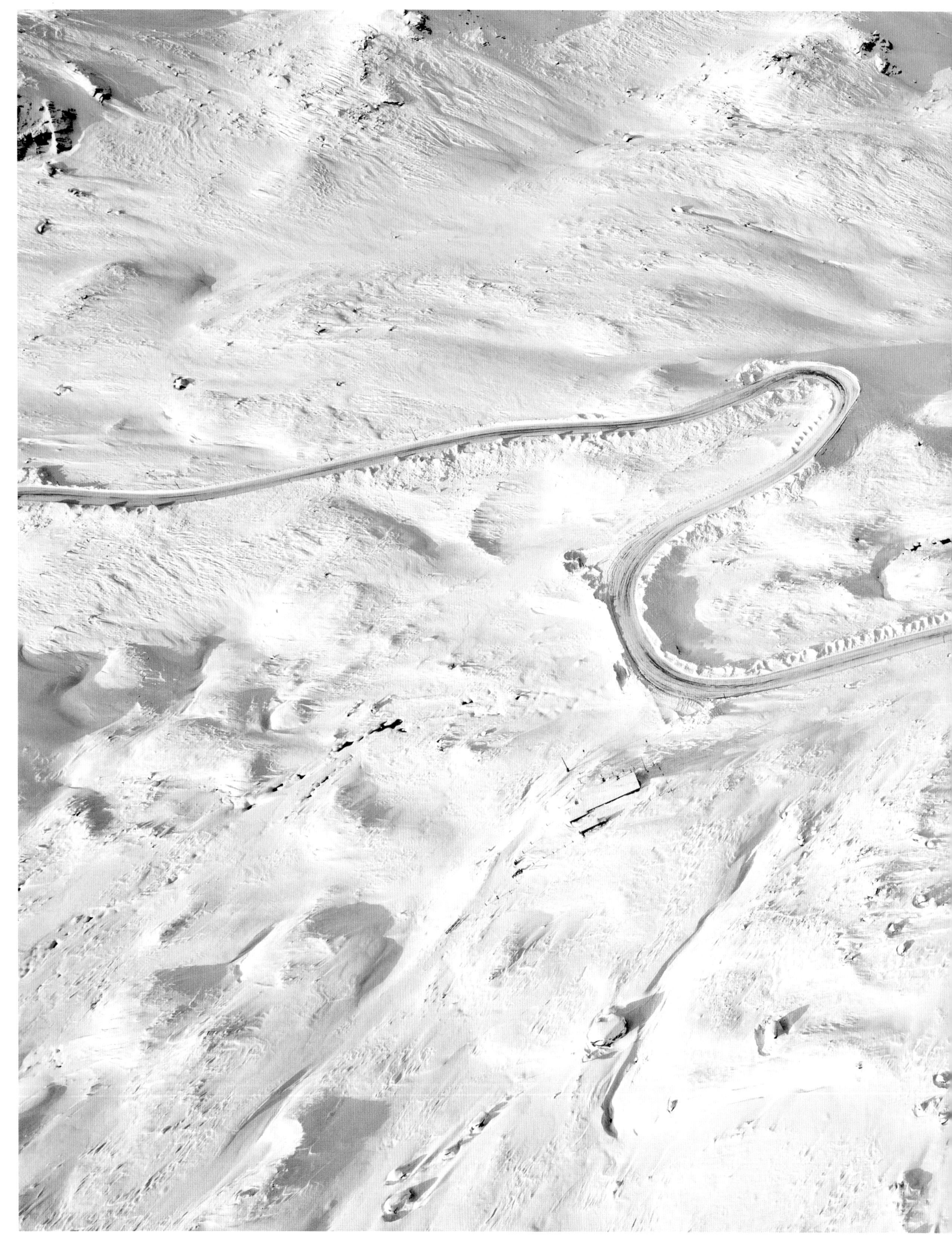

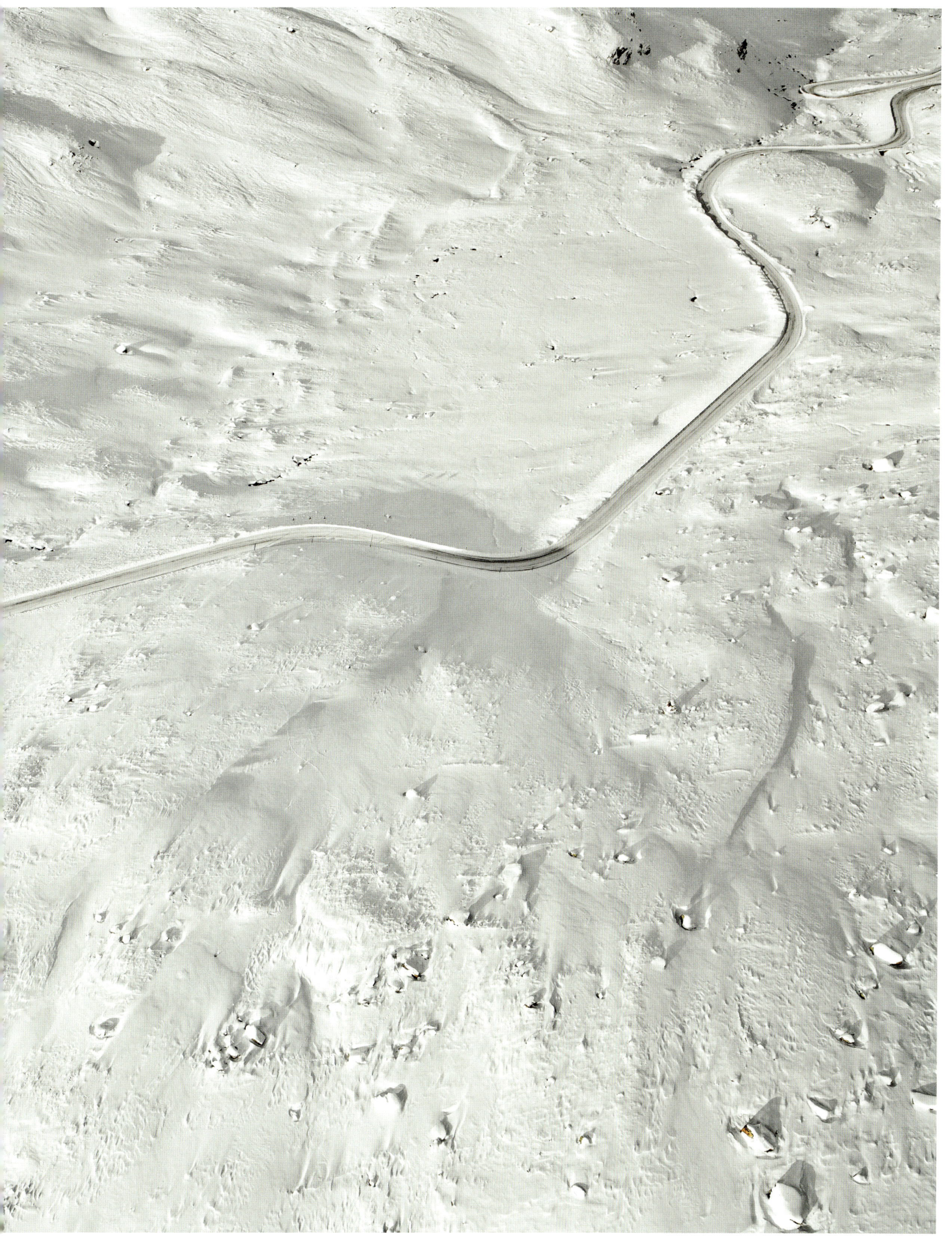

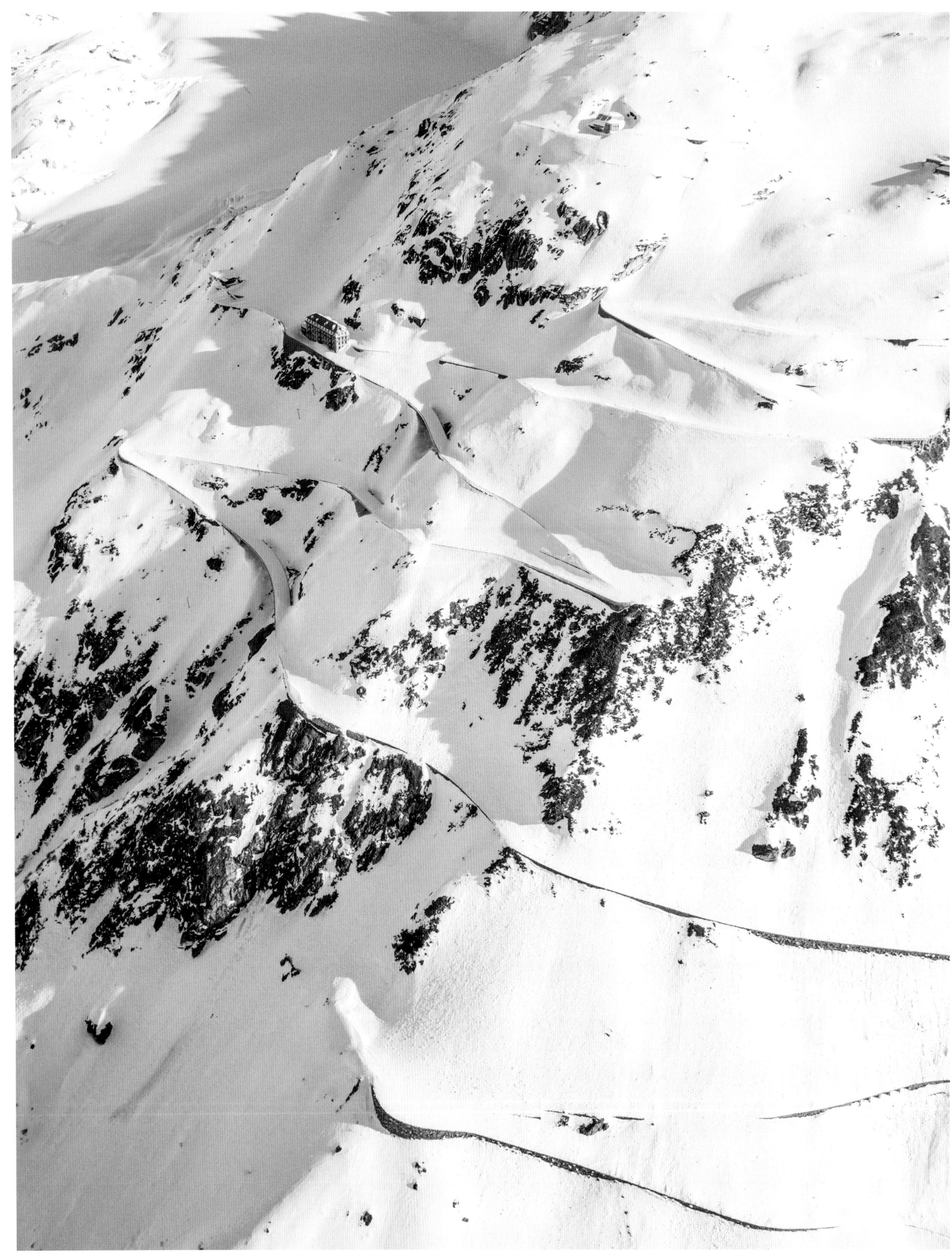

Furkapass

— 2431 Meter
Wintersperre meist von Oktober bis Mai

Von Realp im Kanton Uri führt die Furkapassstraße über den Scheitelpunkt auf 2431 Meter Höhe und mit einzigartigem Blick auf den Rhonegletscher hinunter nach Gletsch im Kanton Wallis. Im Winter ist die Passstraße zwar für Autos geschlossen, sie kann aber als Wanderweg genutzt werden. Der wohl berühmteste Flaneur, der den Furkapass bei Schneetreiben überquerte, war übrigens Johann Wolfgang Goethe im Jahr 1779. Das Hotel Tiefenbach am Wegesrand, in dessen Stube er sich bei einer heißen Ovomaltine hätte aufwärmen können, gab es damals allerdings noch nicht.

— 2,431 metres
Usually closed in winter from October to May

From Realp in the Uri canton, the Furka Pass road leads over the summit at 2,431 metres and allows stunning views to the Rhone Glacier down to Gletsch in the canton of Valais. Although the Pass is closed to traffic in the winter, it can be experienced as a hiking trail. Incidentally, the most famous stroller to traverse the snowy Furka Pass was Johann Wolfgang Goethe in the year 1779. The Hotel Tiefenbach at the roadside, in which he could have warmed up with cocoa, did not exist at this time.

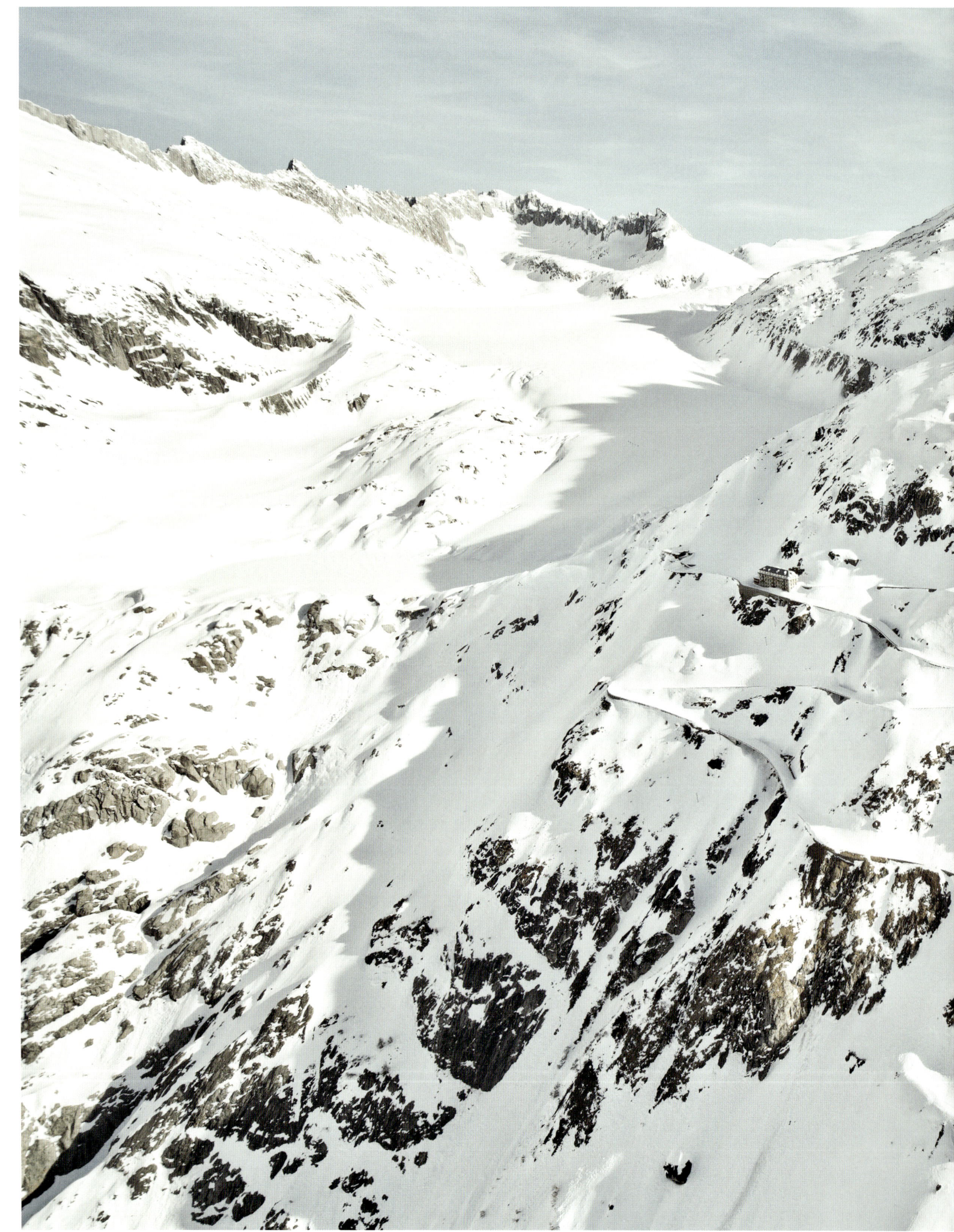

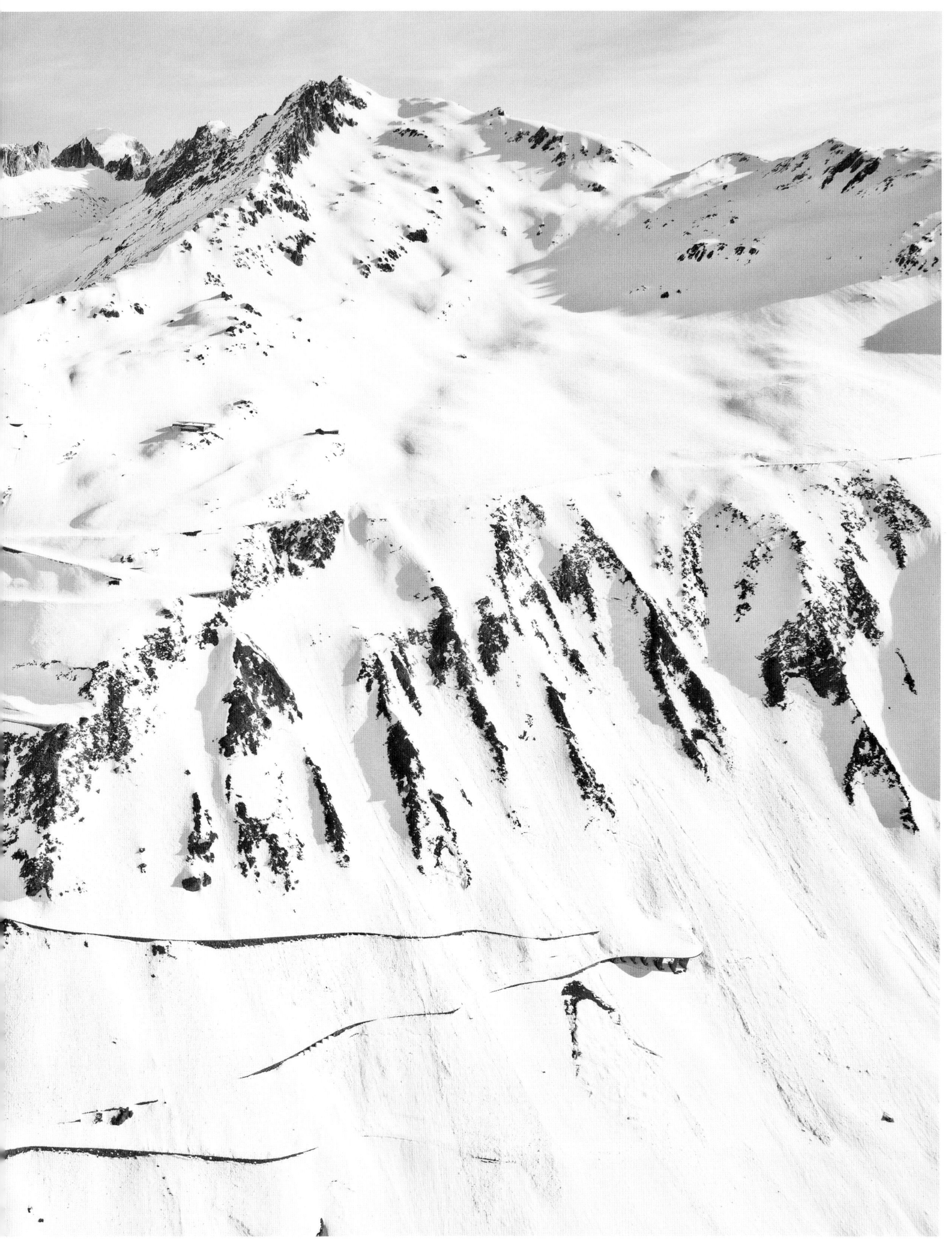

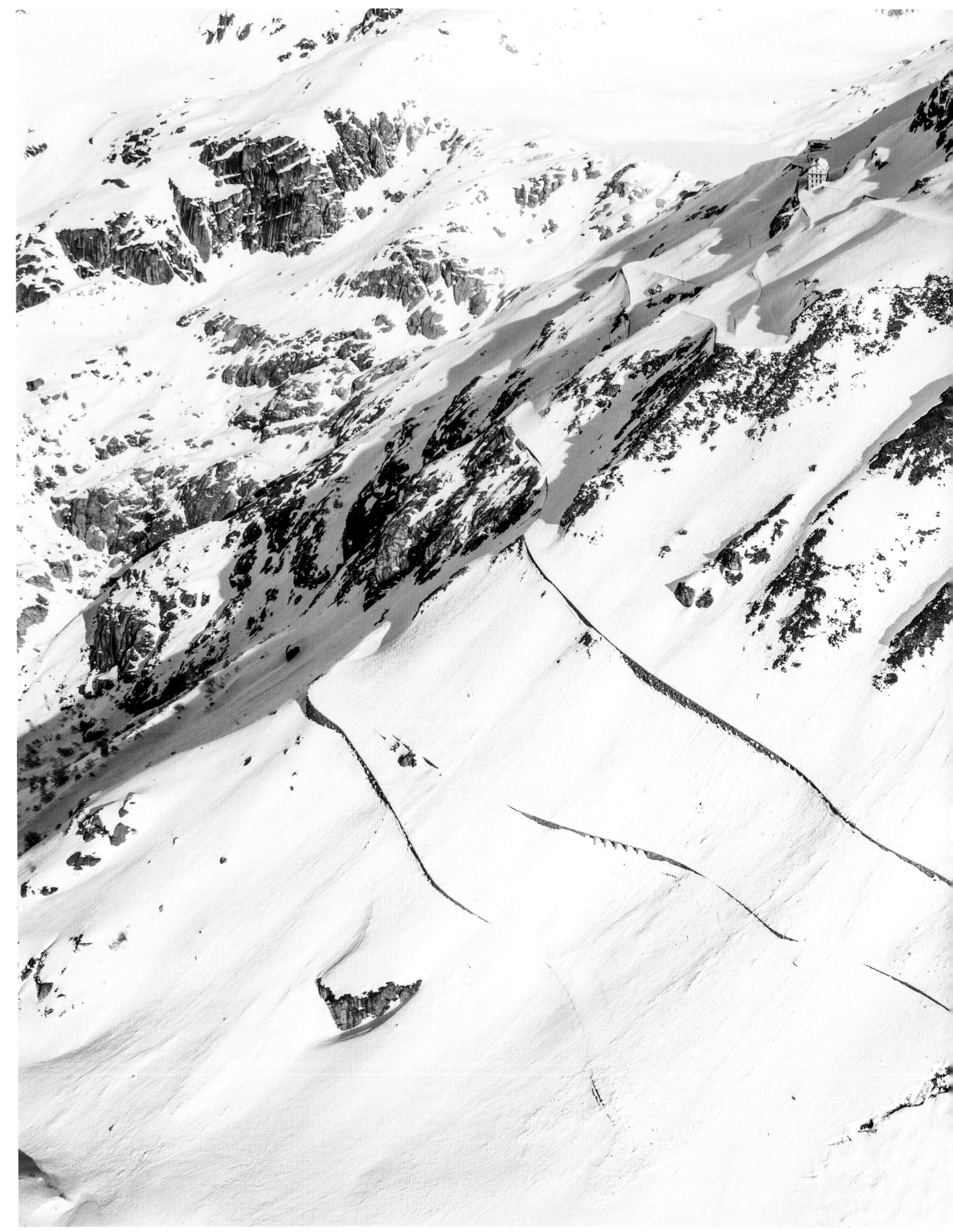

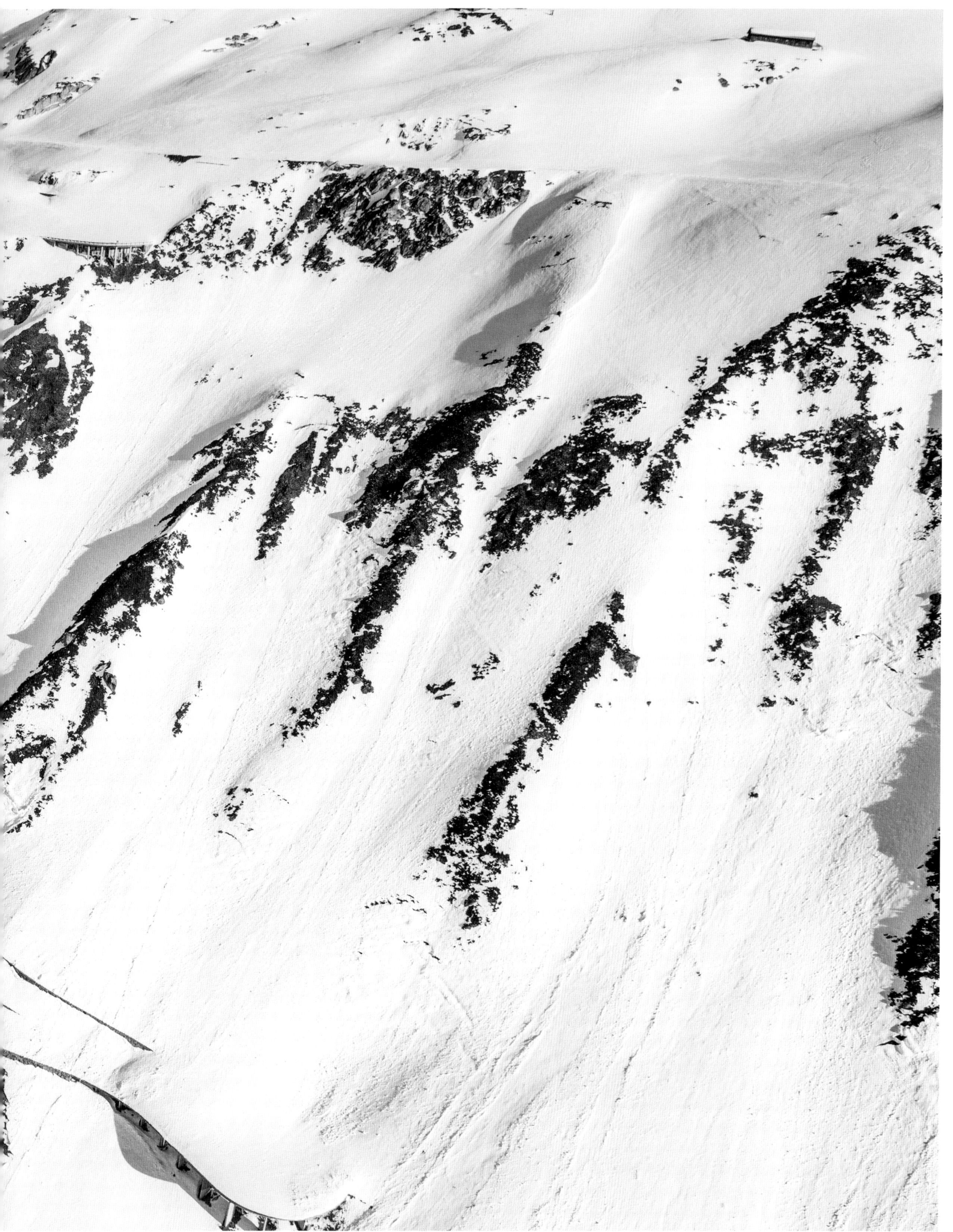

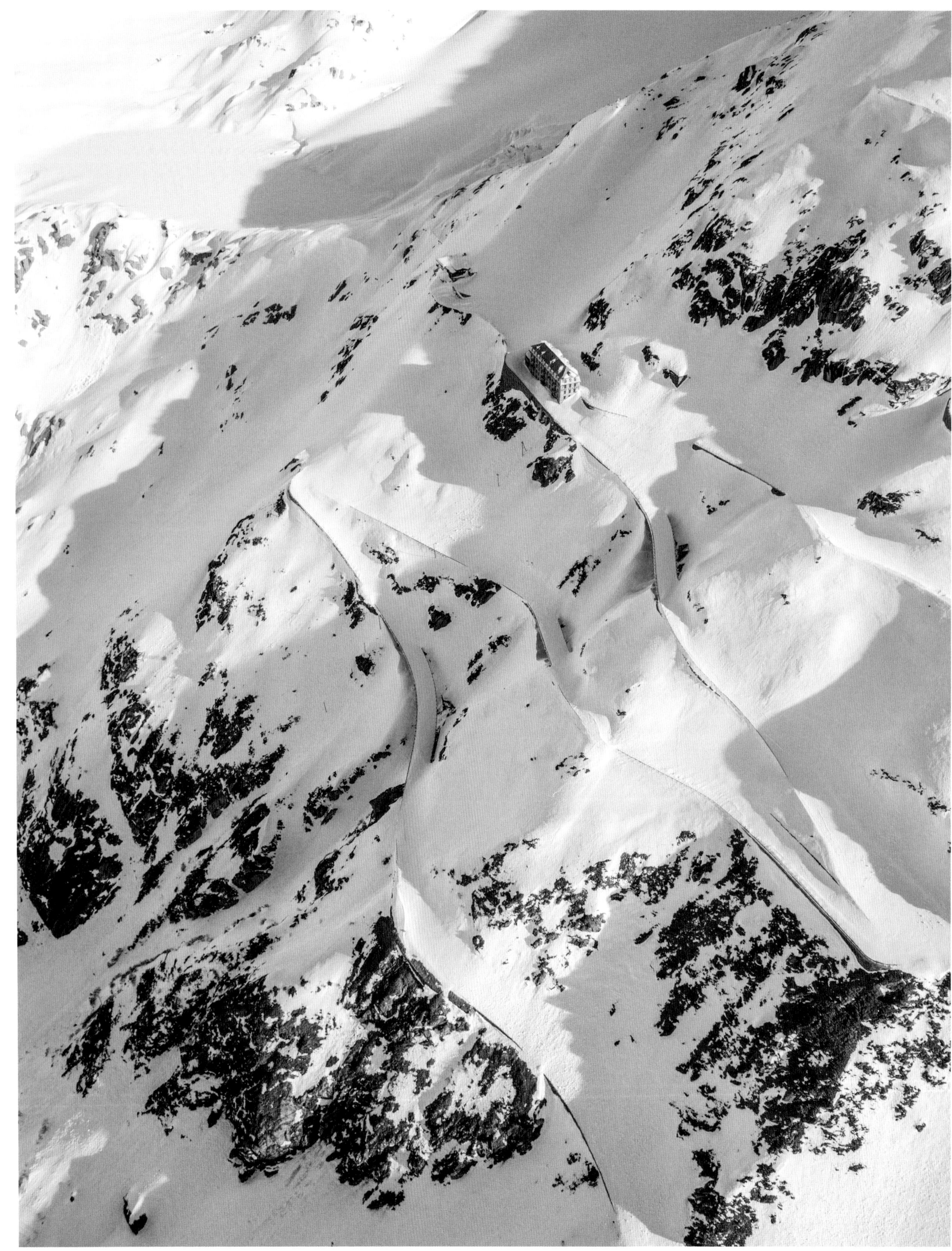

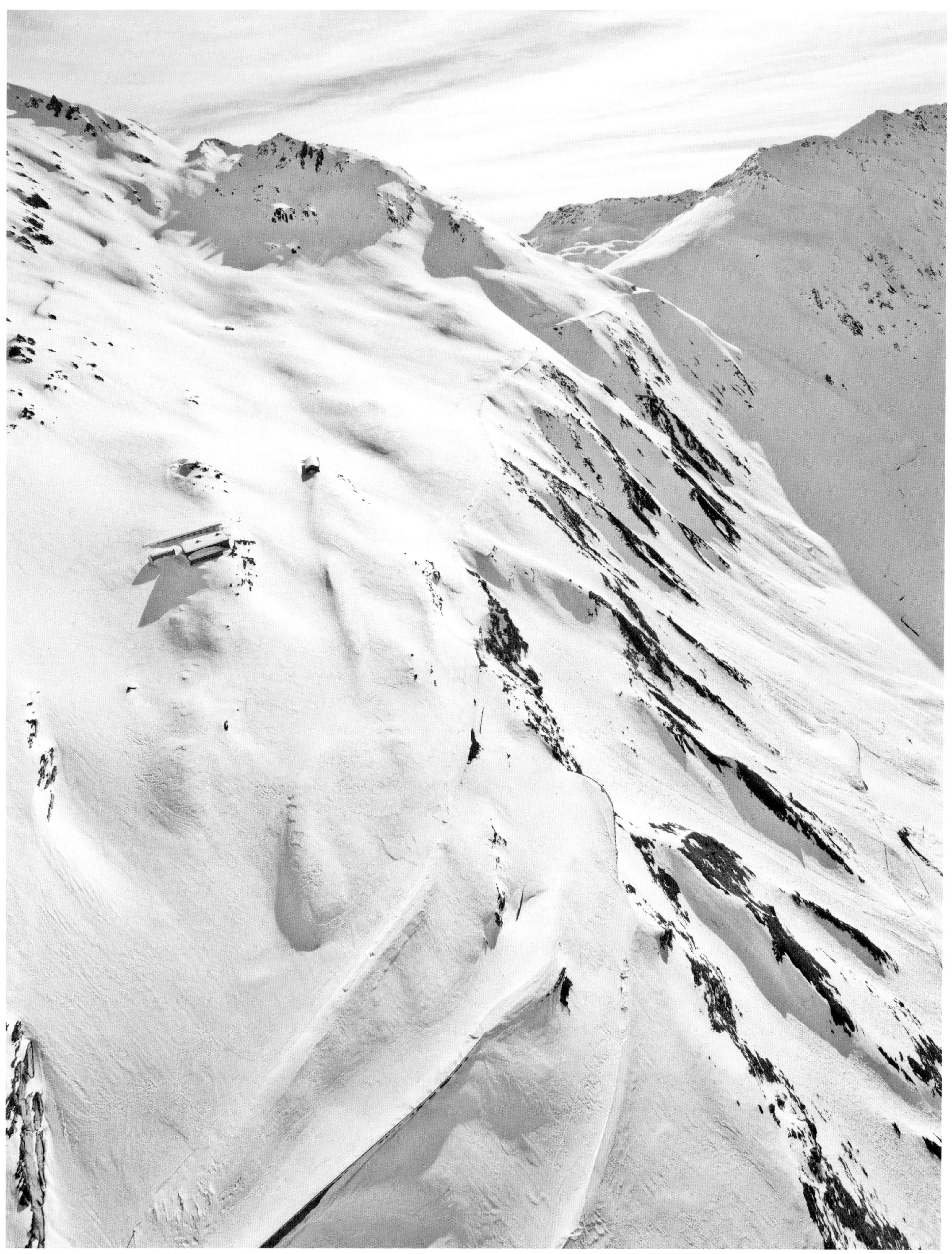

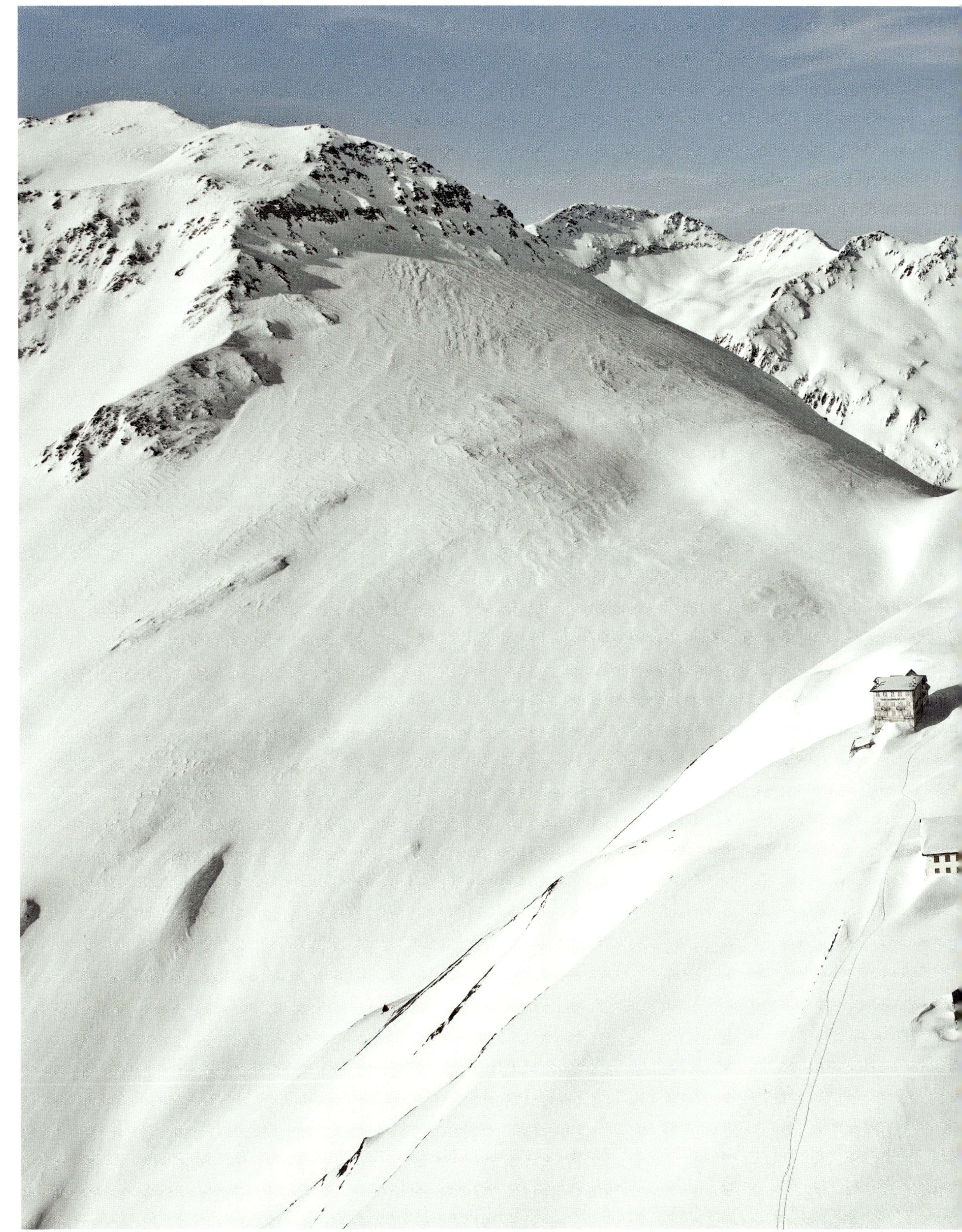

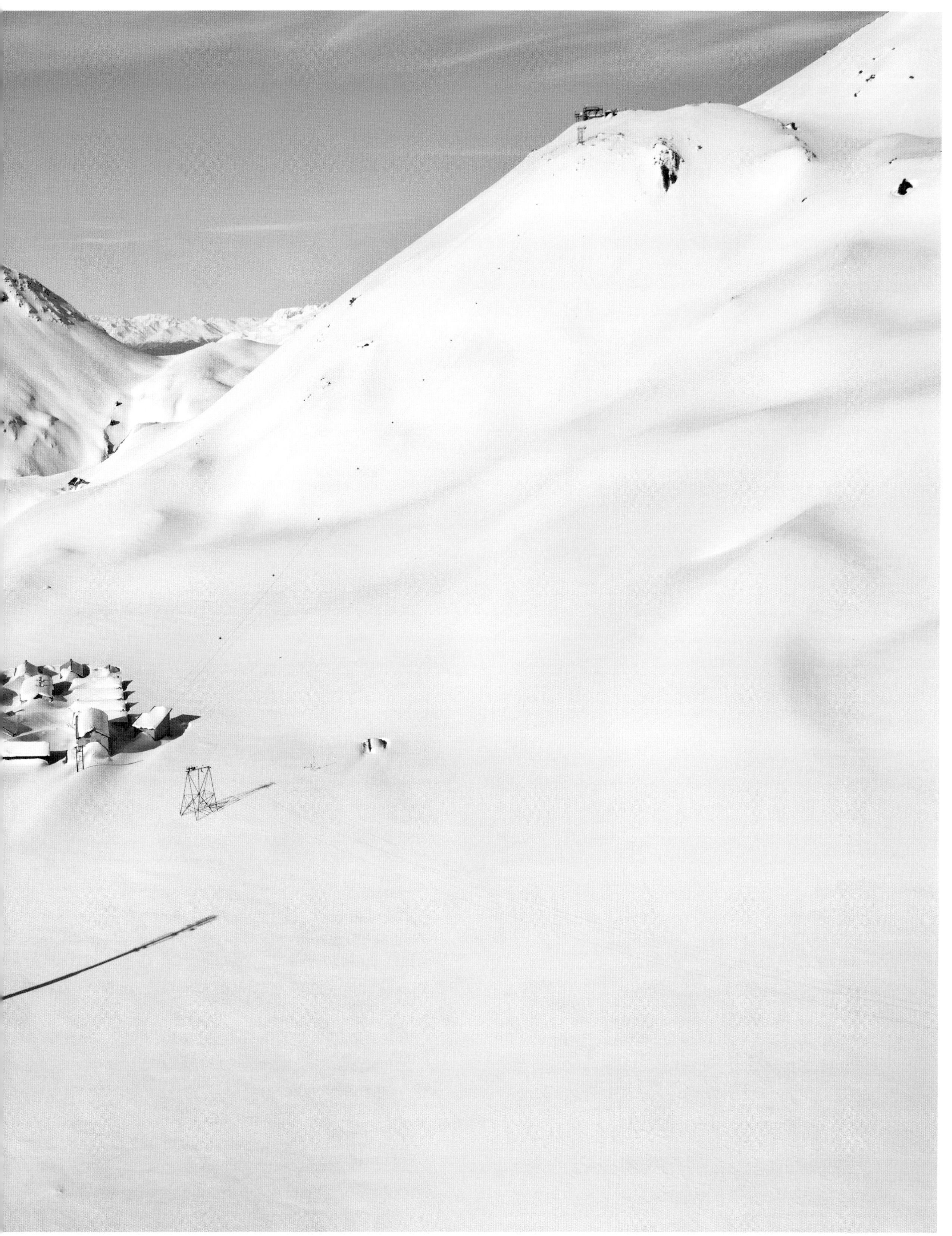

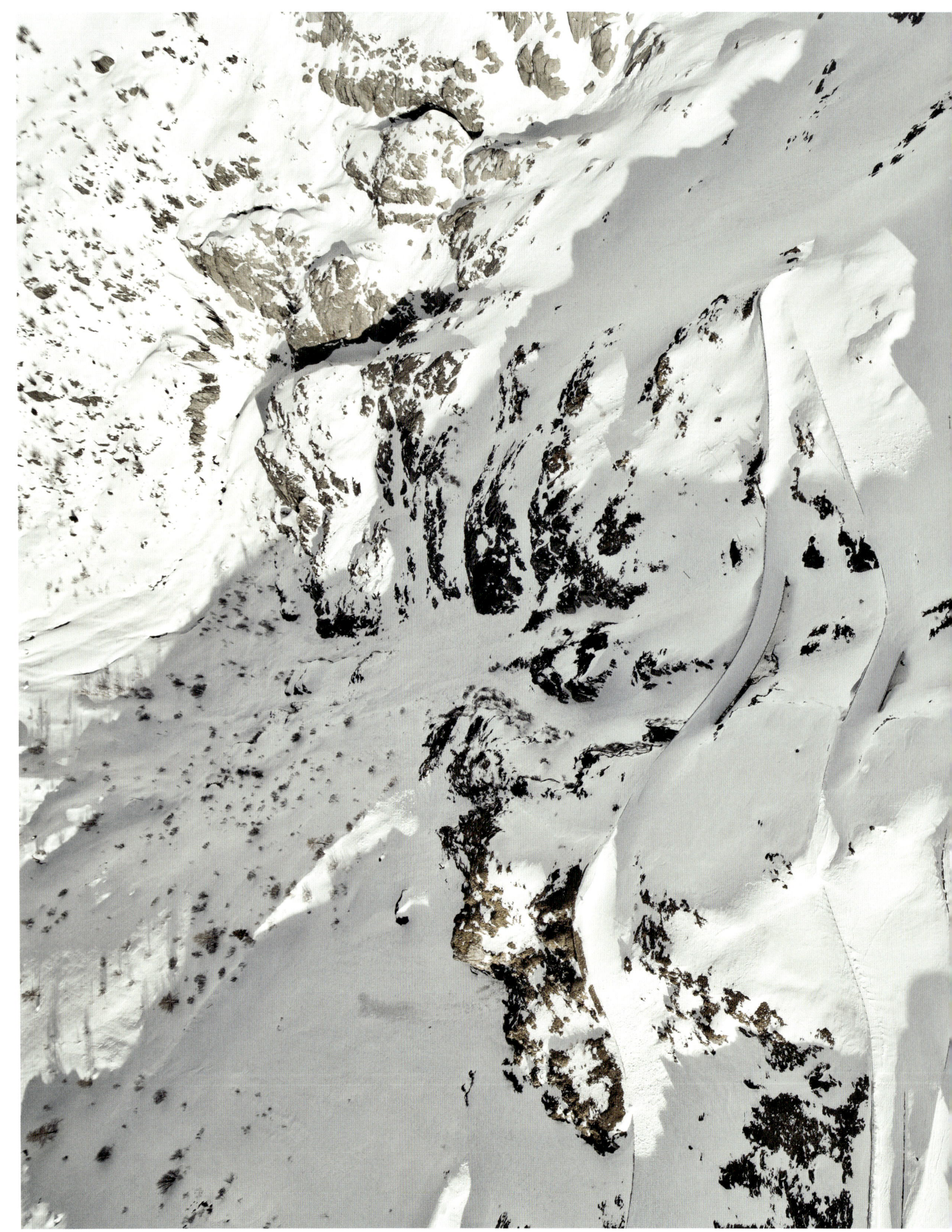

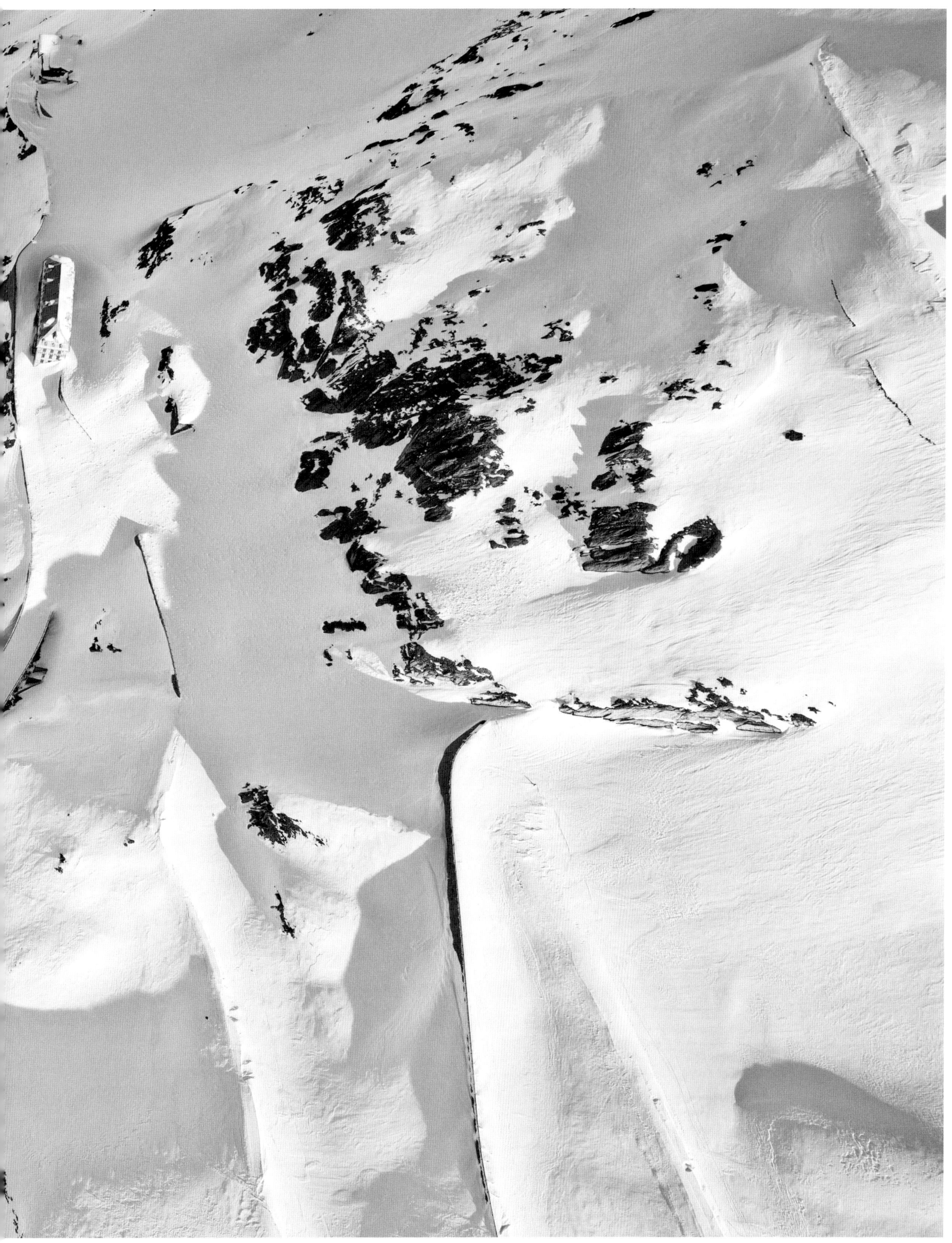

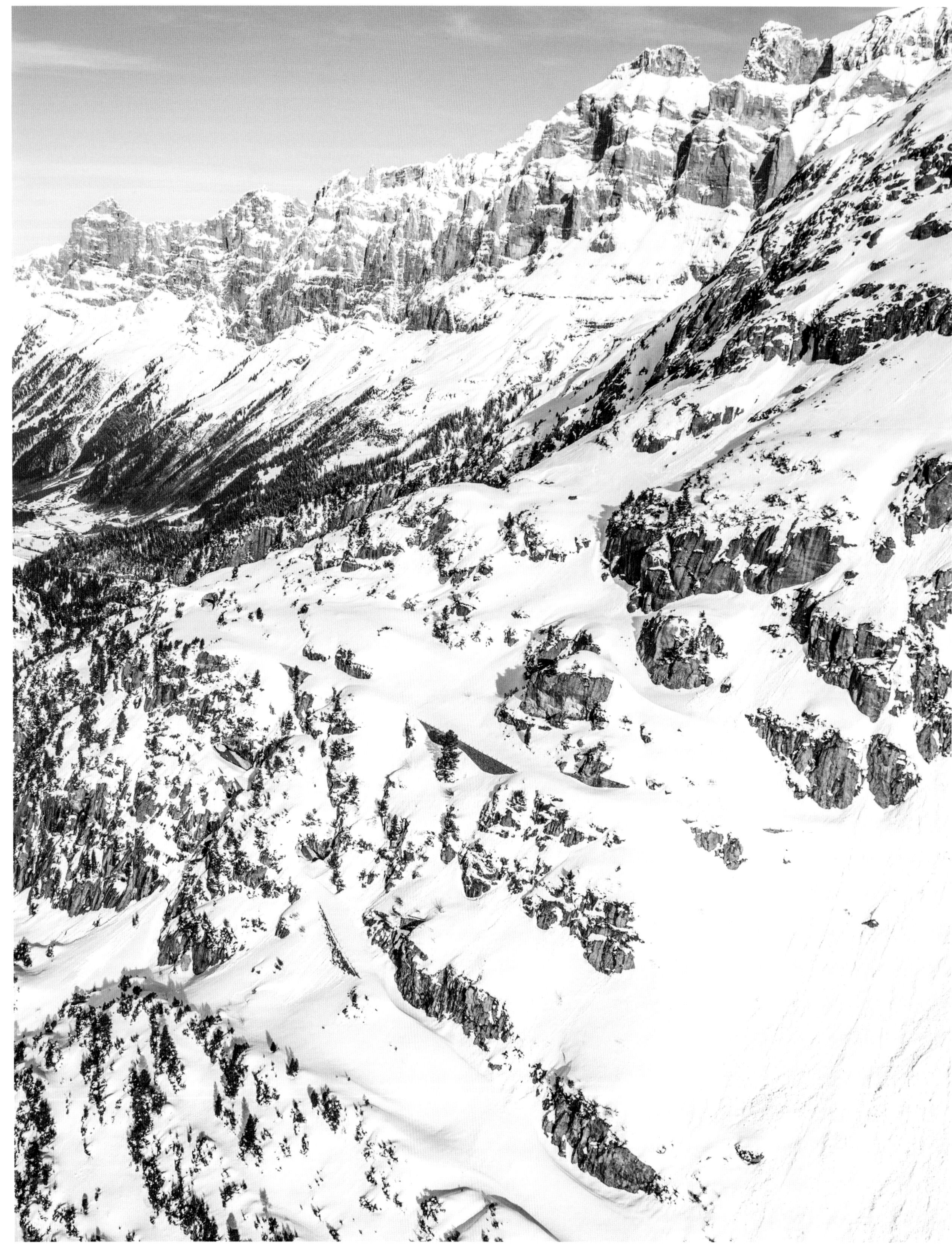

Sustenpass

— 2224 Meter
Wintersperre meist zwischen Oktober und Juni

Die 1946 eröffnete, 45 Kilometer lange Sustenpassstraße verbindet die Orte Wassen im Kanton Uri und Innertkirchen im Kanton Bern und gehört im Sommer zu den meistbefahrenen Pässen der Schweiz – gilt sie doch als Meisterstück der Straßenbaukunst. Kaum eine Schweizer Bergstrecke ist vielseitiger und abwechslungsreicher als der Sustenpass. Ab dem frühen Winter, wenn die Passstraße unter einer tiefen Schneedecke verschwunden ist, locken zur Entschädigung anspruchsvolle Skitouren in die Berge rund um das Sustenhorn.

— 2,224 metres
Usually closed in winter between October and June

Opened in 1946, the 45-kilometre-long Susten Pass road links the towns of Wassen in the canton of Uri with Innertkirchen in the Bern canton. In summer, it is one of the busiest passes in Switzerland and is regarded as a masterpiece of road construction. Very few other Swiss mountain routes are as interesting and varied as the Susten Pass. From early winter when the road to the pass is buried under a blanket of deep snow, challenging ski tours in the mountains around the Sustenhorn prove a popular alternative.

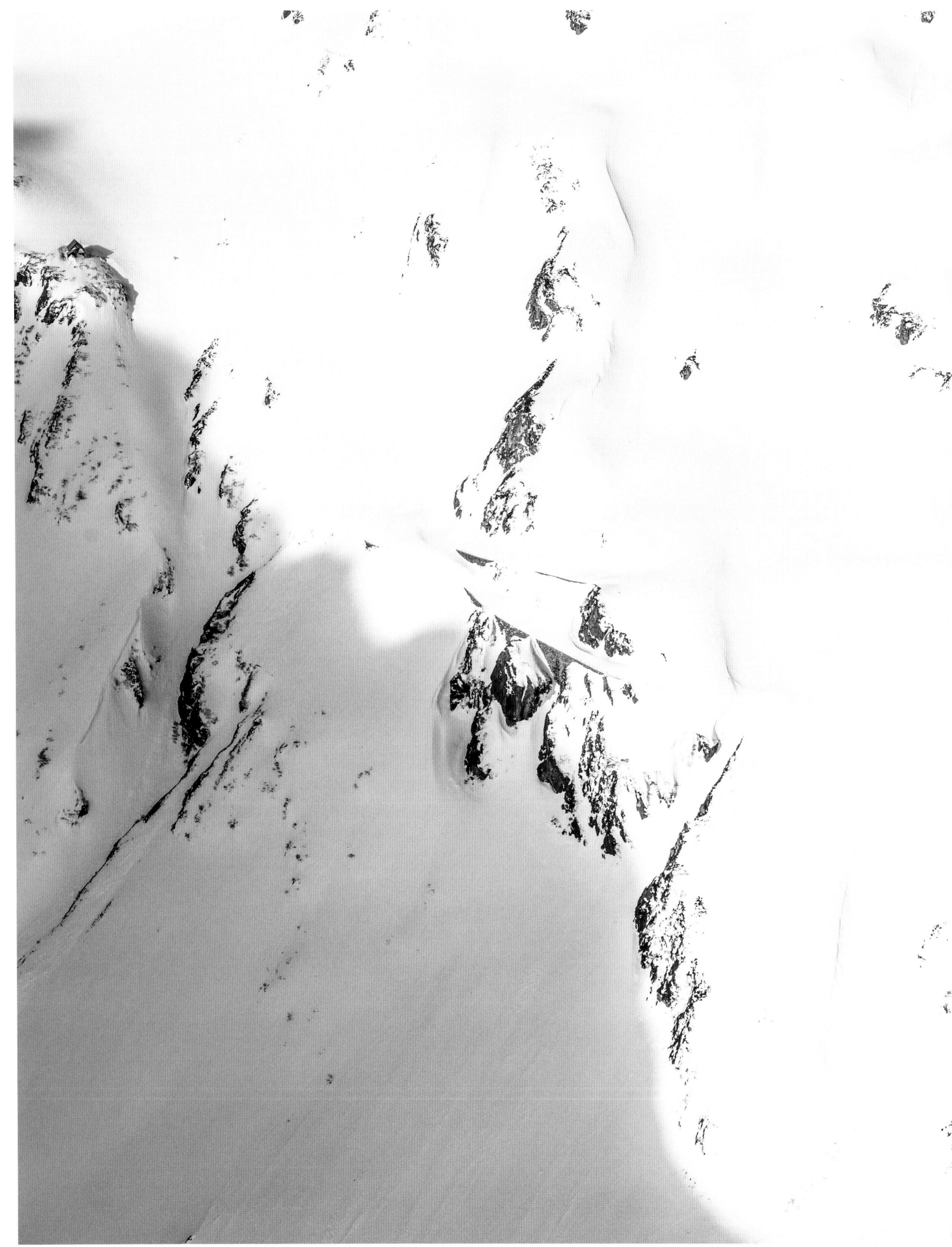

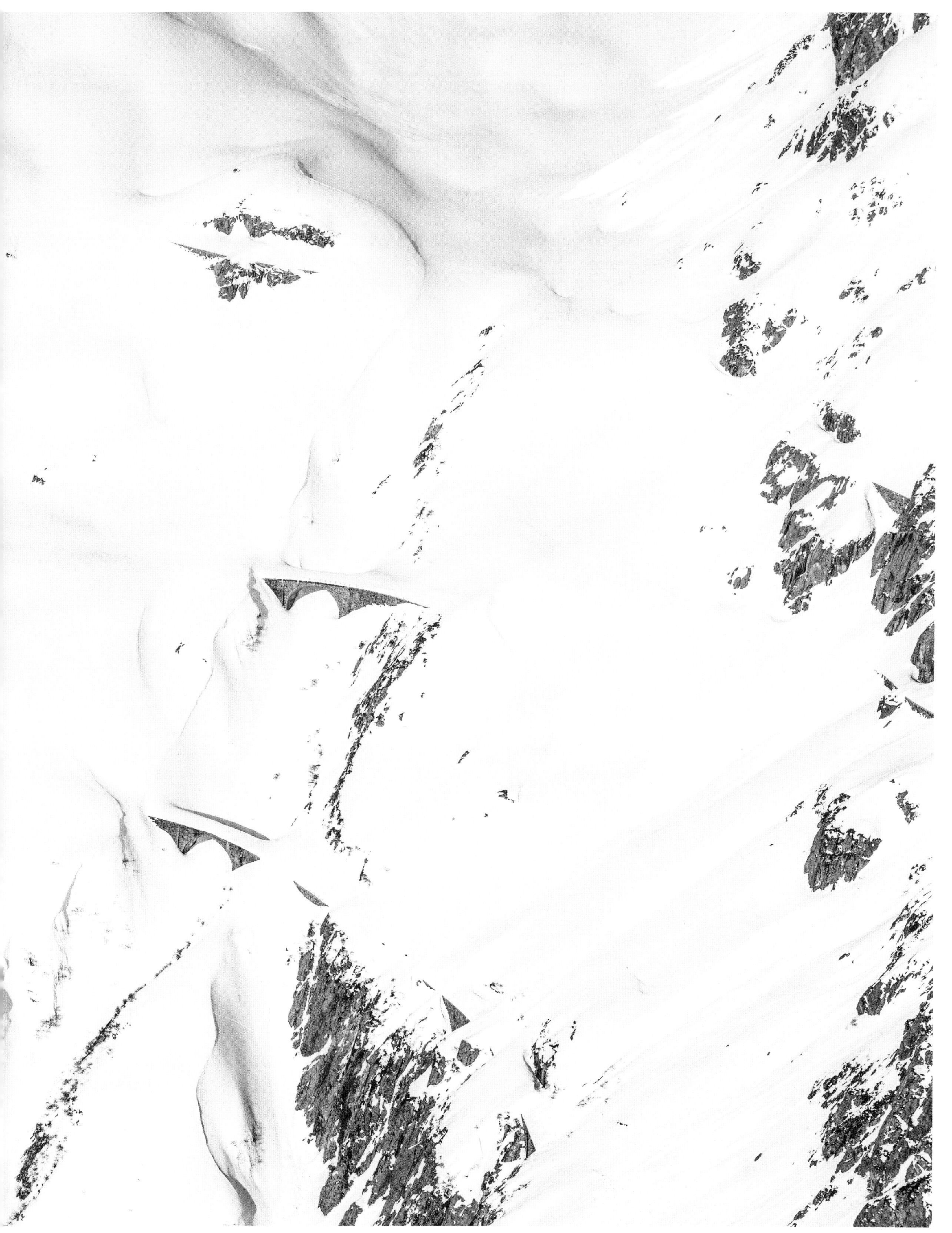

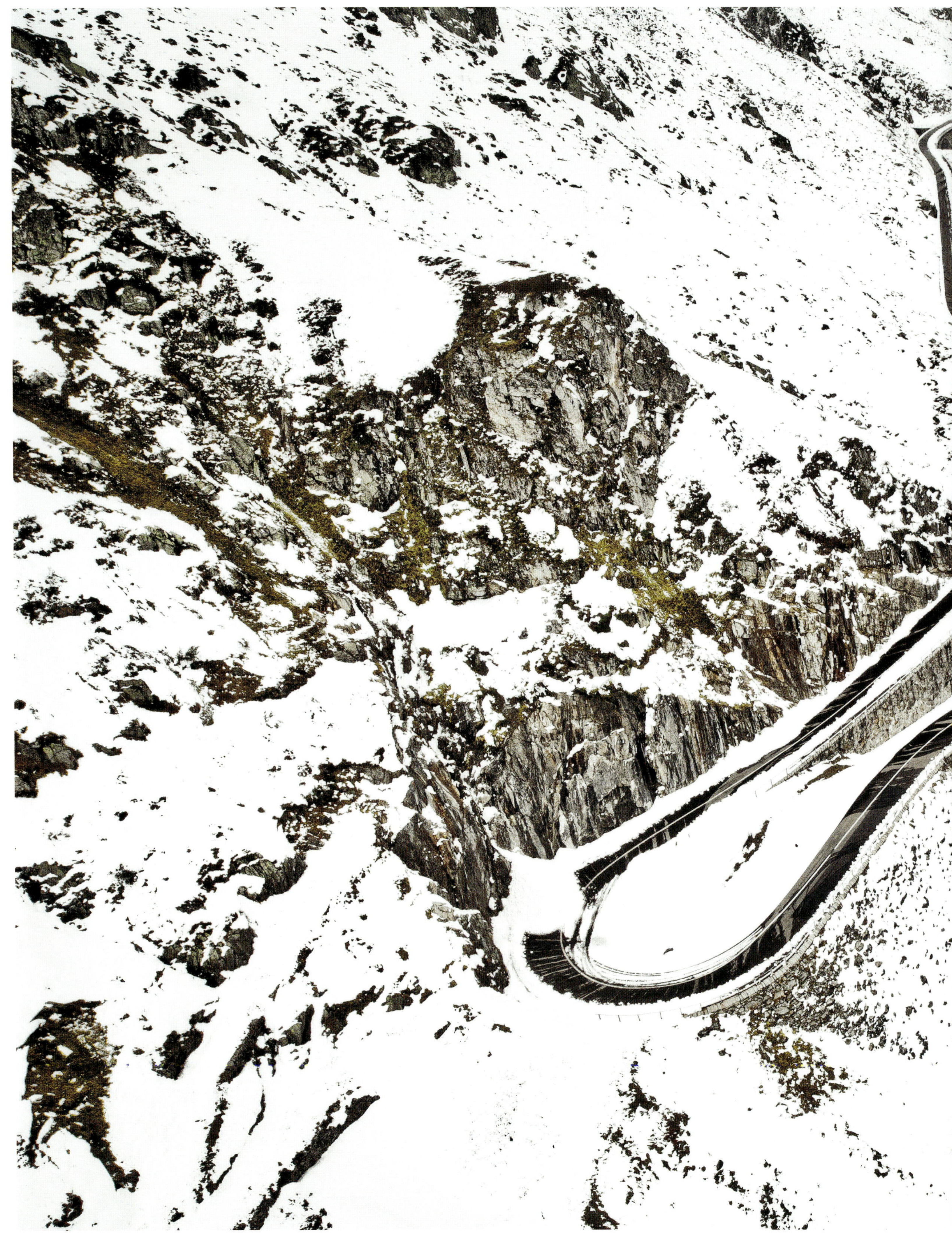

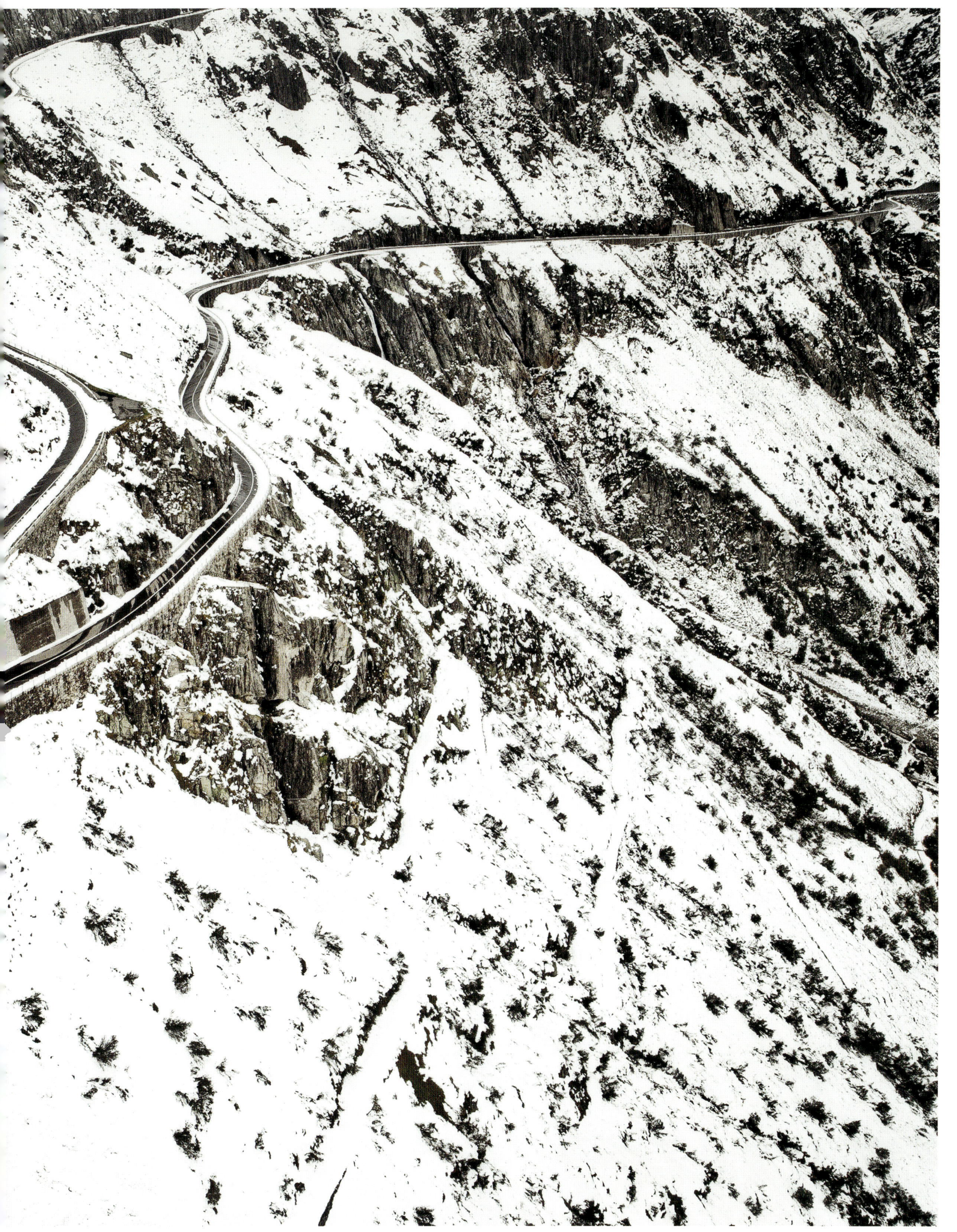

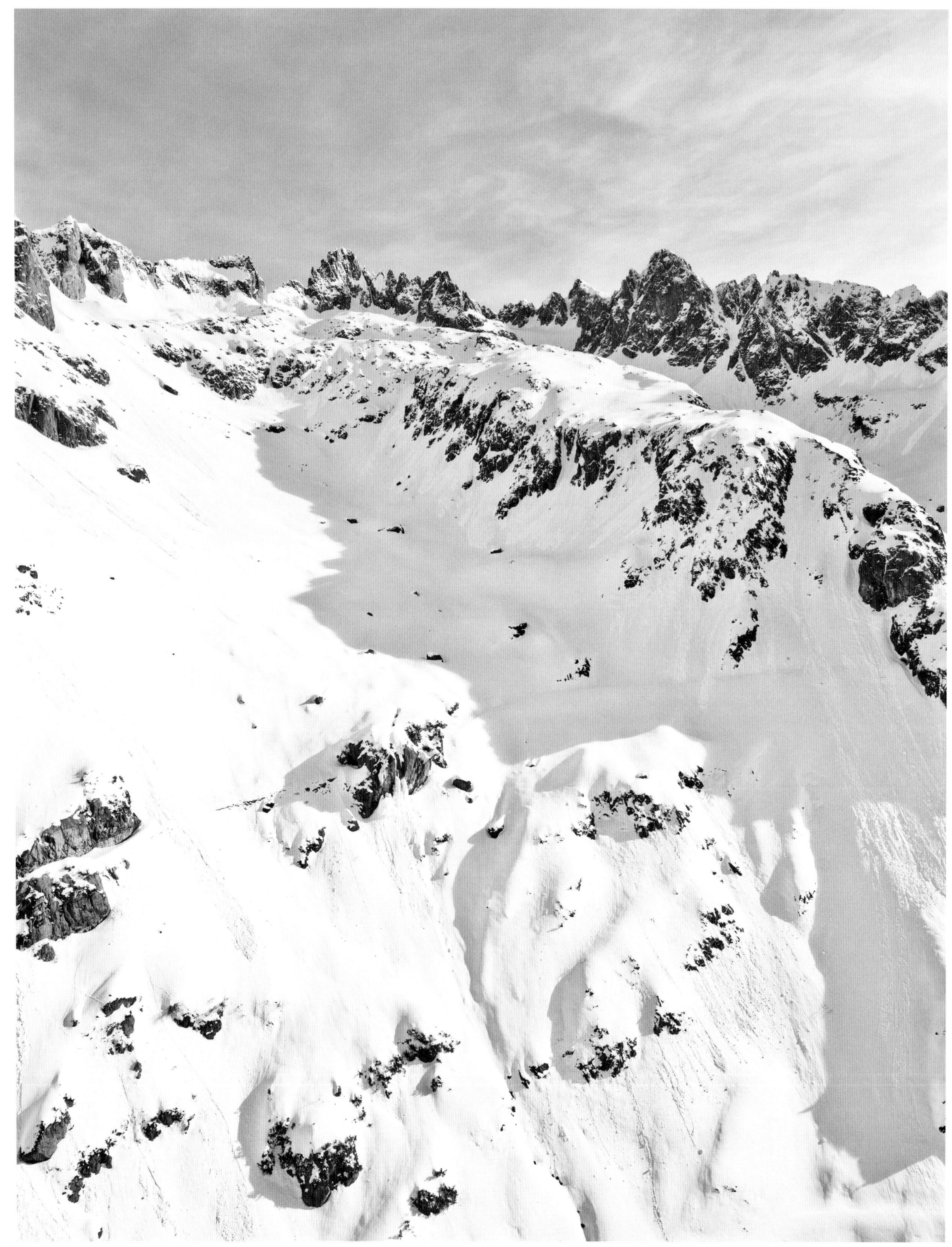

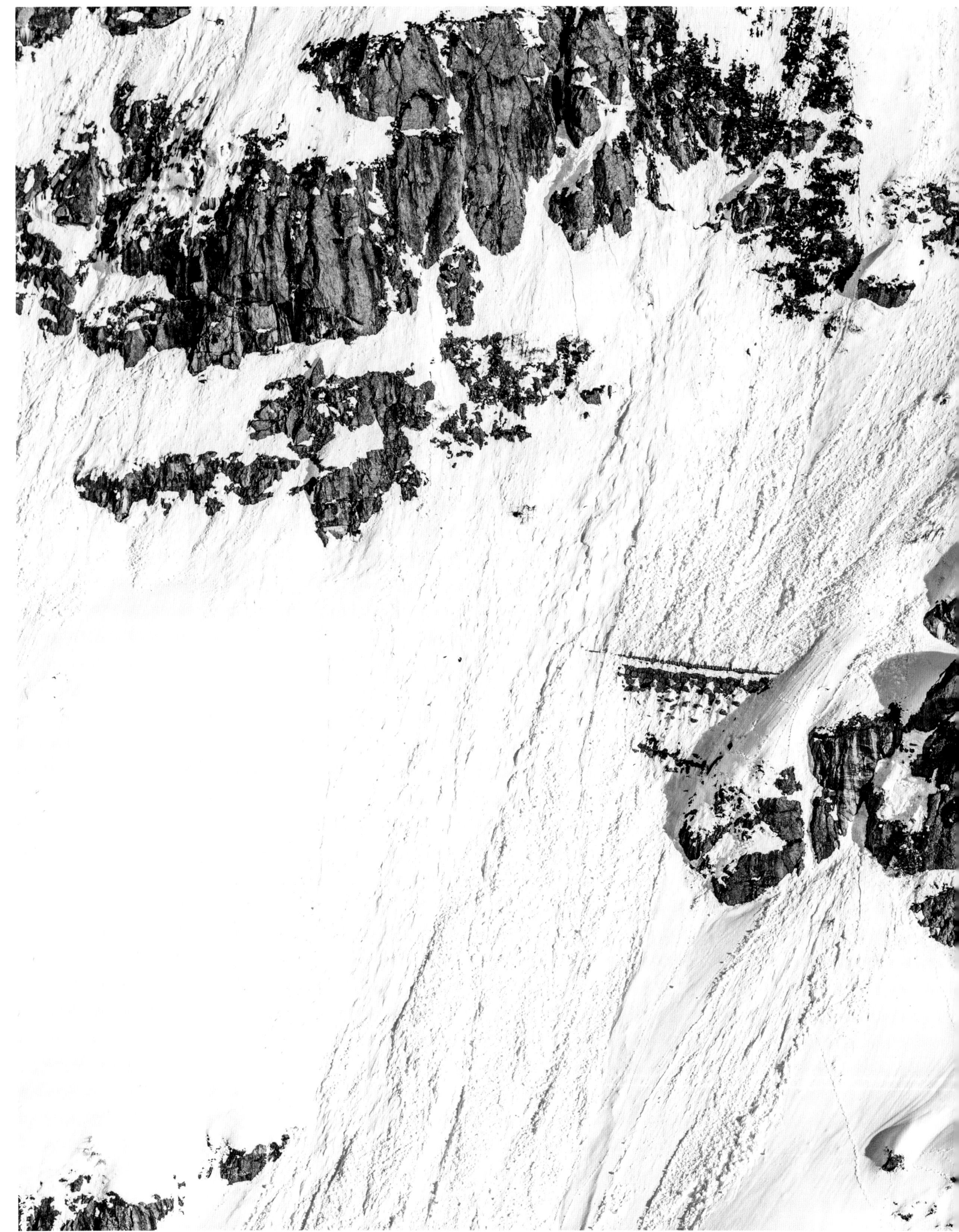

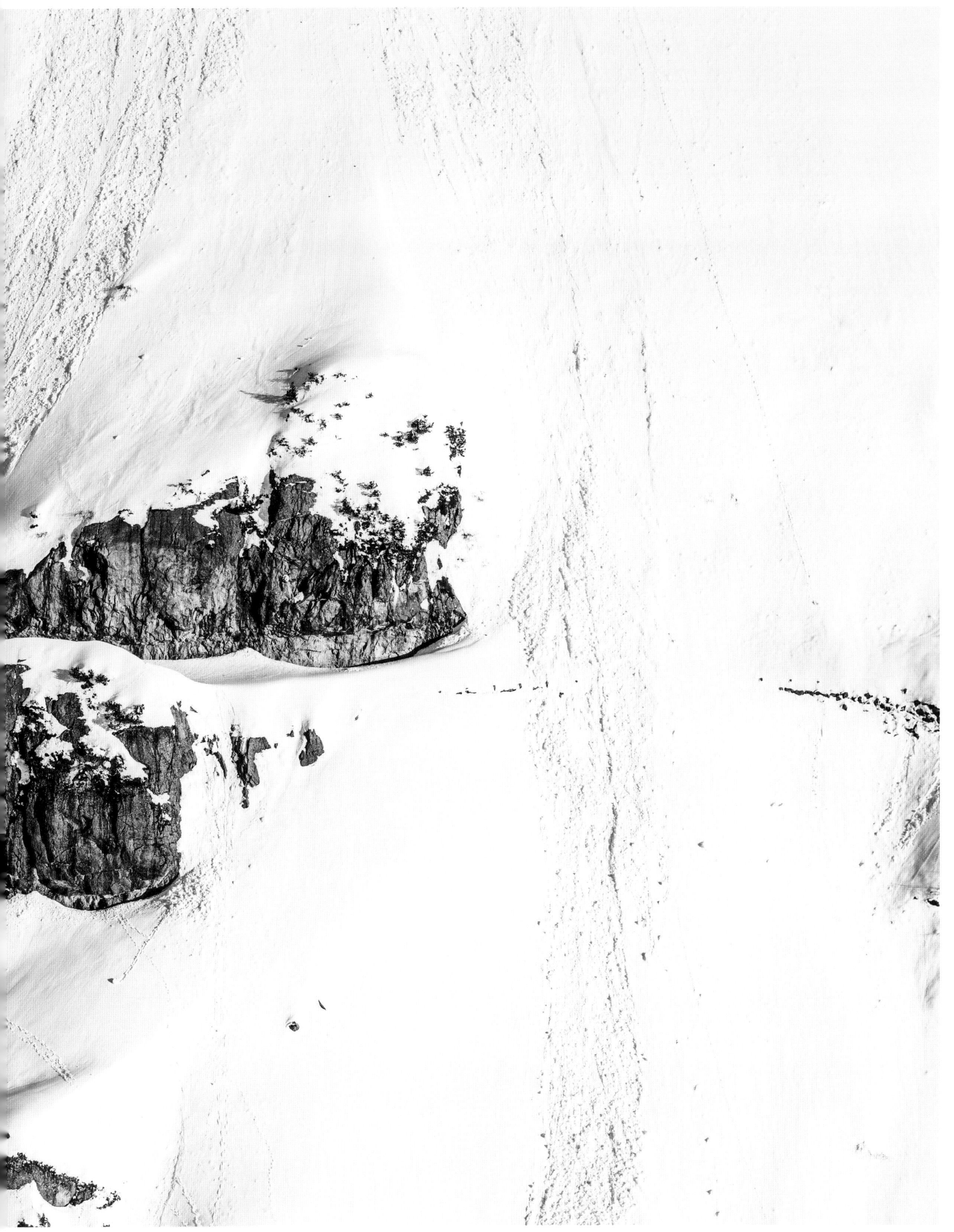

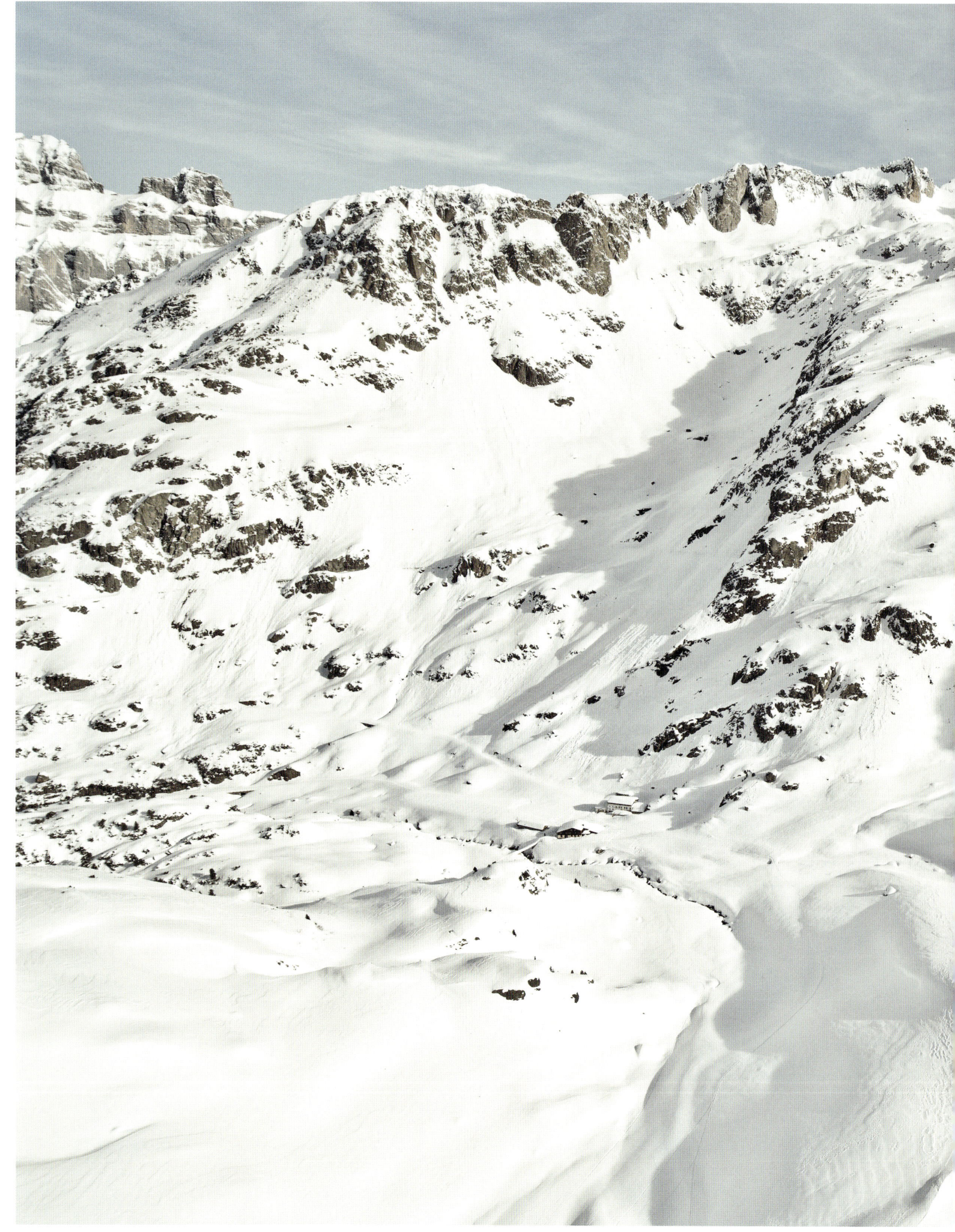

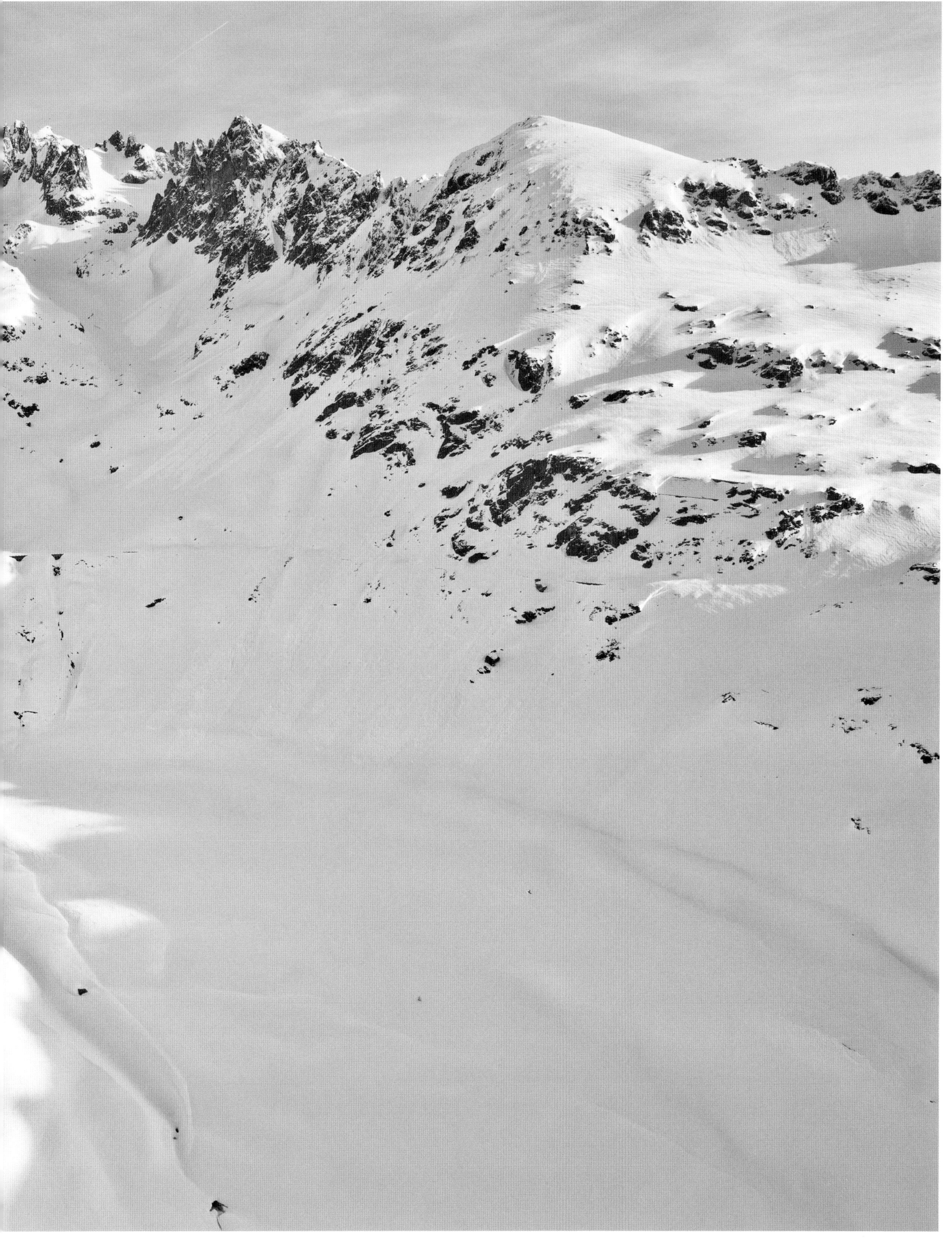

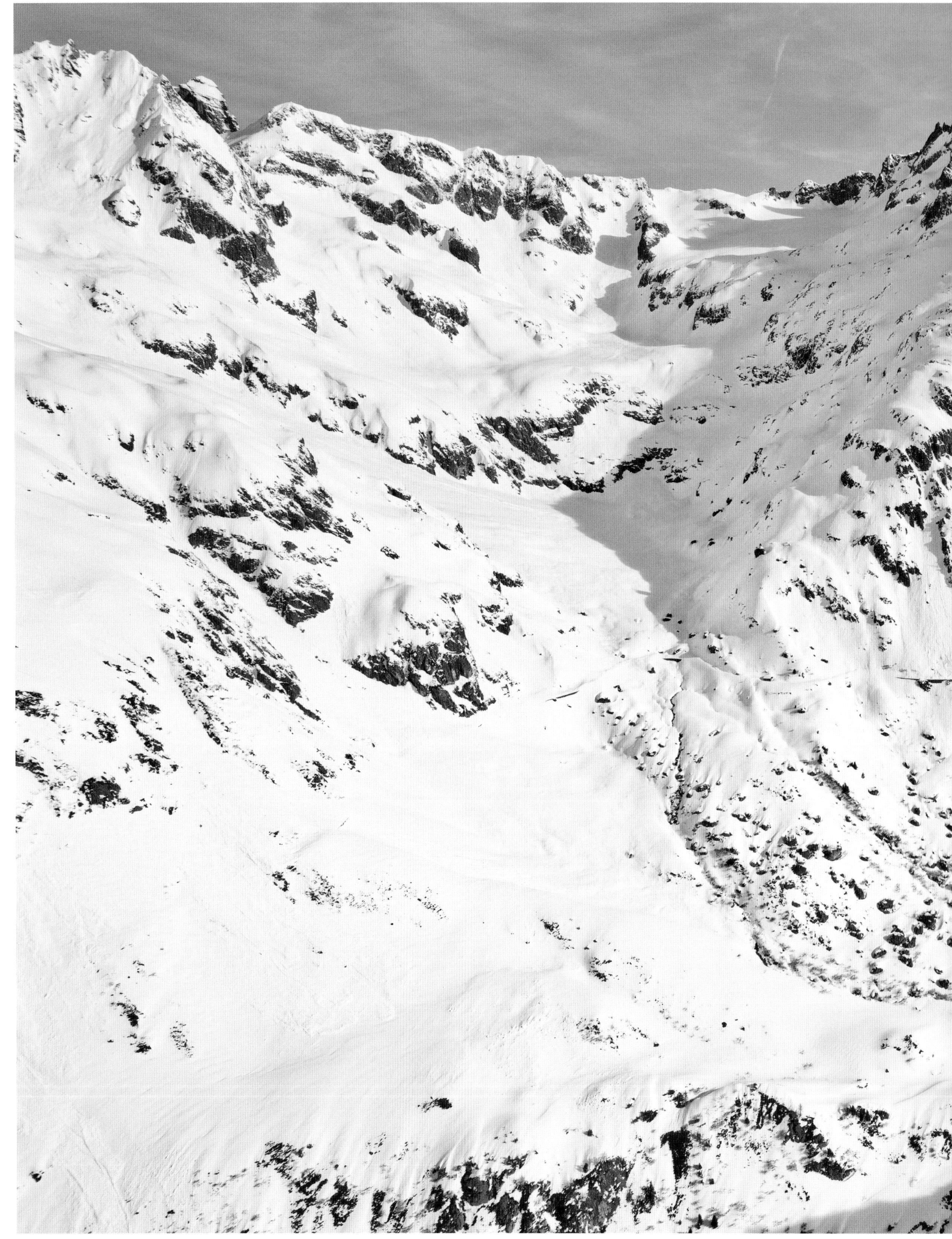

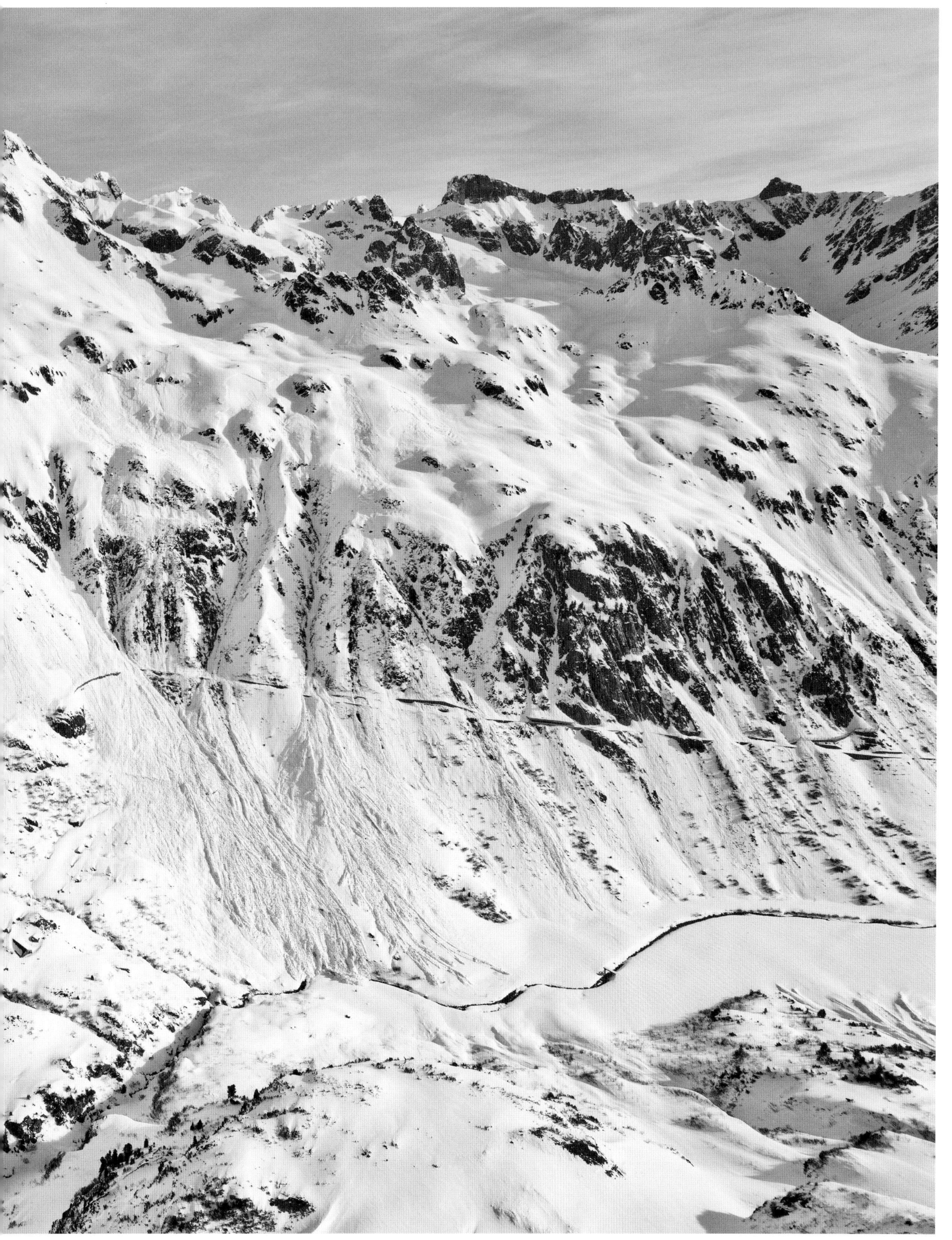

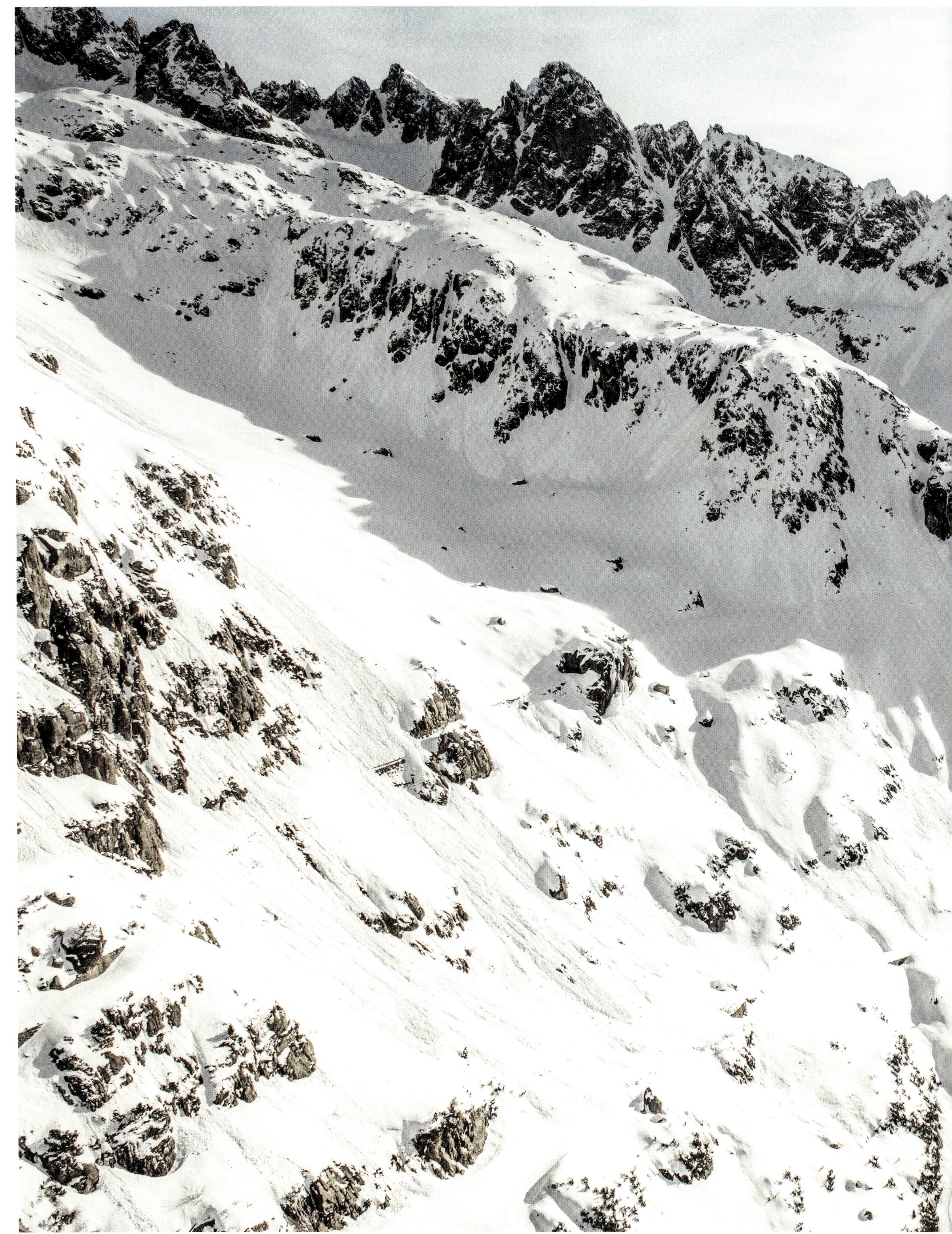

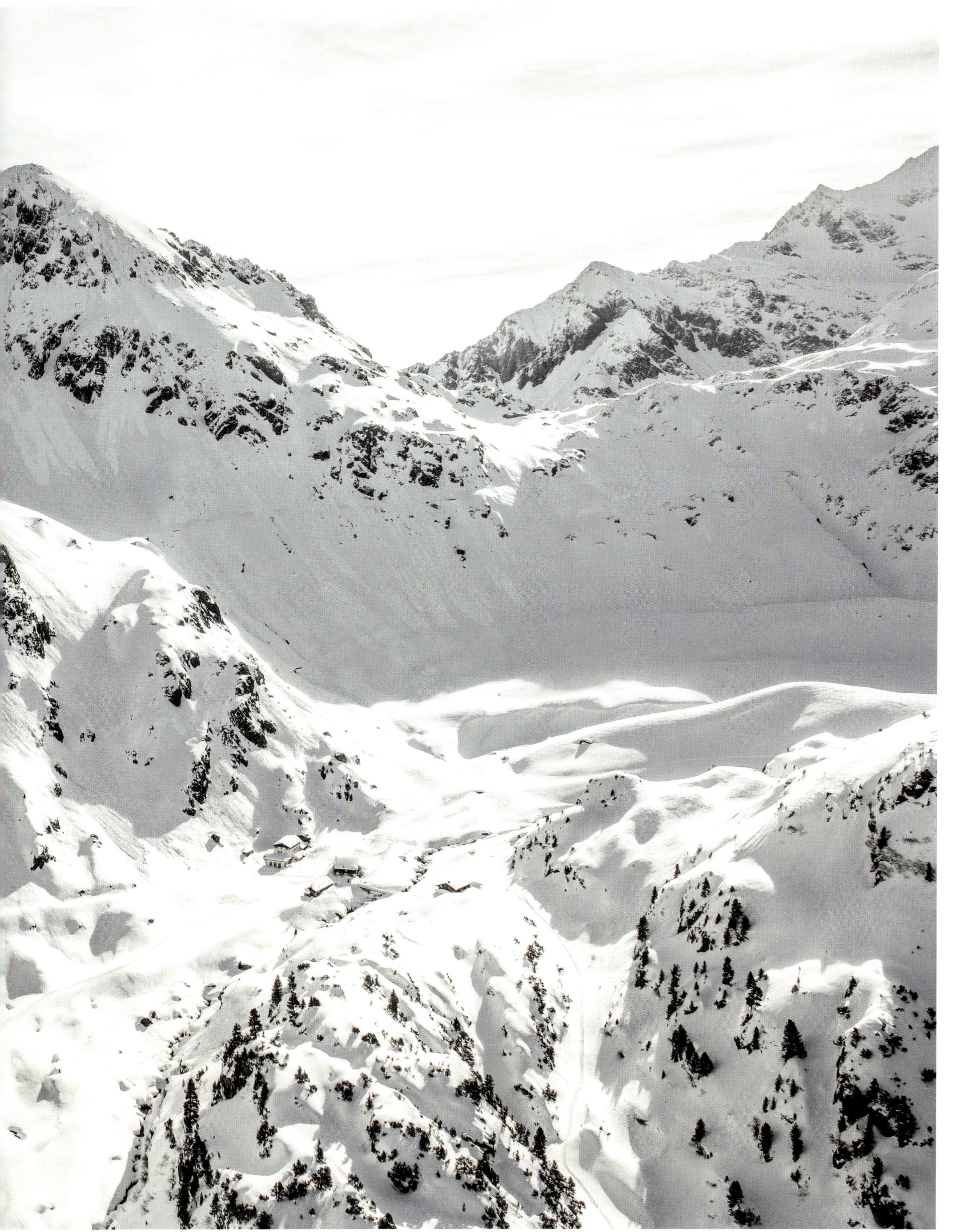

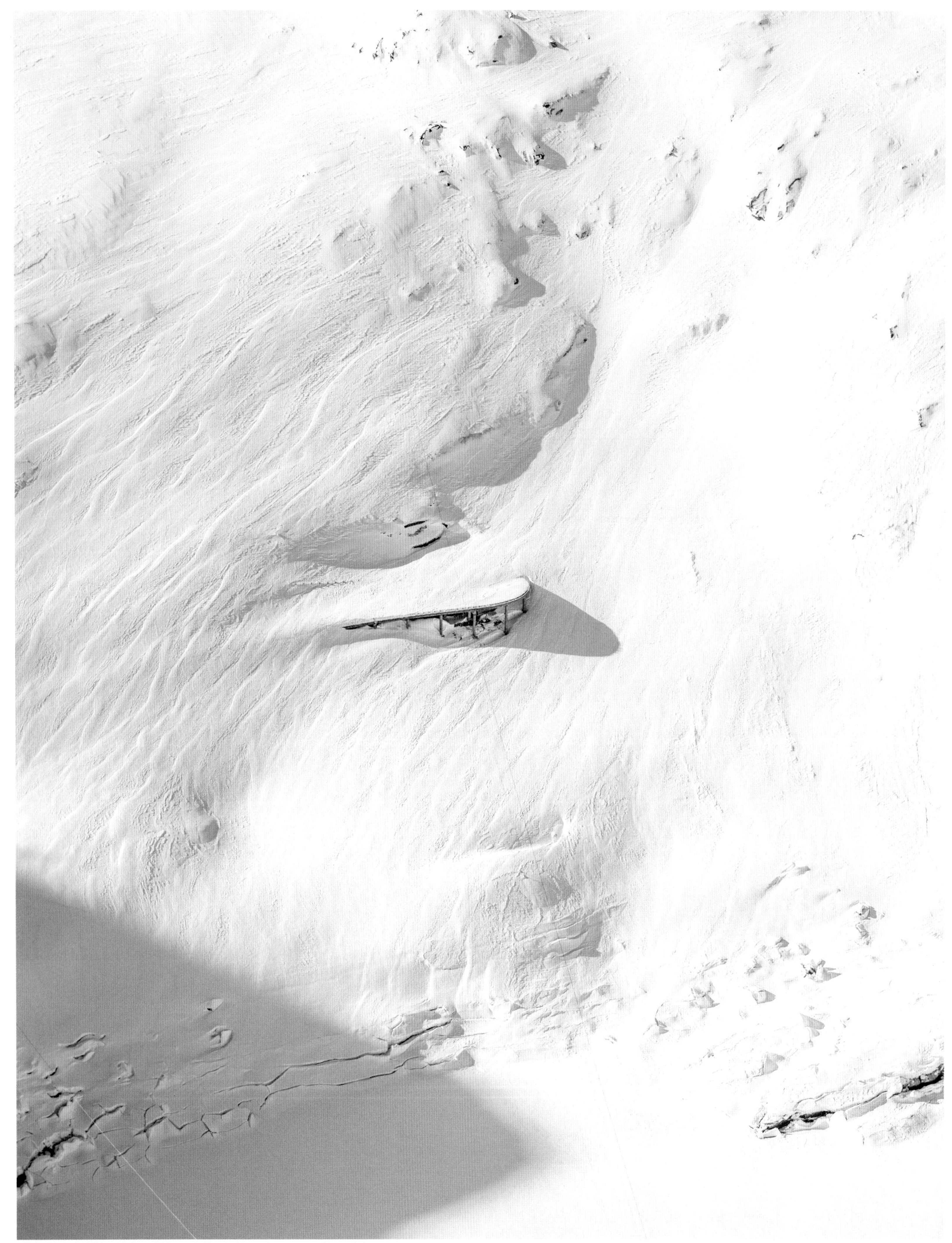

Grimselpass

— 2165 Meter
Wintersperre meist von Oktober bis Mai

Einst ein alter Säumerpfad, auf dem ausdauernde Lastenträger von Schnaps und Wein bis Gold und Silber Waren aller Art über die Alpen transportierten, verbindet seit 1894 eine moderne Hochalpenstraße die Ortschaft Innertkirchen im Berner Oberland mit Gletsch im Oberwallis. Im Winter führt ein anspruchsvoller Wanderweg mit atemberaubendem Alpenpanorama von Oberwald über die Rhonequelle hinauf zur Passhöhe. Wer auch bei Eis und Schnee nicht auf sein Drift-Vergnügen verzichten mag, darf von dort aus auch mit dem Schlitten zurück ins Tal sausen. Zwischen Dezember und April kann man zudem im Grimsel Hospiz wunderbar gemütliche Stunden vorm prasselnden Kamin verbringen.

— 2,165 metres
Usually closed in winter from October to May

Formerly an old mule track on which the sure-footed hauliers transported everything from schnapps and wine to gold and silver goods over the Alps. By 1894 it had become a modern high Alpine road connecting the towns of Innertkirchen in the Bernese Oberland with Gletsch in Upper Valais. In winter, a challenging path rewards hikers with breathtaking panoramic views from Oberwald to the source of the Rhone and up to the pass. Those who are keen to satisfy their speed fever on ice and snow can jump on a sled and swoop back down into the valley. Between December and April, visitors can also enjoy some warm and cosy hours in front of a crackling fire at the Grimsel Hospice.

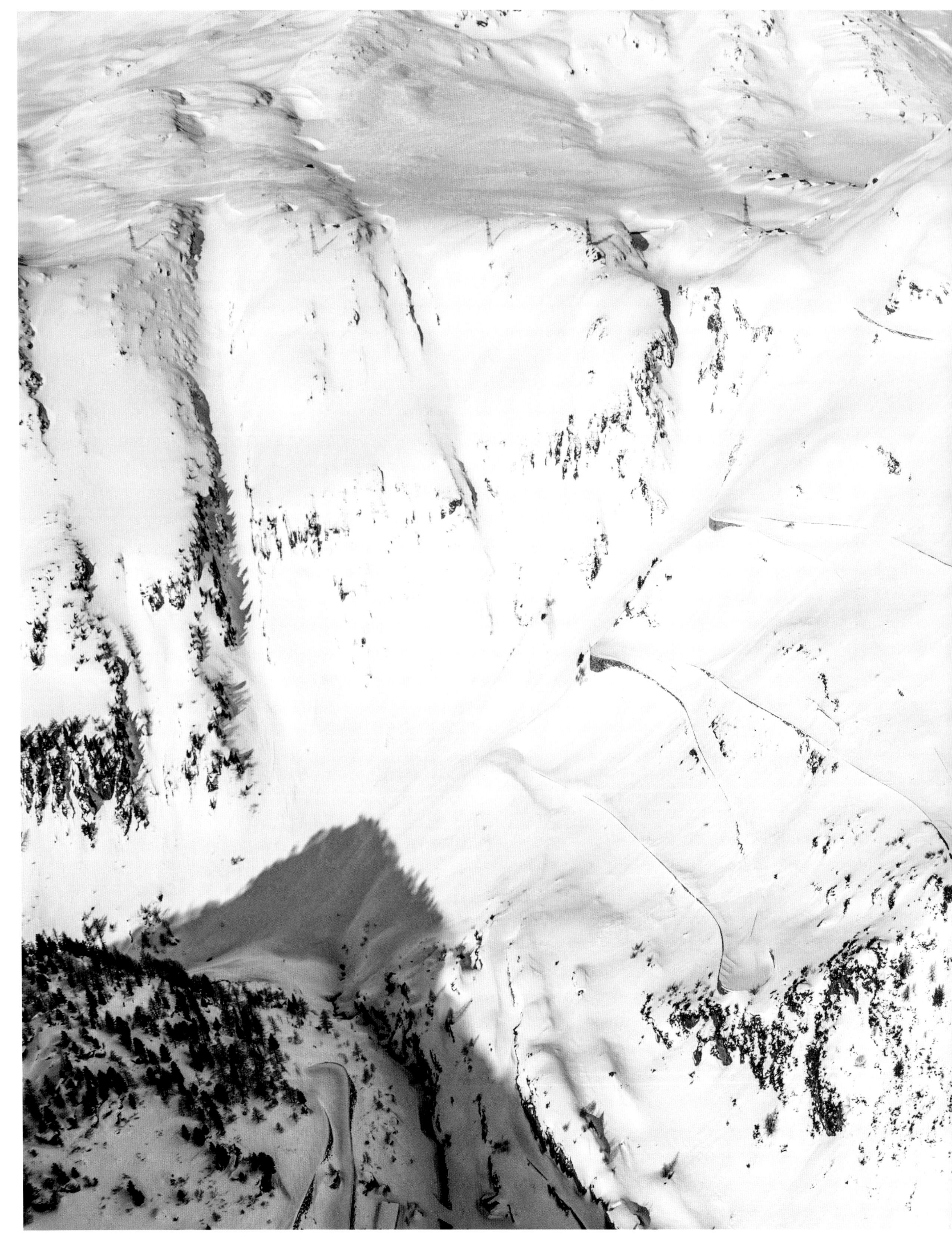

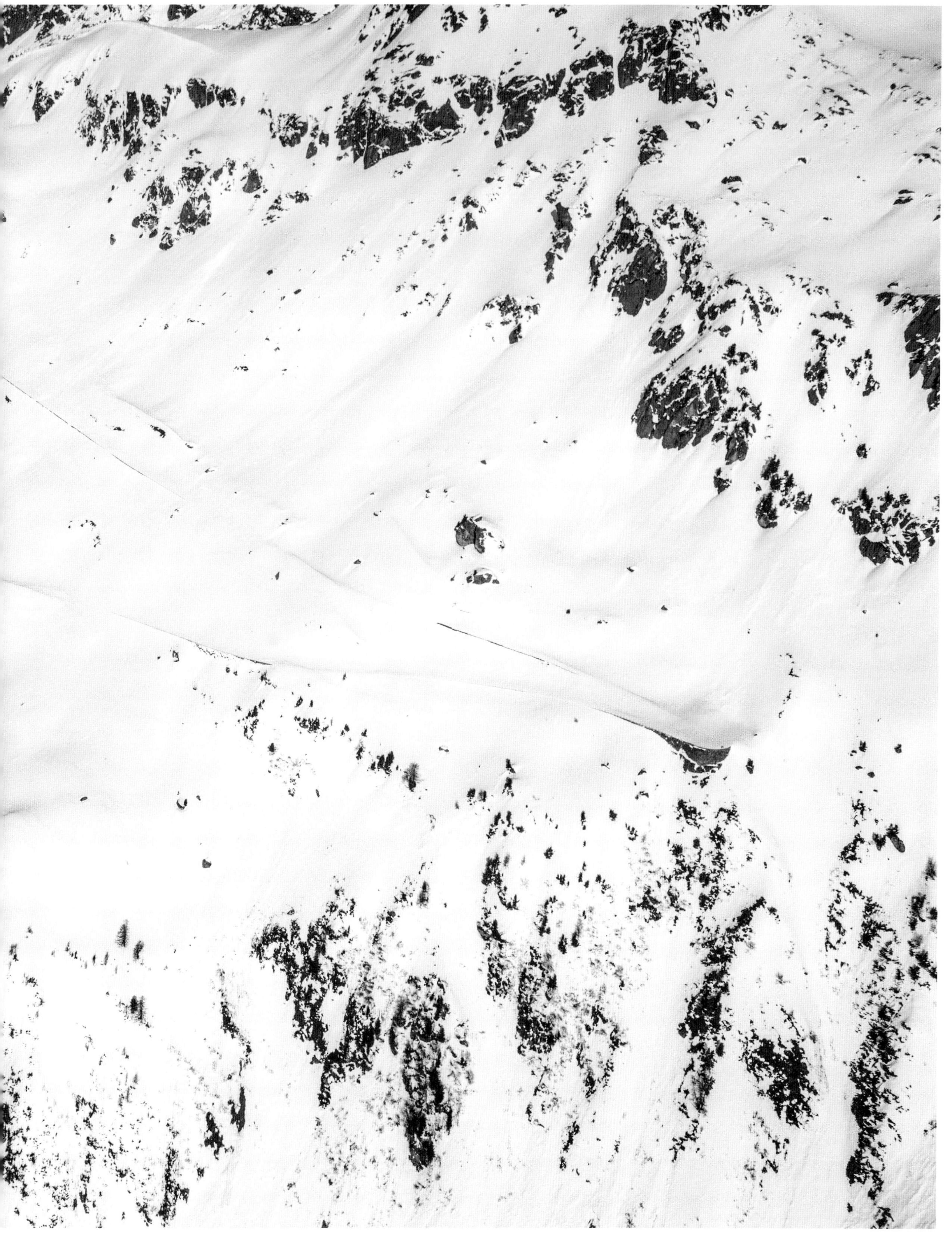

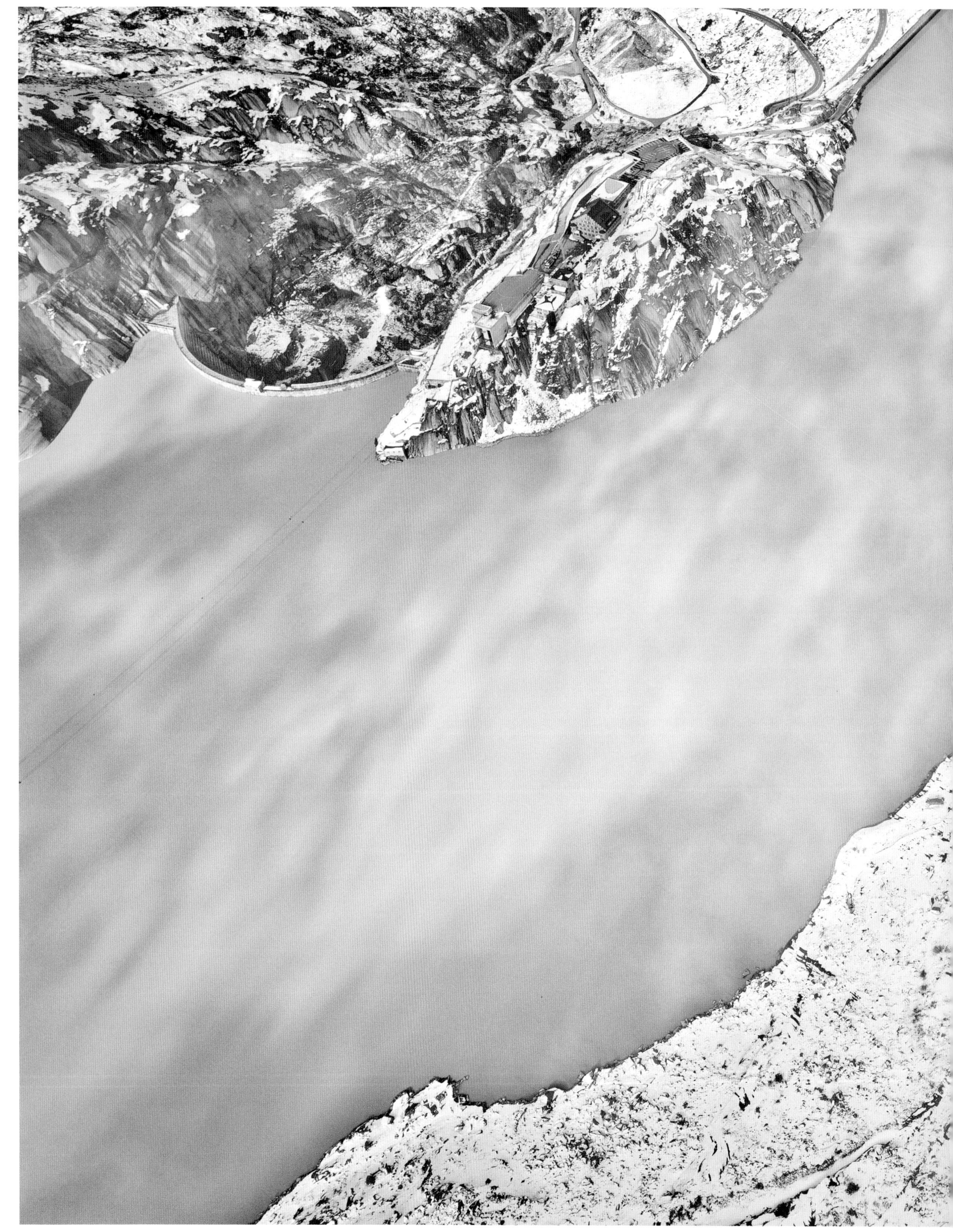

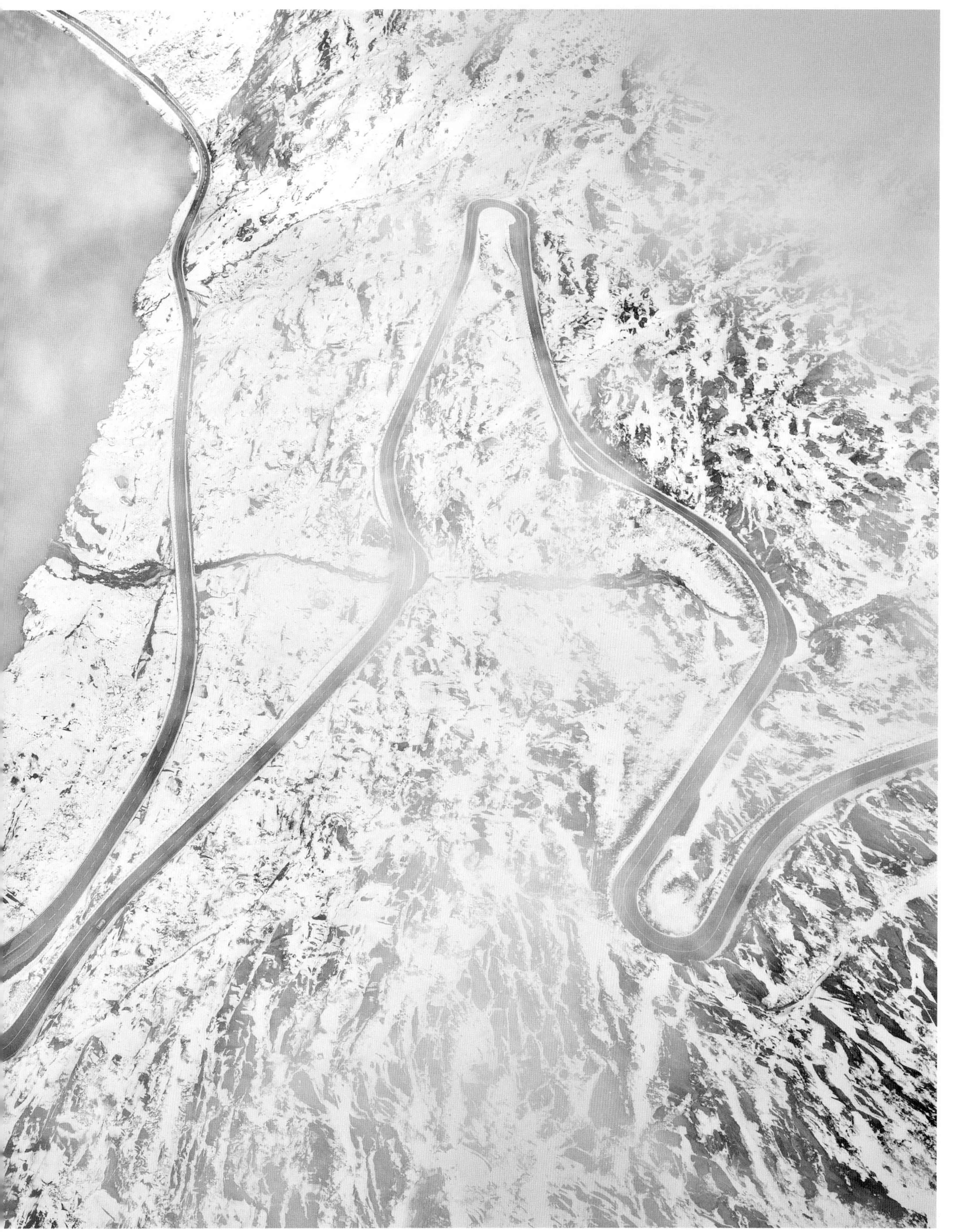

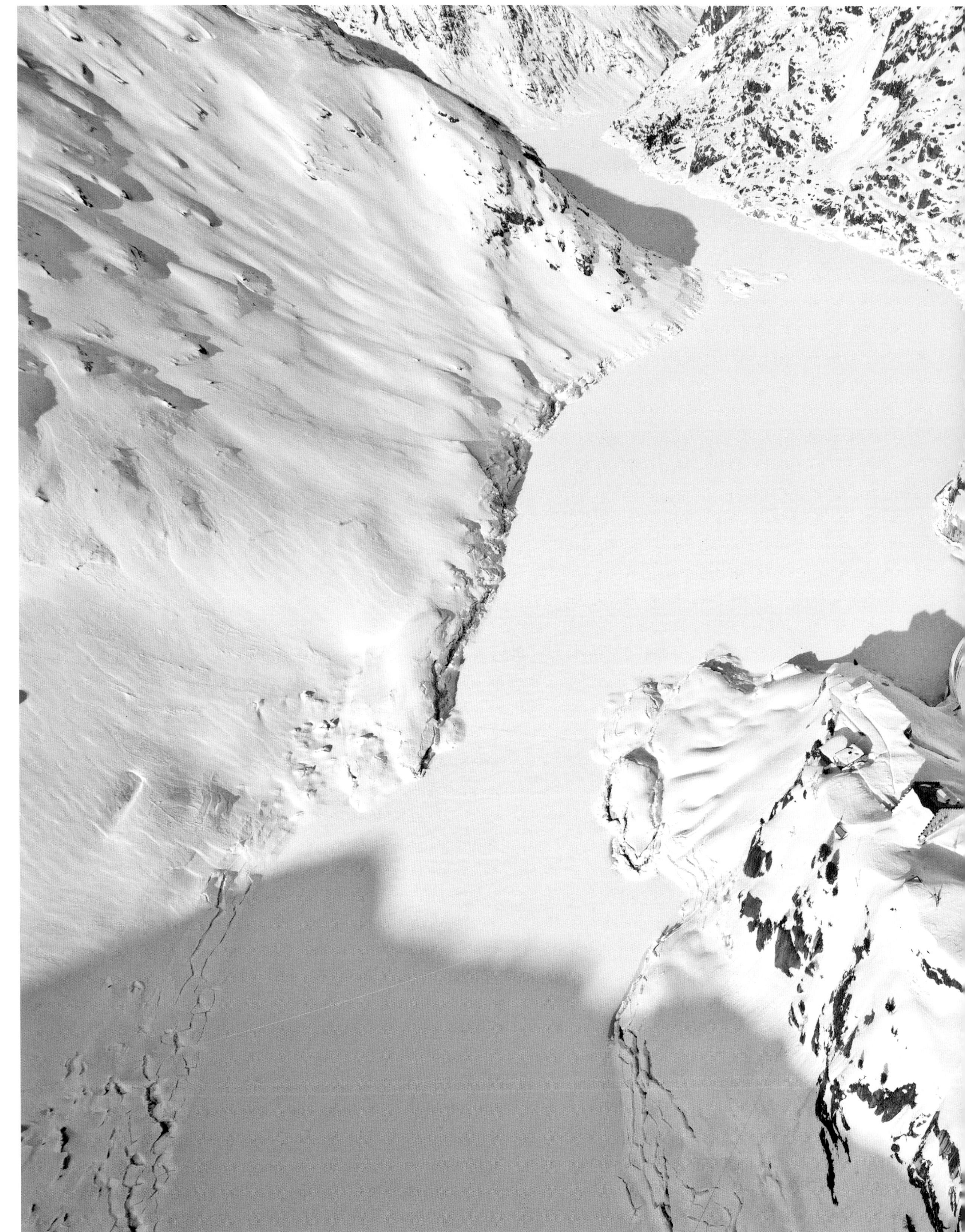

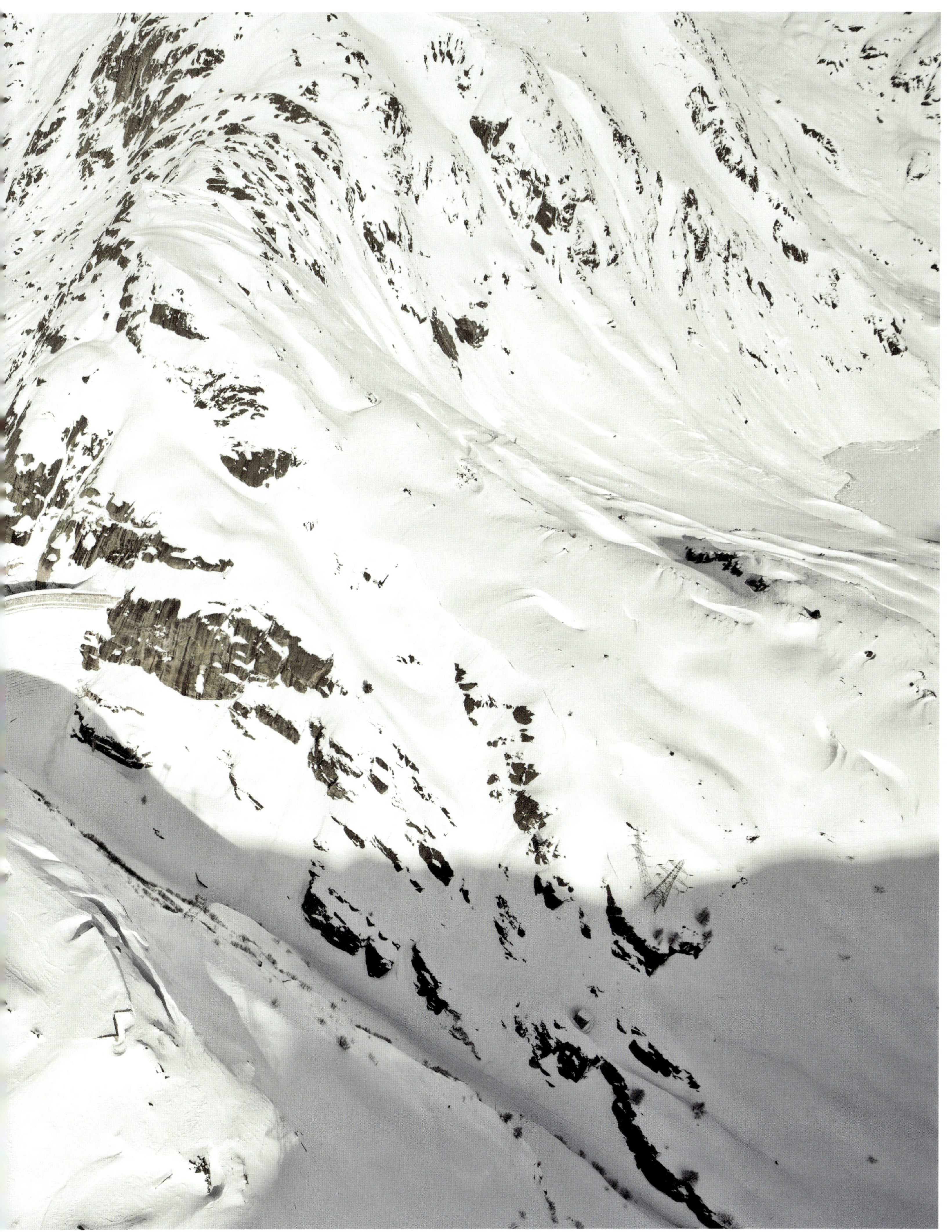

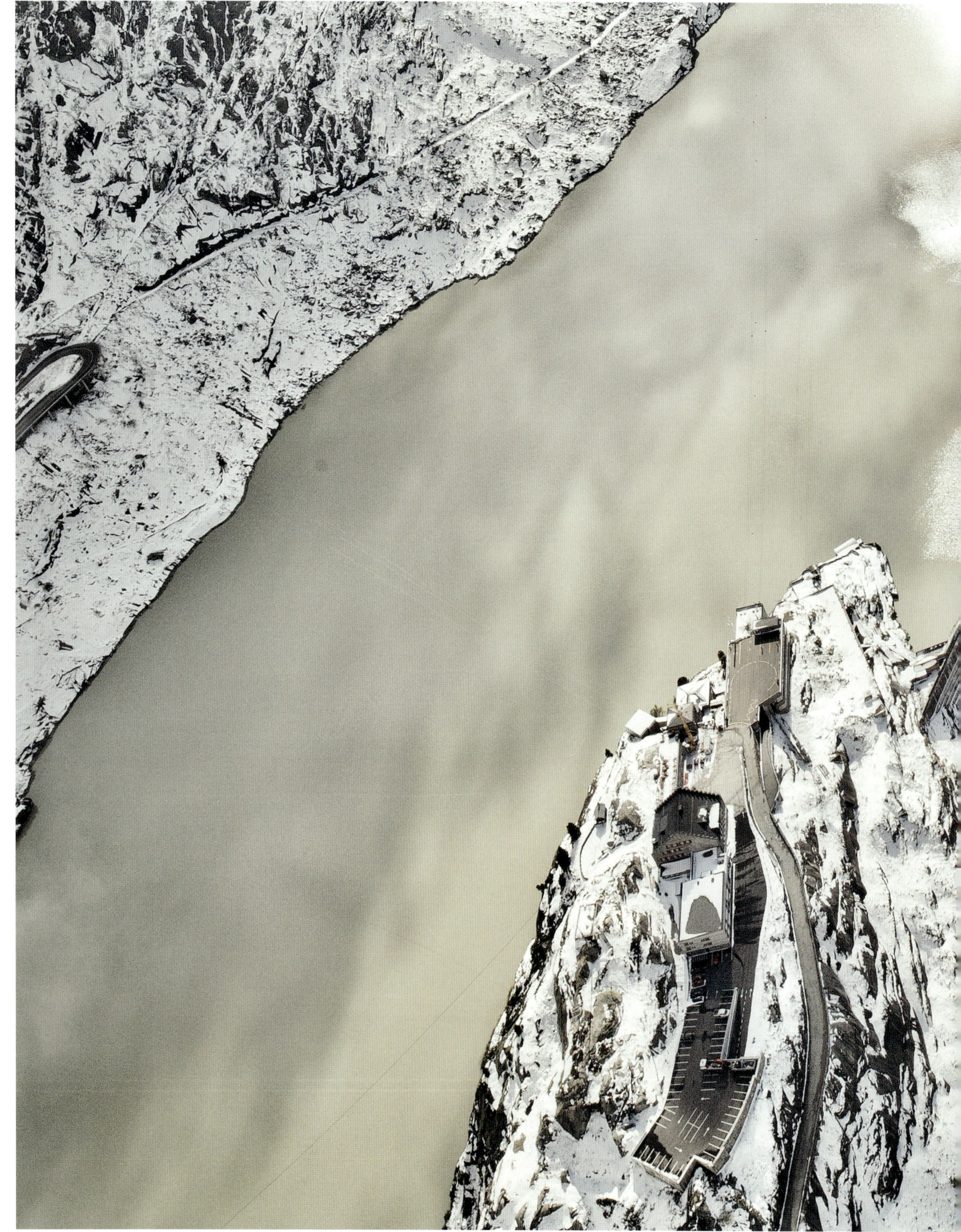

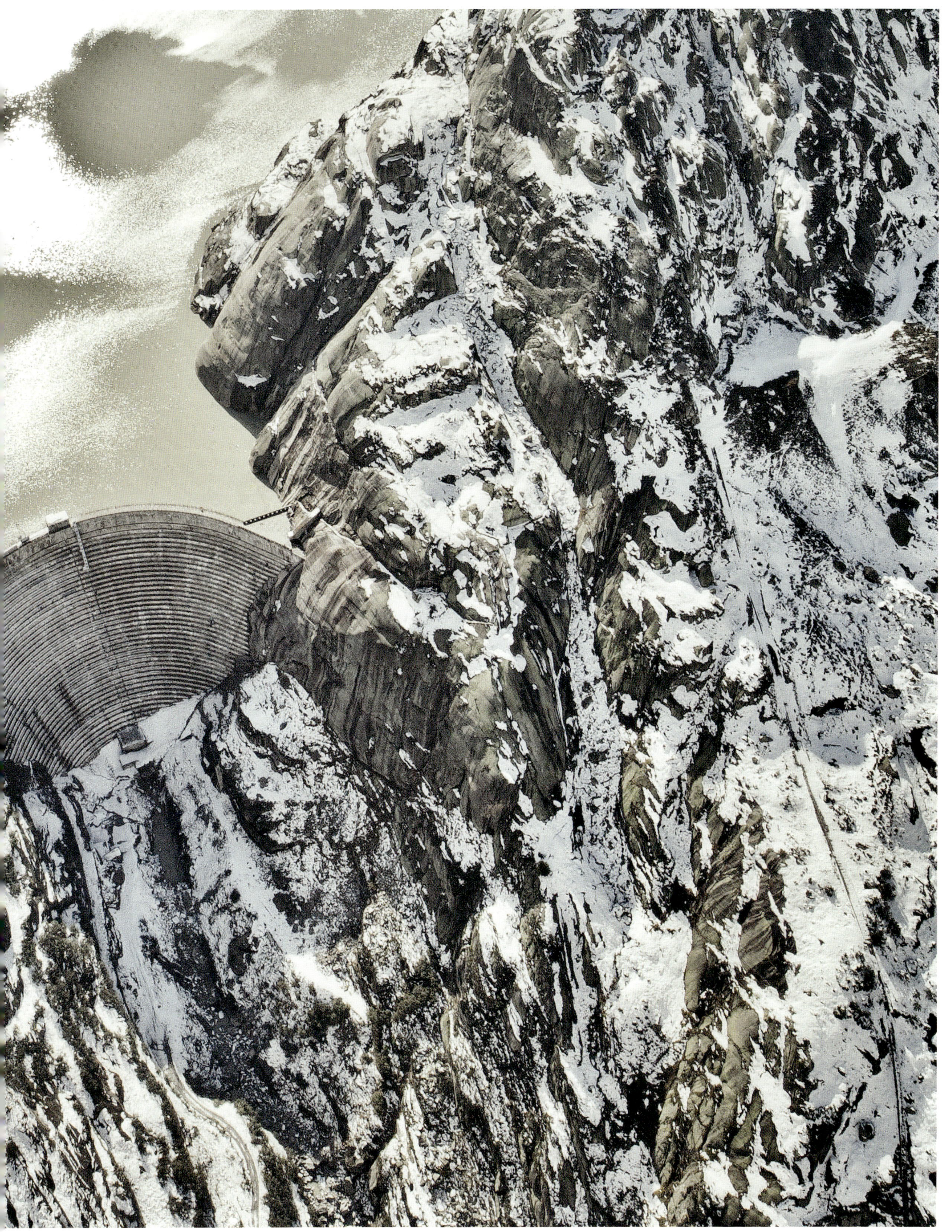

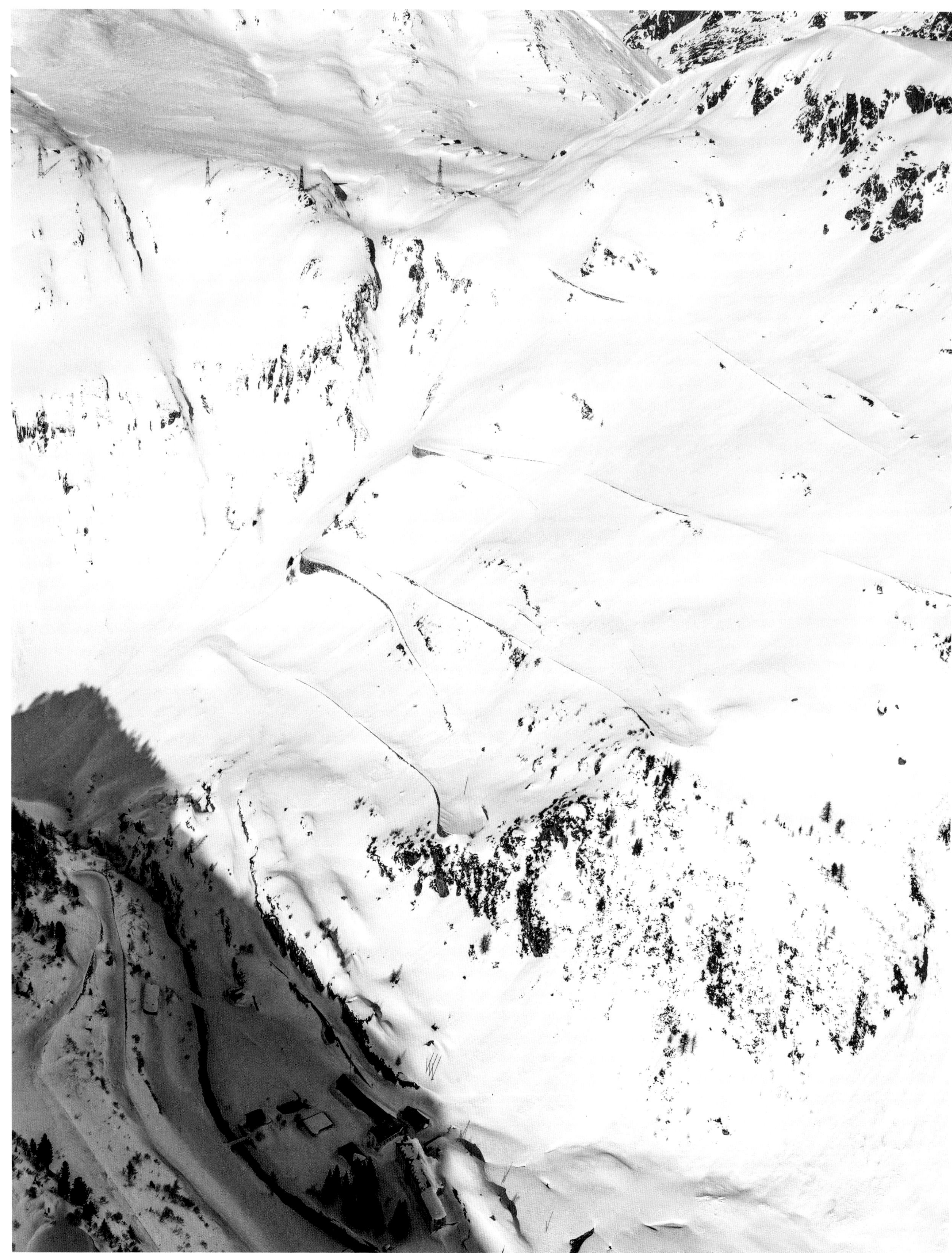

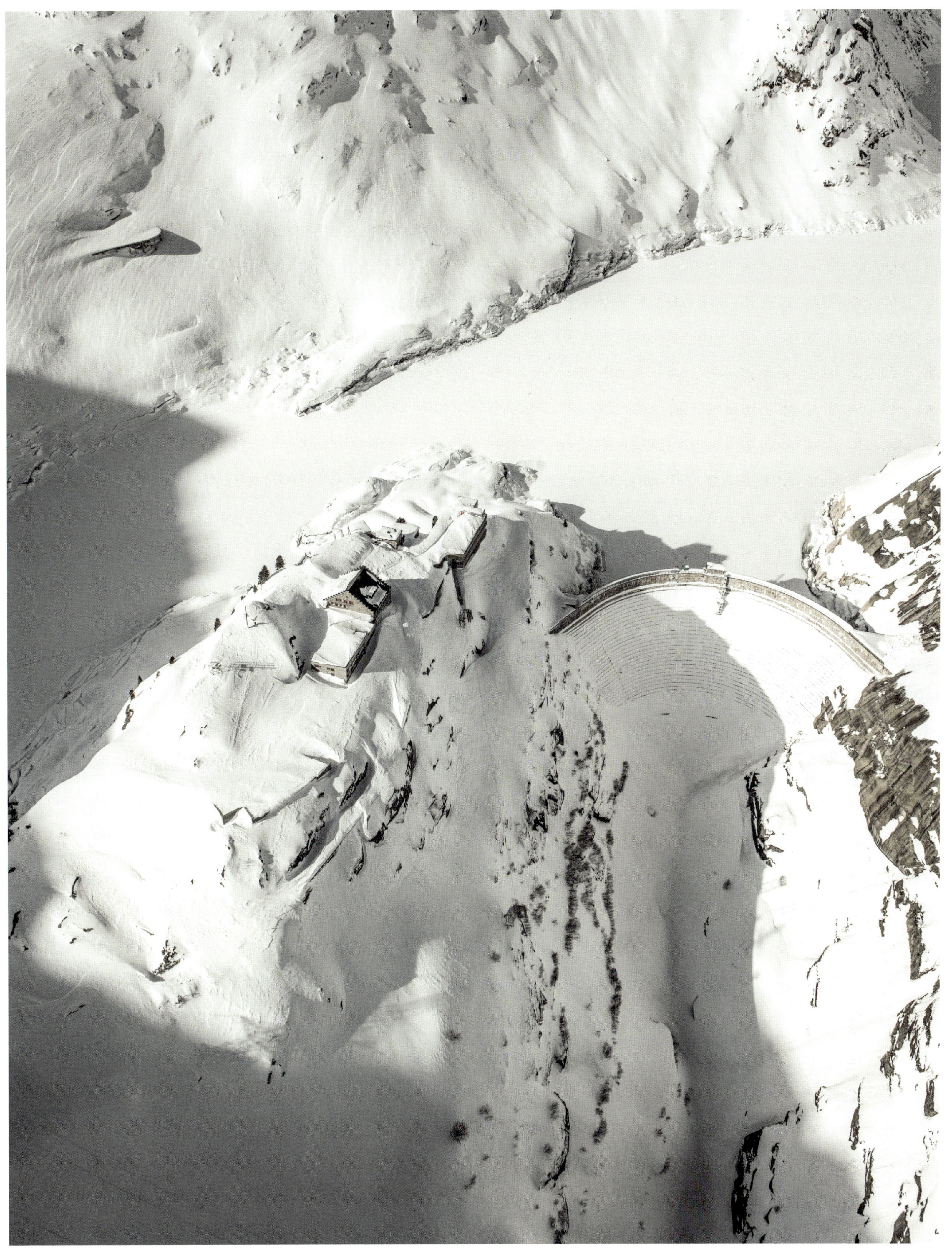

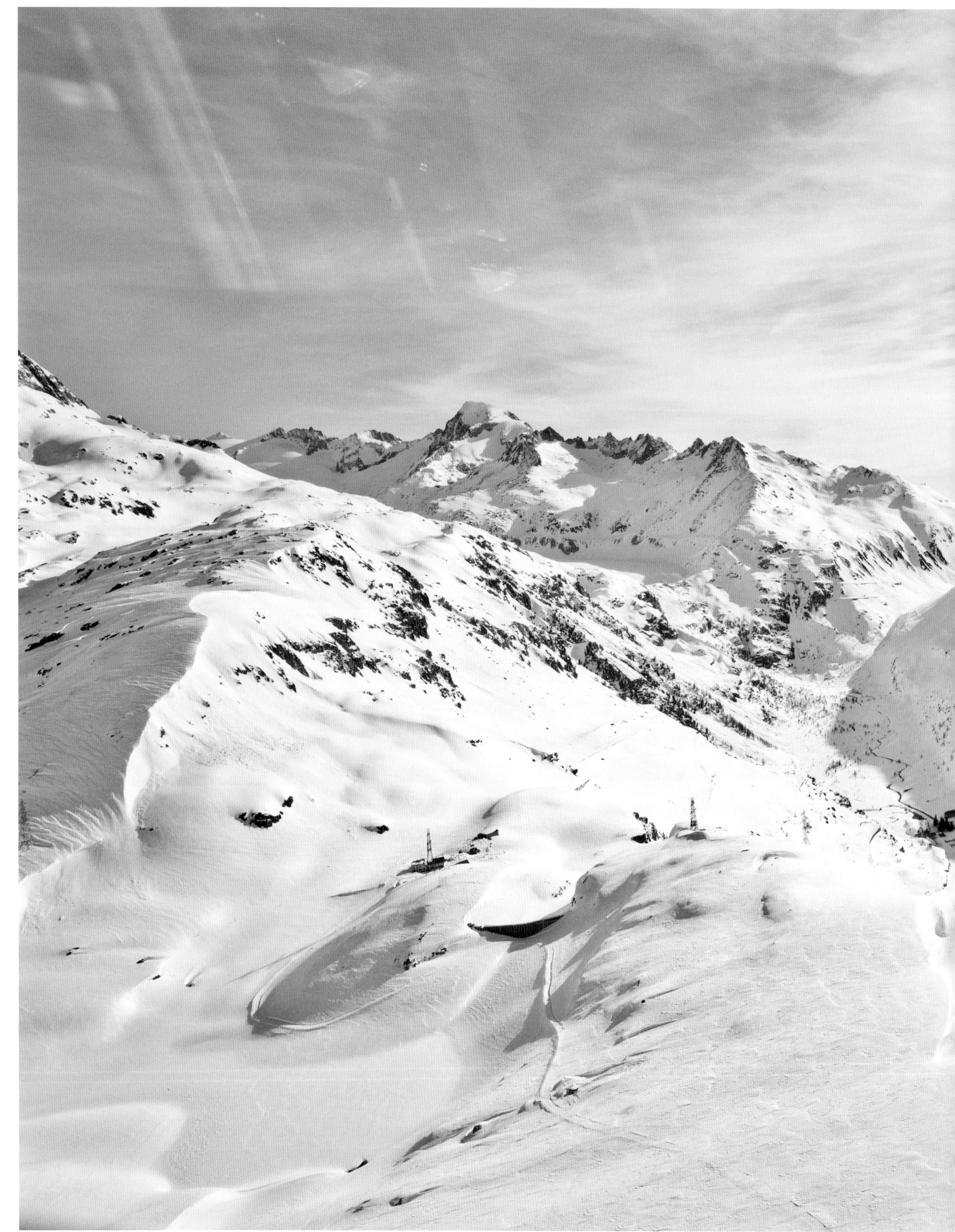

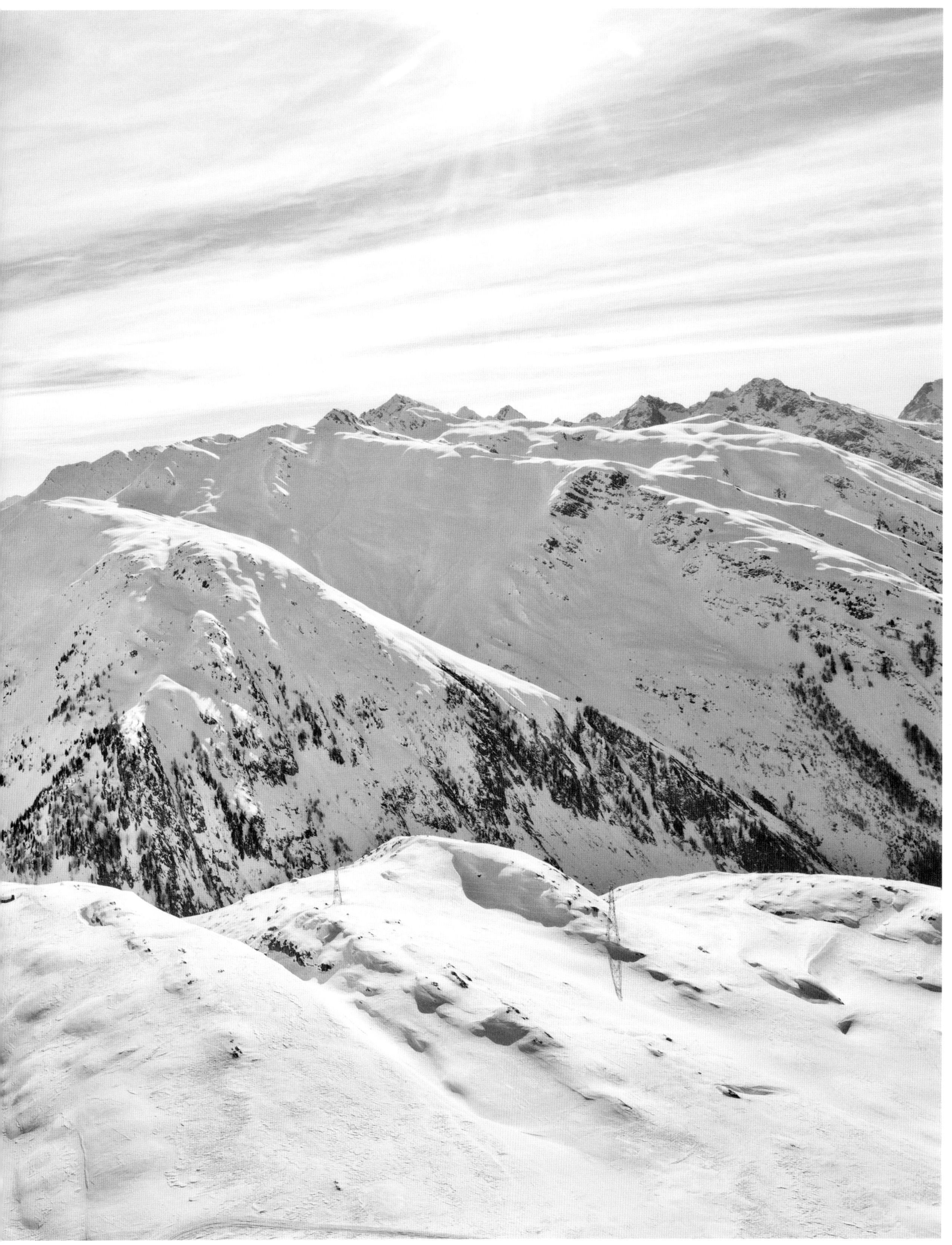

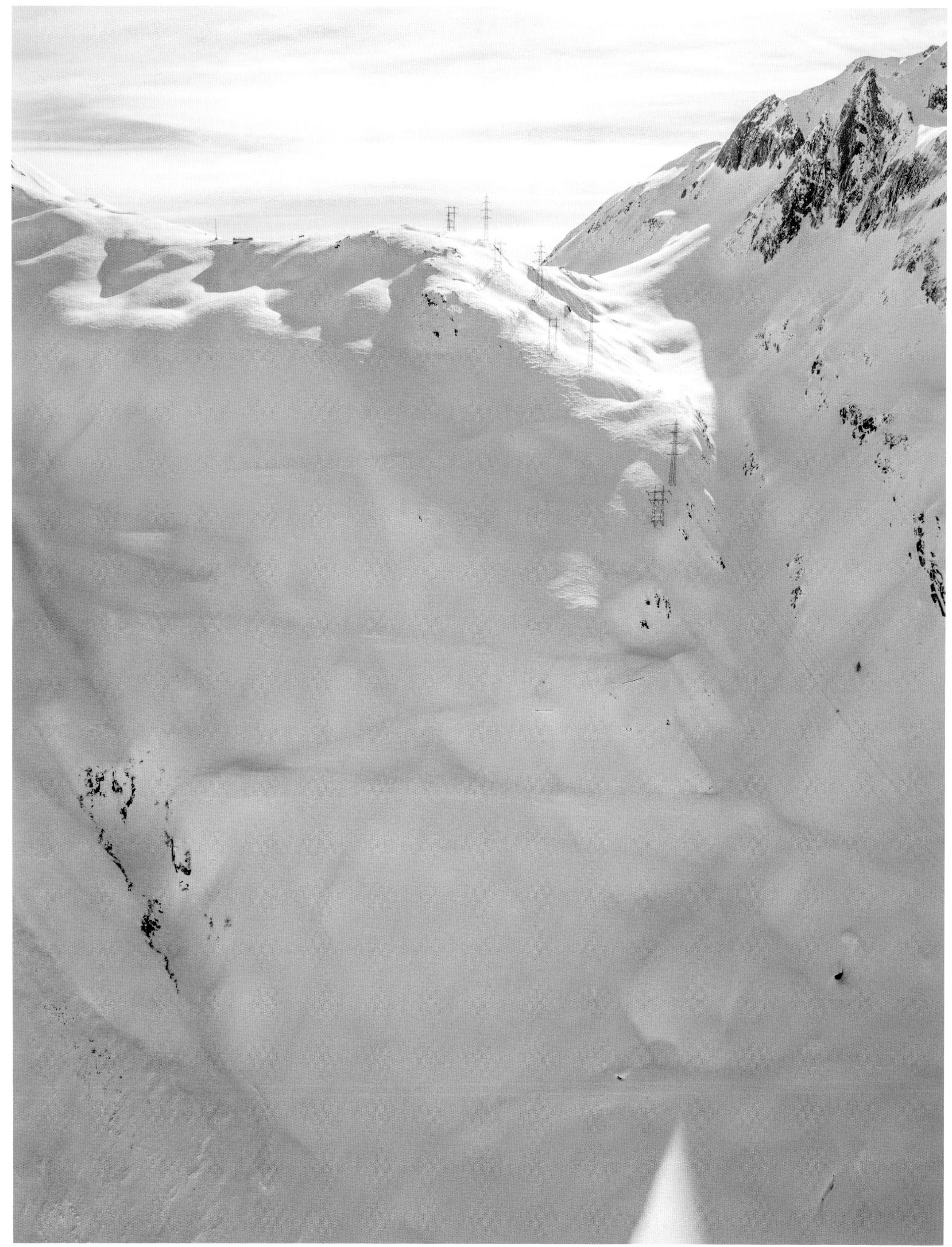

Nufenenpass

— 2478 Meter
Wintersperre meist von Oktober bis Juni

Der Nufenenpass ist mit einer Höhe von 2478 Metern die höchste befahrbare Passstraße der Schweiz. Entsprechend früh fällt die Verbindung zwischen dem Rhonetal im Kanton Wallis und dem Tessin meist in den Winterschlaf. Doch auch im Hochsommer kann einen am Pass durchaus ein Schneesturm kalt erwischen. Es lohnt sich deshalb, im Voraus den Wetterbericht zu studieren und die Schneeketten stets dabei zu haben. Das Bedrettotal auf der Südseite der Passstraße gilt zudem als eine der schneereichsten Regionen der Schweiz.

— 2,478 metres
Usually closed in winter from October to June

At an altitude of 2,478 metres, the Nufenen Pass is the highest navigable pass in Switzerland. For this reason, the route connecting the Rhone Valley in the Valais canton and Ticino tends to go into early hibernation. But even in midsummer, a snowfall at the pass can take motorists by surprise. Hence, it pays to study the weather report in advance and always have snow chains at the ready. Moreover, the Val Bedretto on the southern side of the pass road is regarded as one of the snowiest regions in Switzerland.

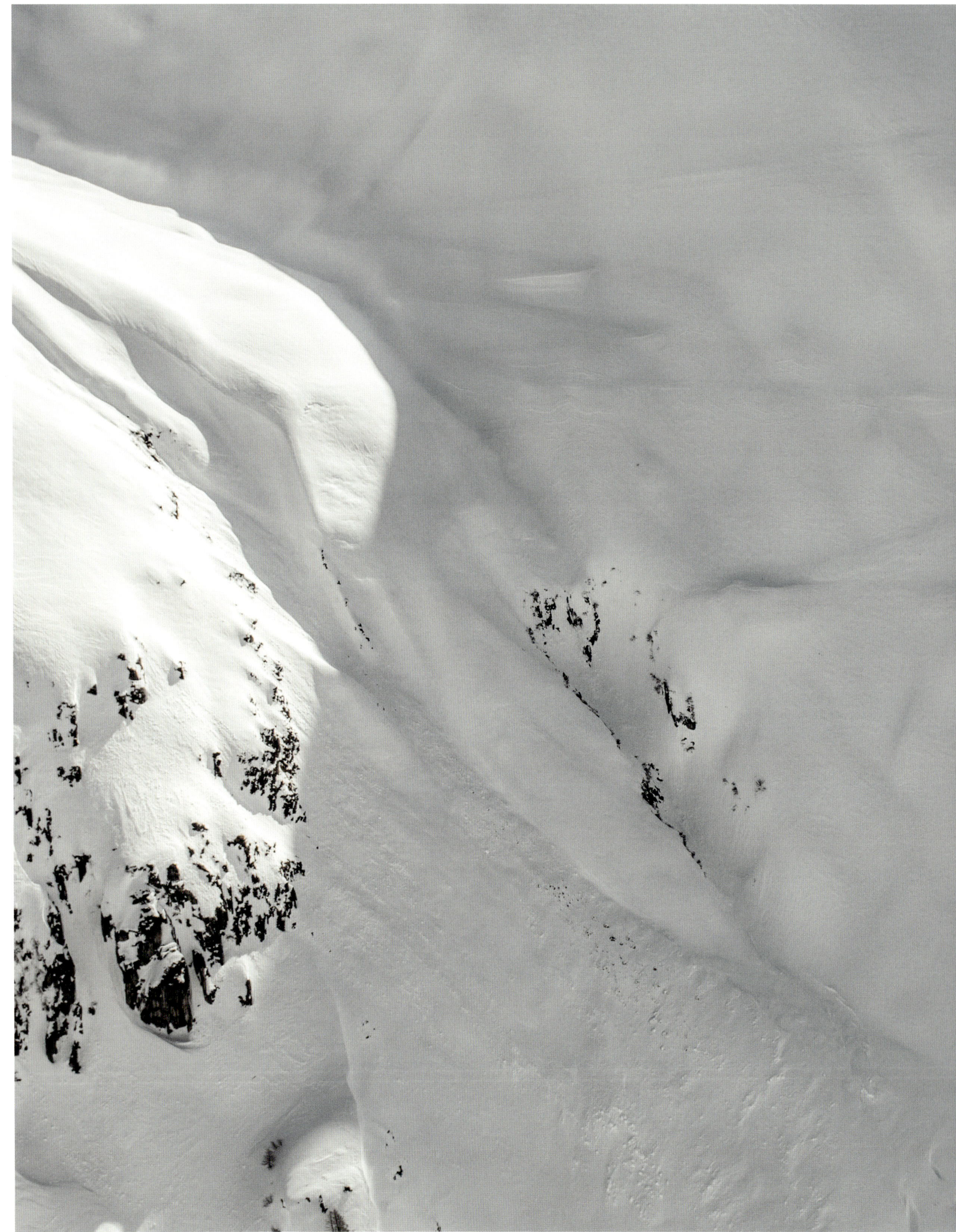

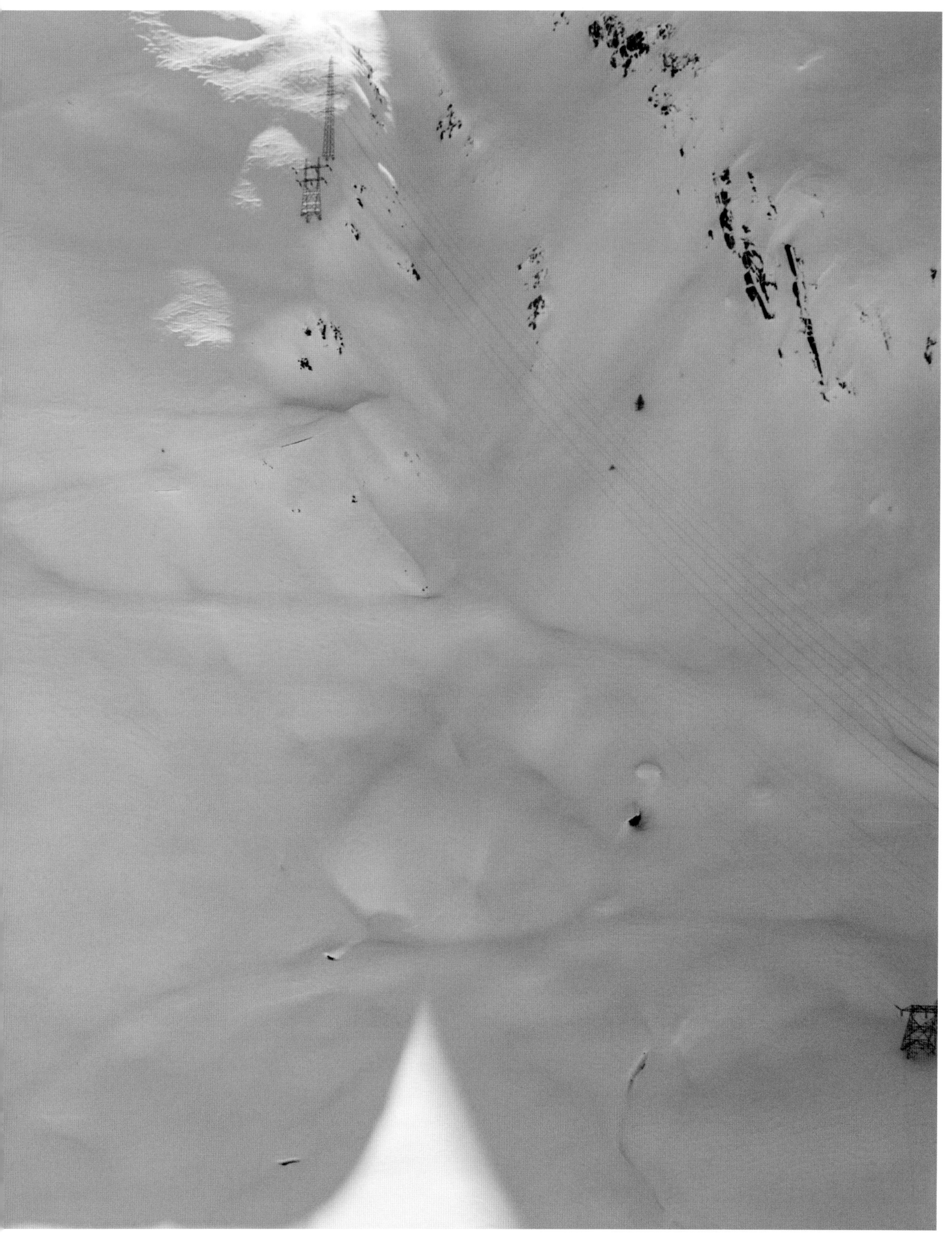

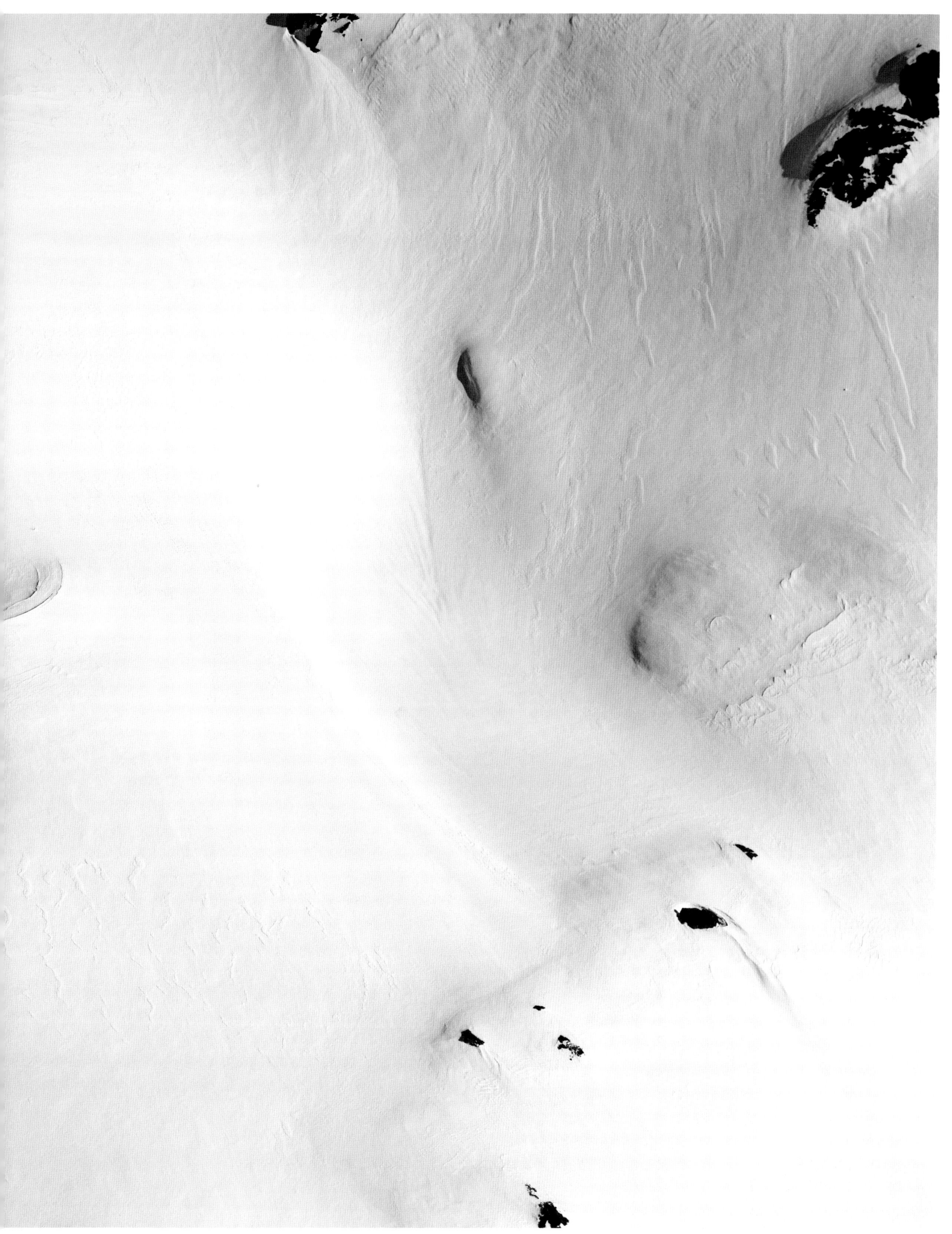

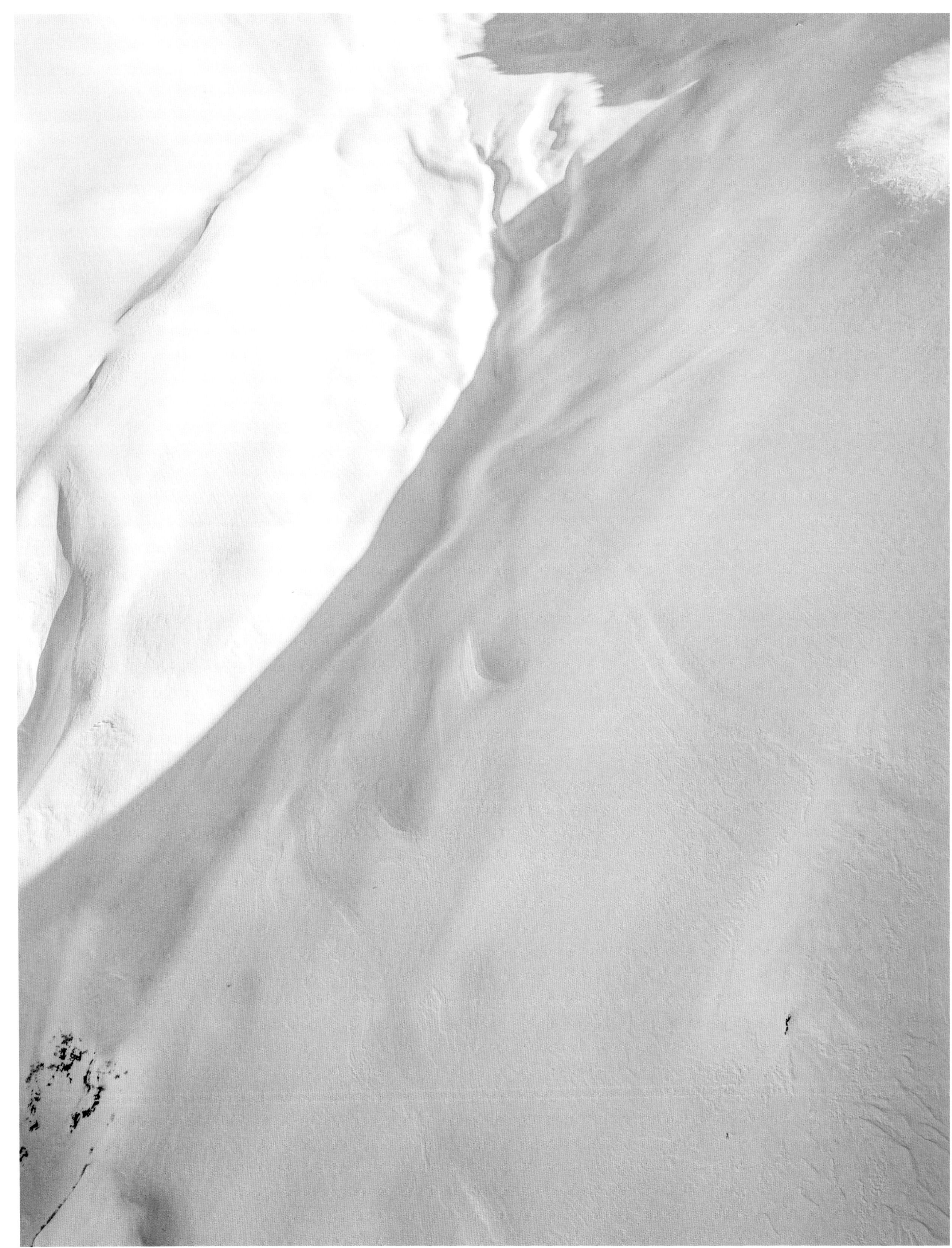

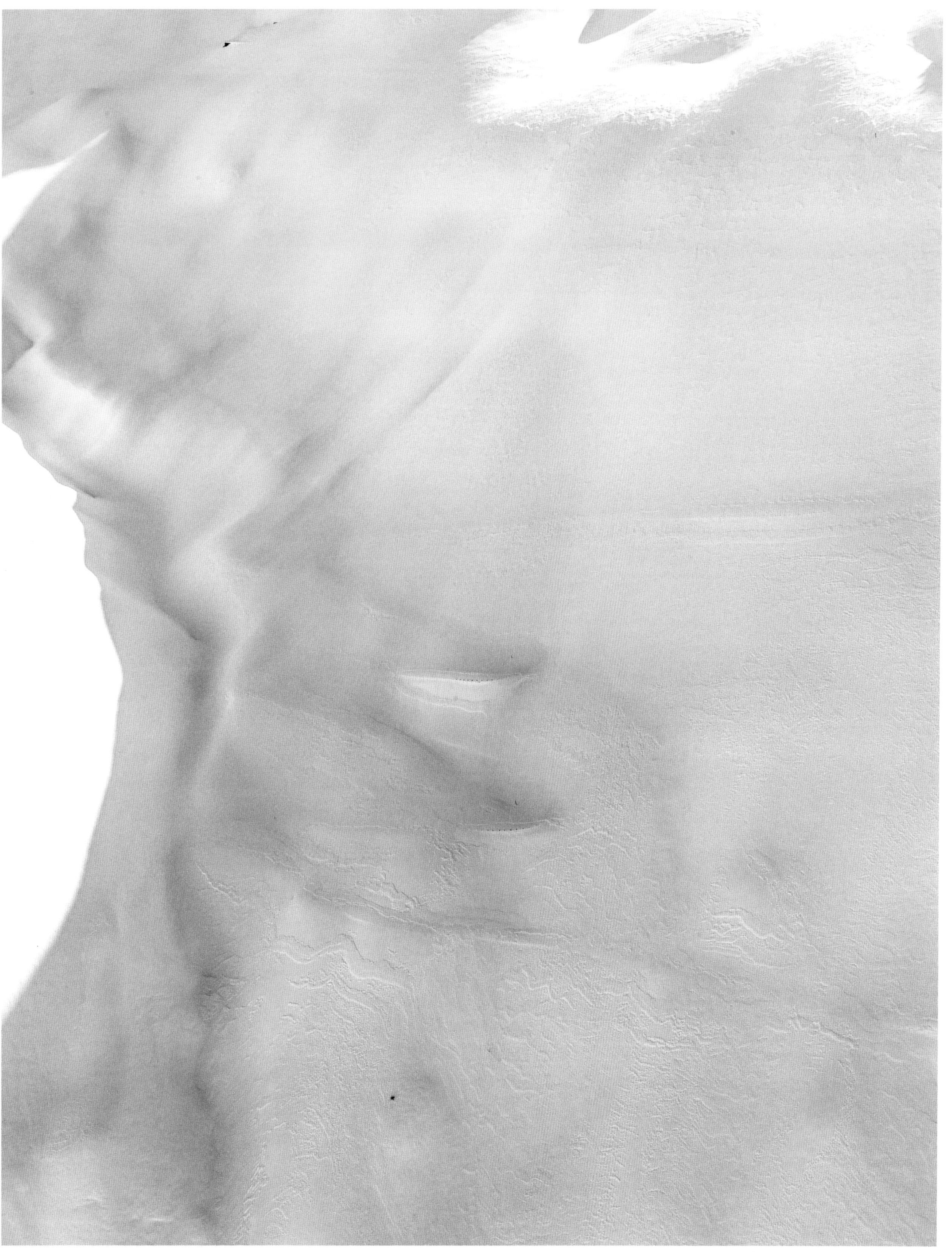

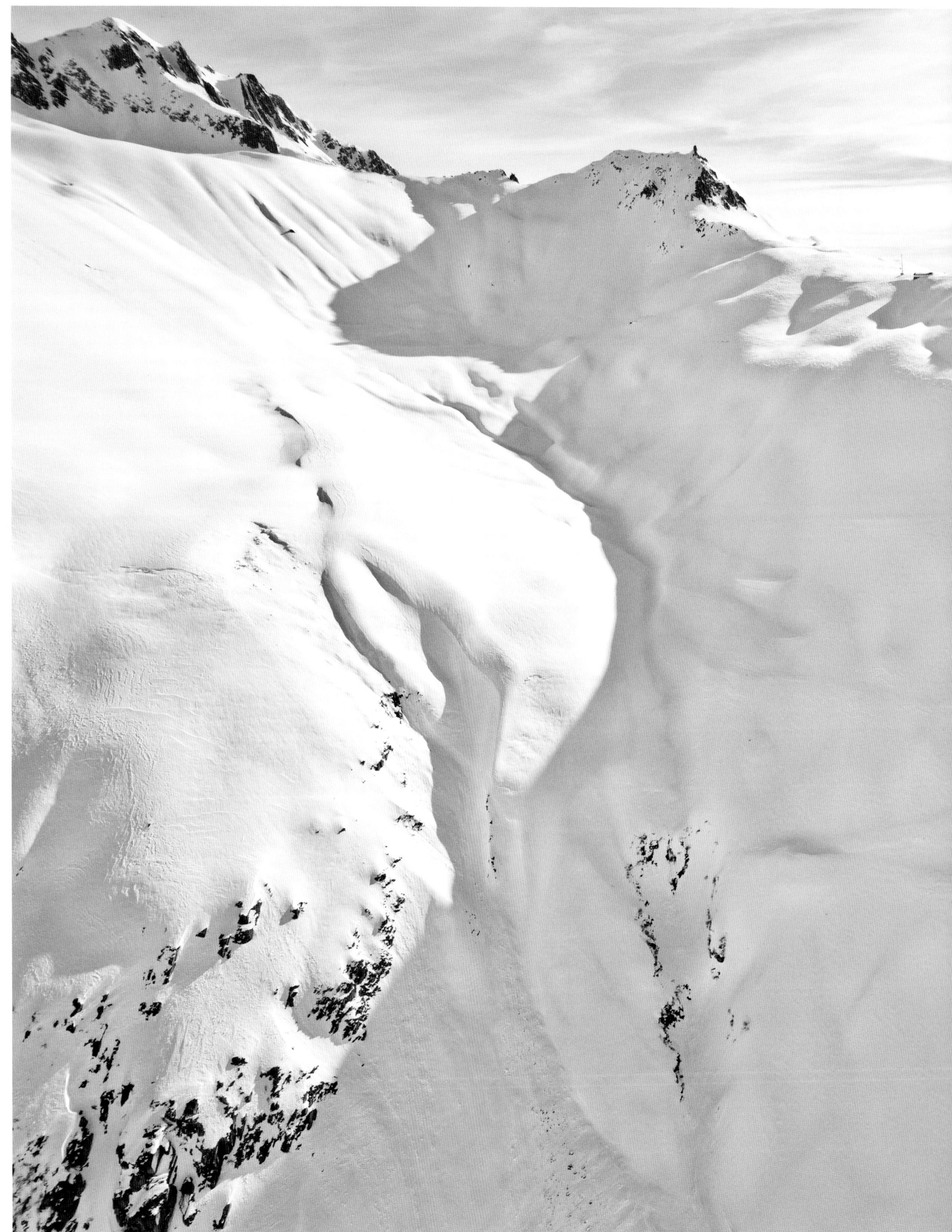

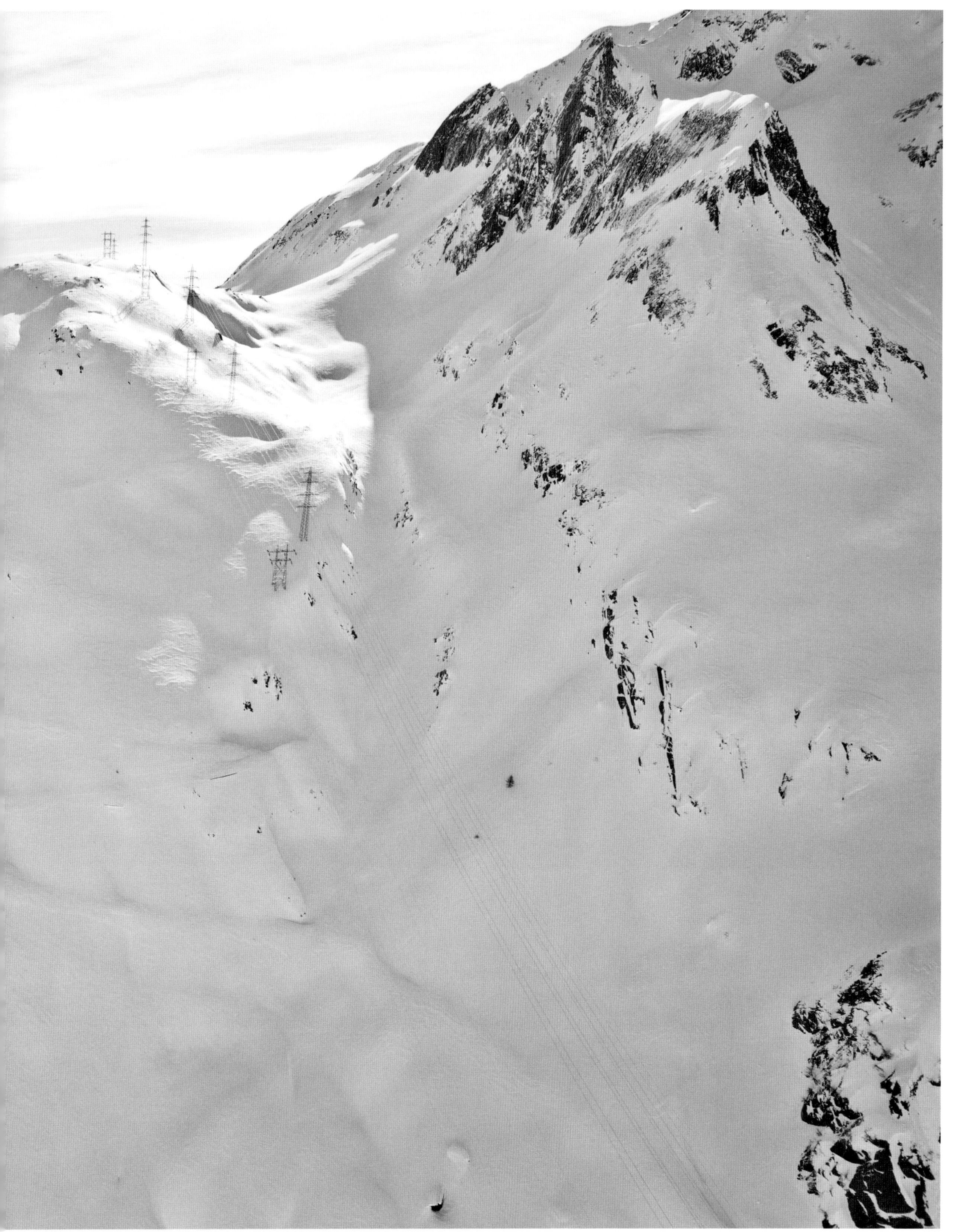

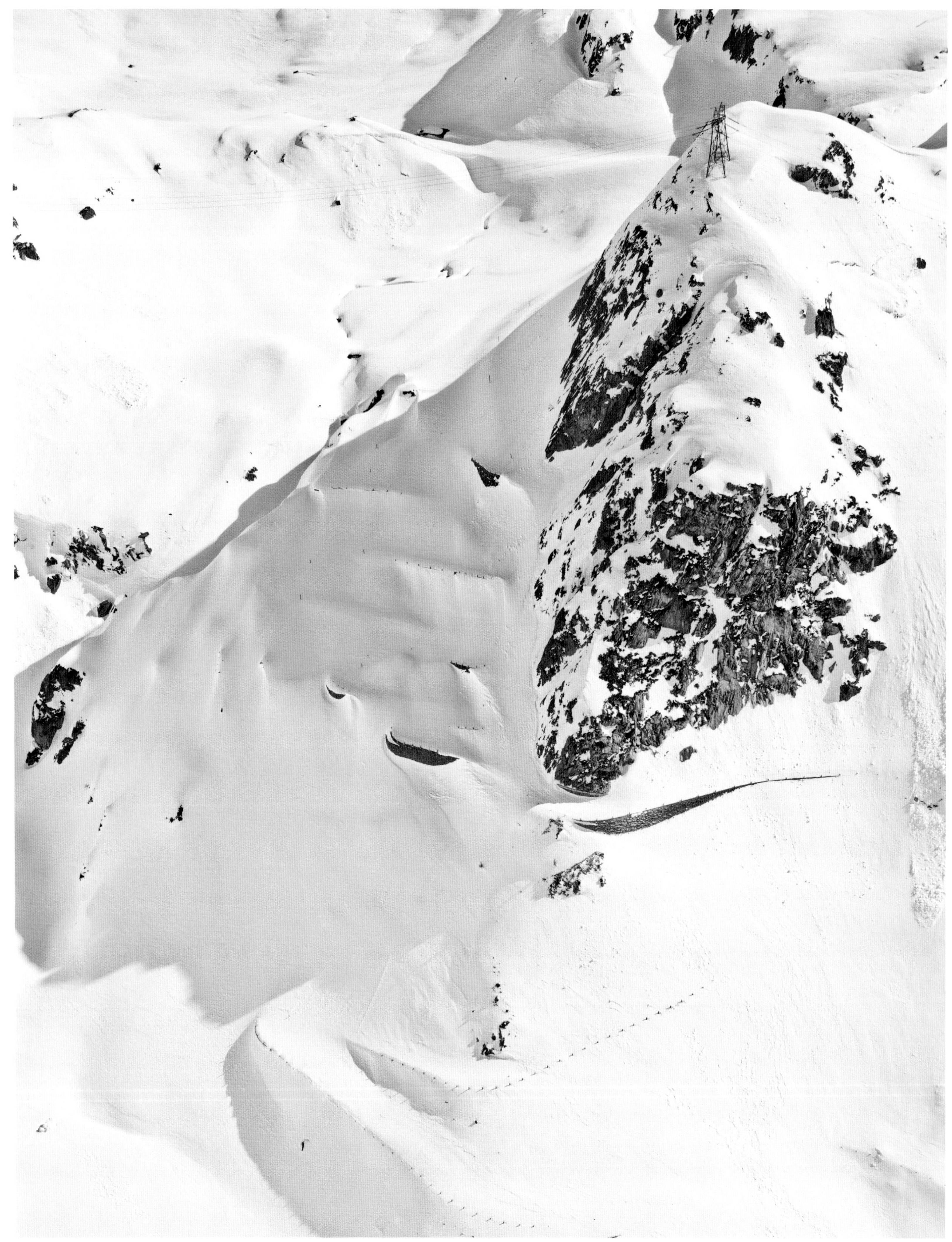

St. Gotthardpass

— 2106 Meter
Wintersperre meist von November bis Mai

Der St. Gotthardpass wird seit der Steinzeit begangen und war seit dem Mittelalter eine der wichtigsten Nord-Süd-Verbindungen der Alpen. Von Göschenen im Kanton Uri führt die Straße durch die schroffe Schöllenen-Schlucht und hinauf nach Andermatt und zur Passhöhe, von wo sich die alte Südrampe, die Tremola, in 24 gepflasterten Kehren hinab in Richtung Airolo im Tessin windet. Natürlich ist die Bahnfahrt durch den neu eröffneten Gotthard-Basistunnel viel komfortabler – warme Füße bekommt man aber auch beim Bremsen und Kuppeln am Berg.

— 2,106 metres
Usually closed in winter from November to May

The Gotthard Pass has been used since the Stone Age, and in the Middle Ages was one of the Alps' most important routes between the north and the south. From Göschenen in the canton of Uri, the road leads through the craggy Schöllenen Gorge and up to Andermatt and the summit of the pass, from where the old southern flank, the Tremola, winds through 24 paved hairpins towards Airolo in Ticino. Of course, the ride through the newly-opened Gotthard Base Tunnel is much more comfortable – but you can also warm your feet working the brakes and clutch on the mountain.

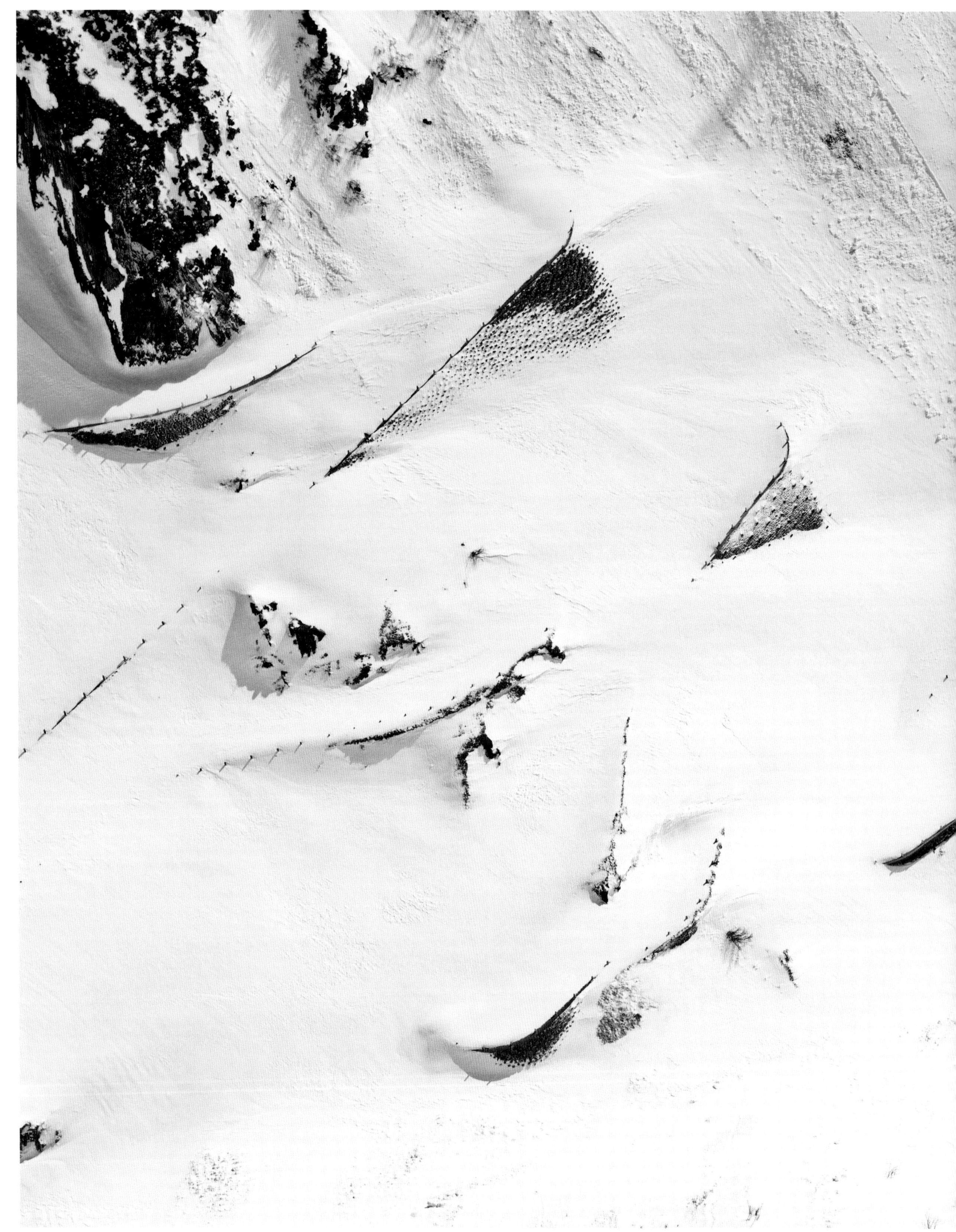

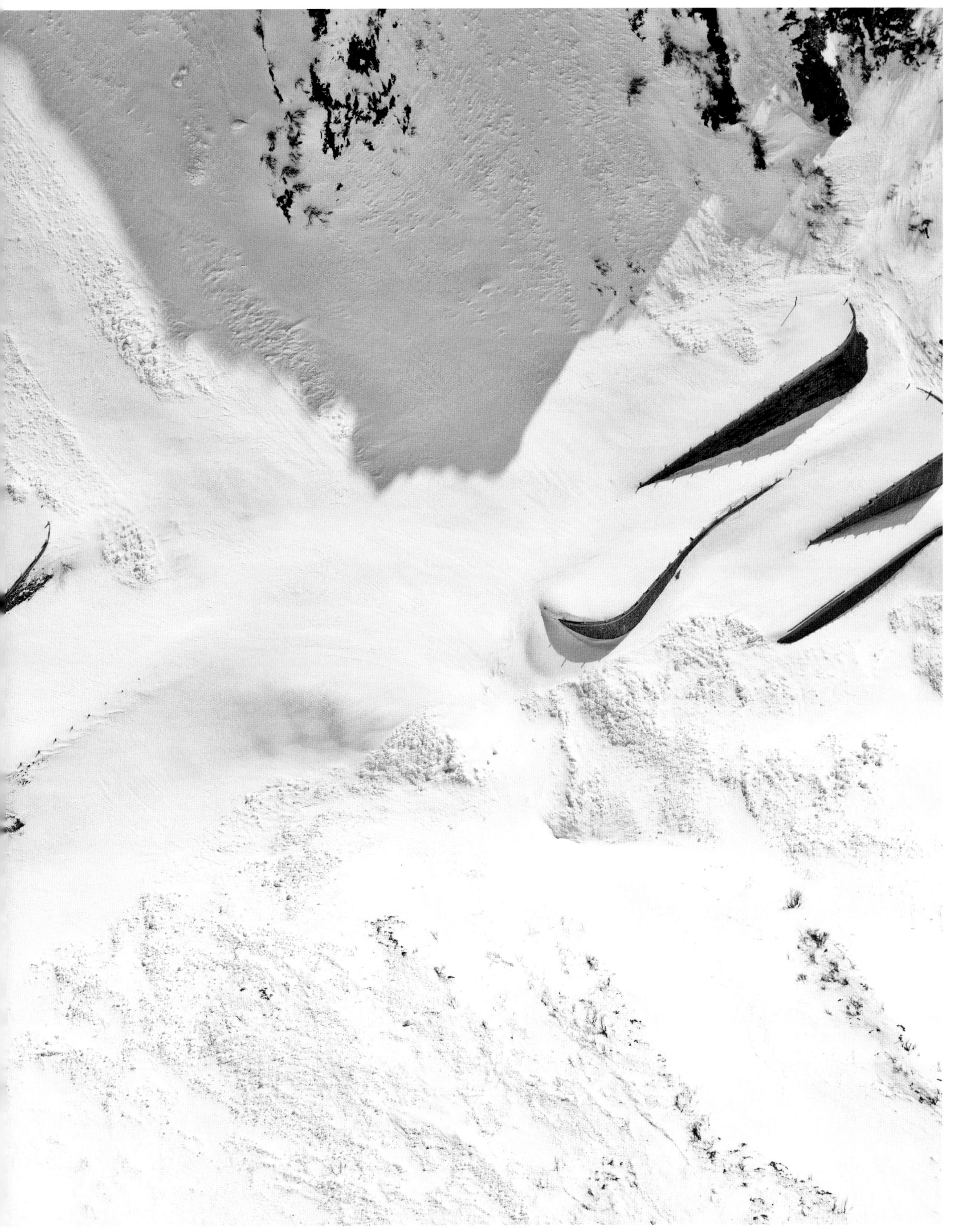

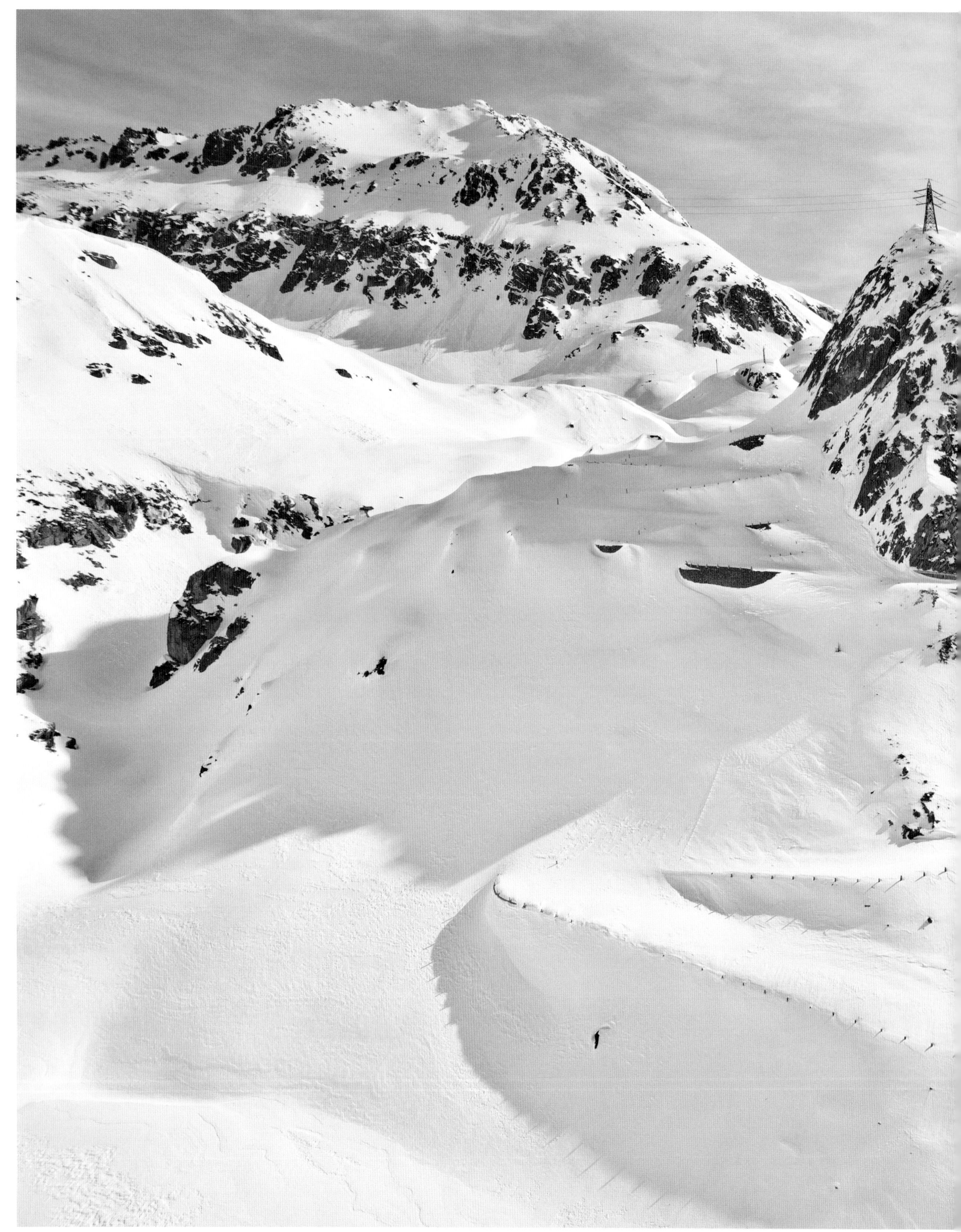

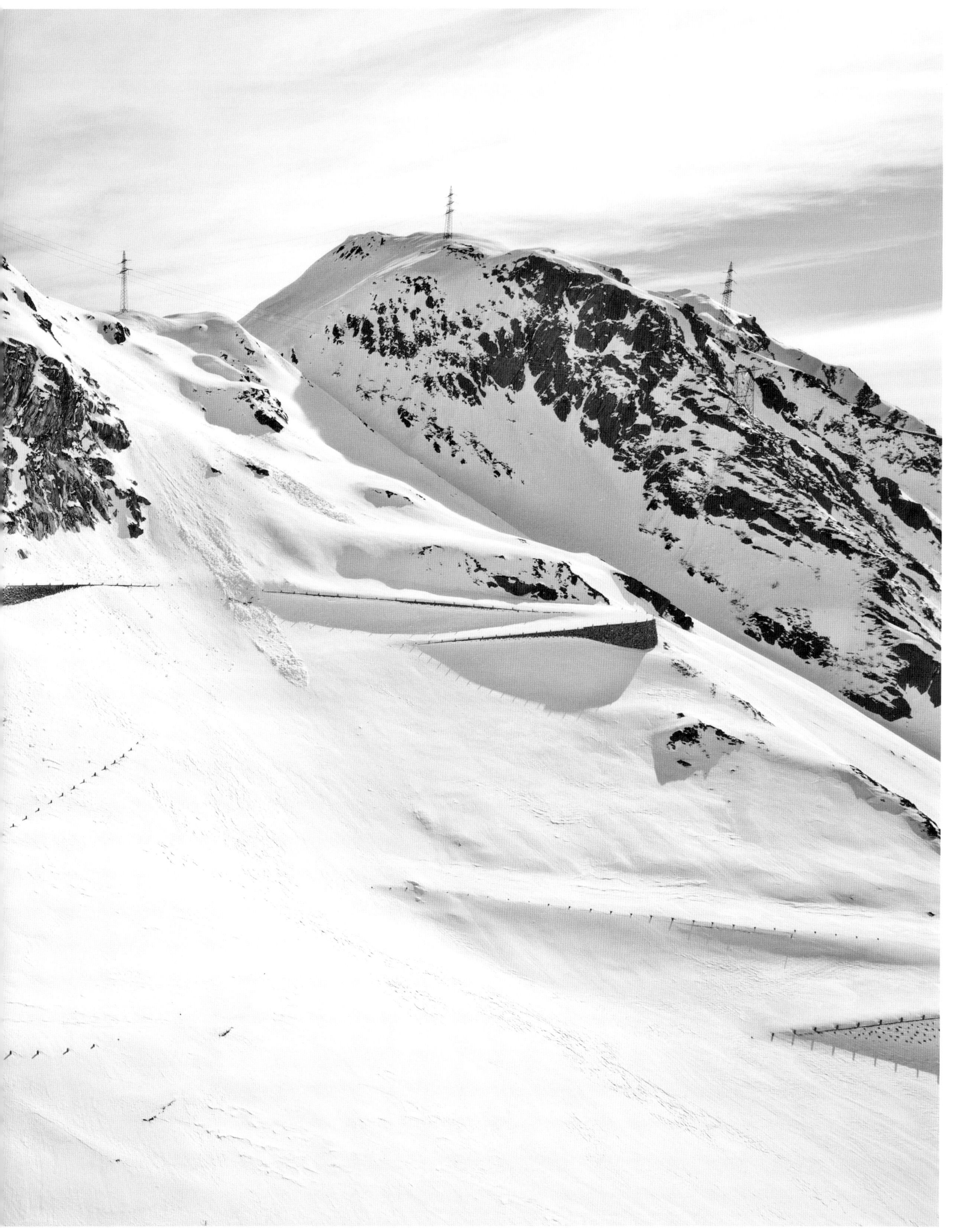

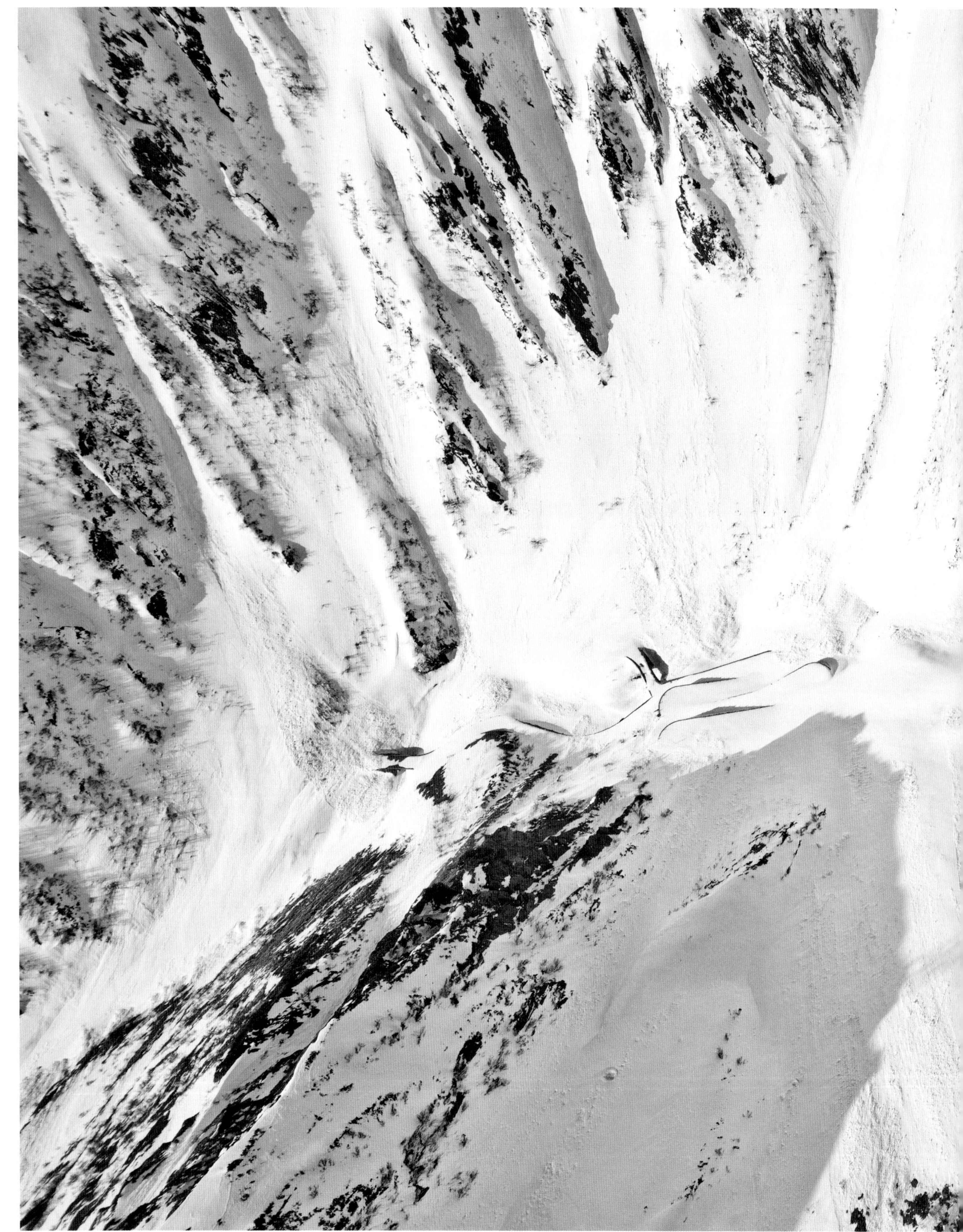

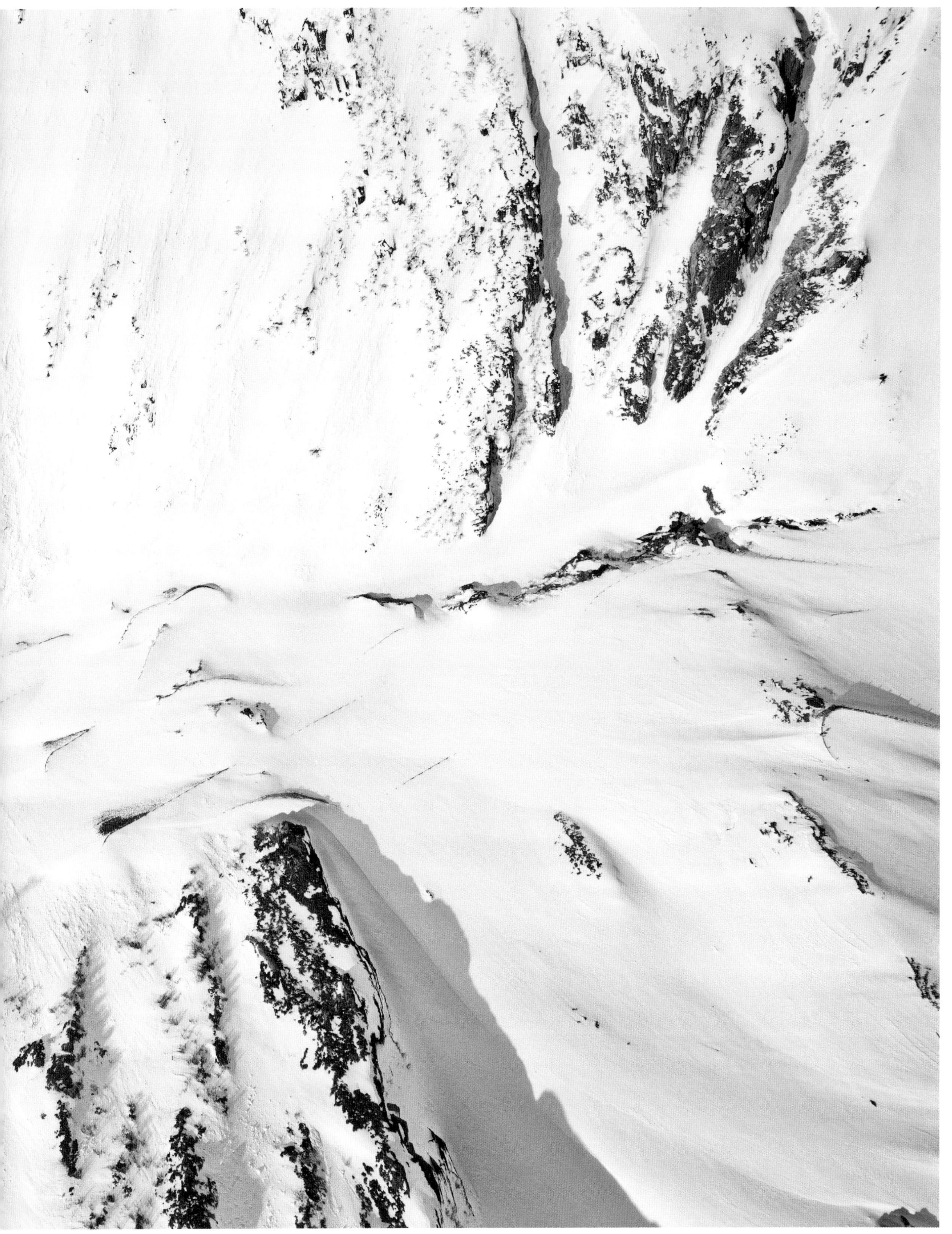

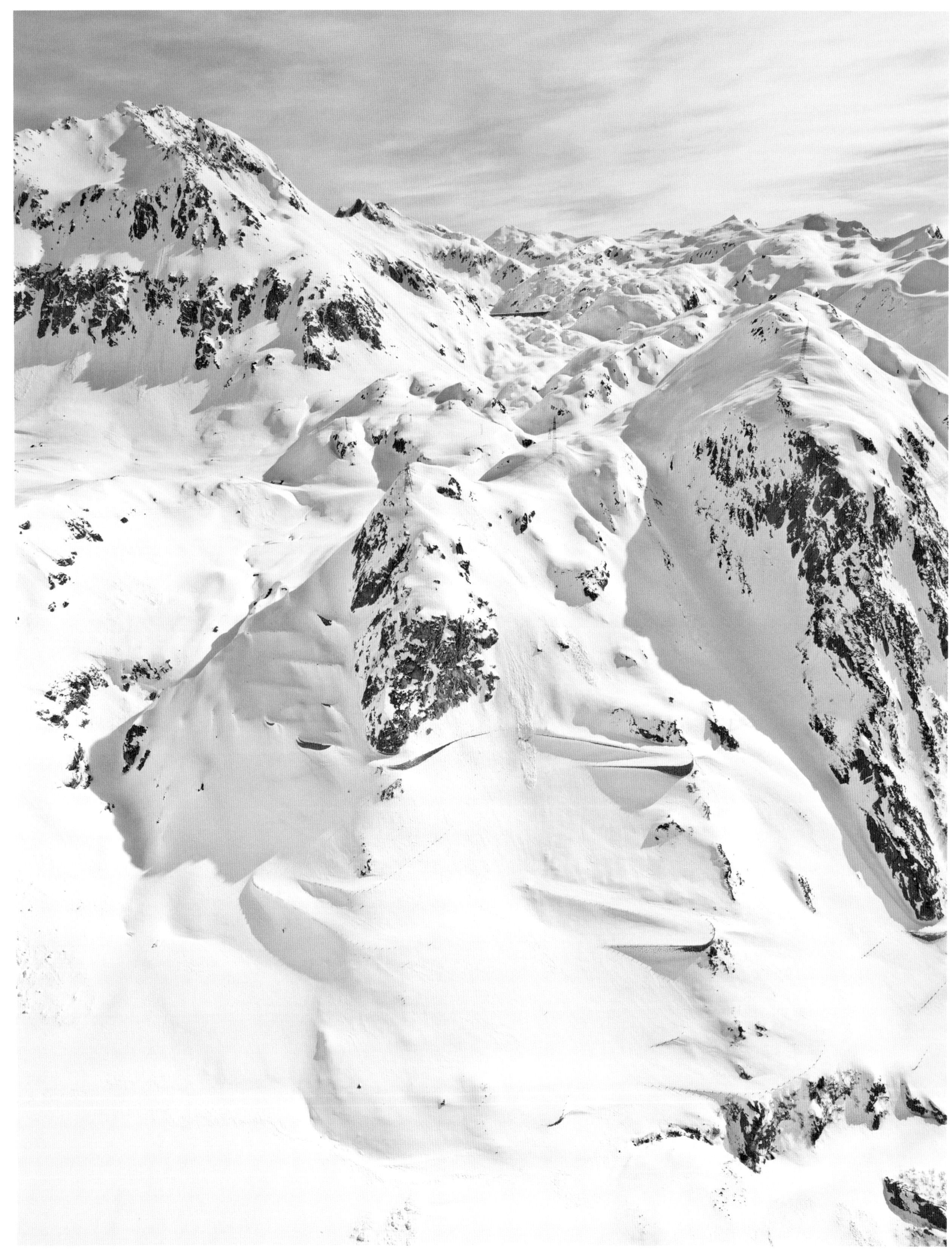

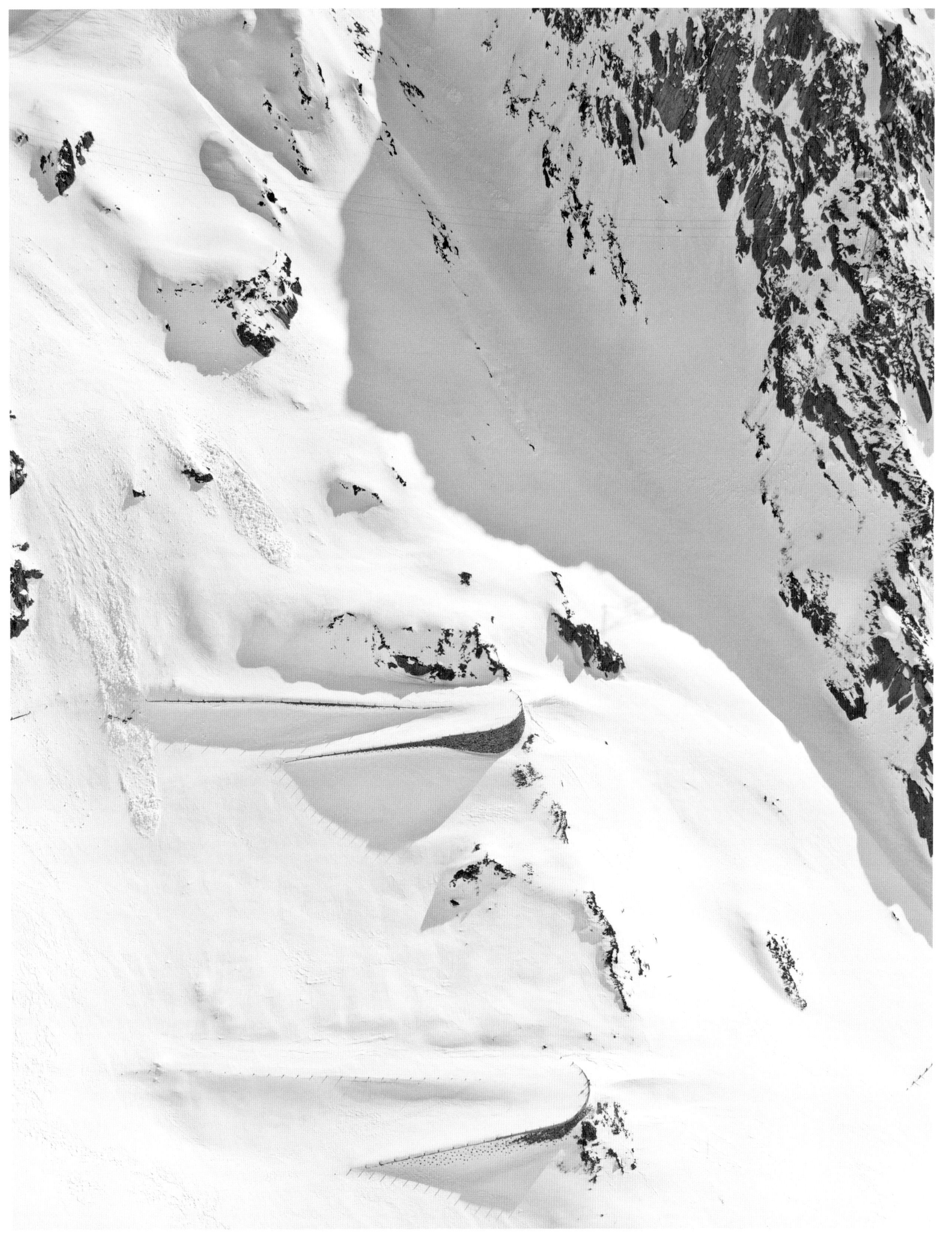

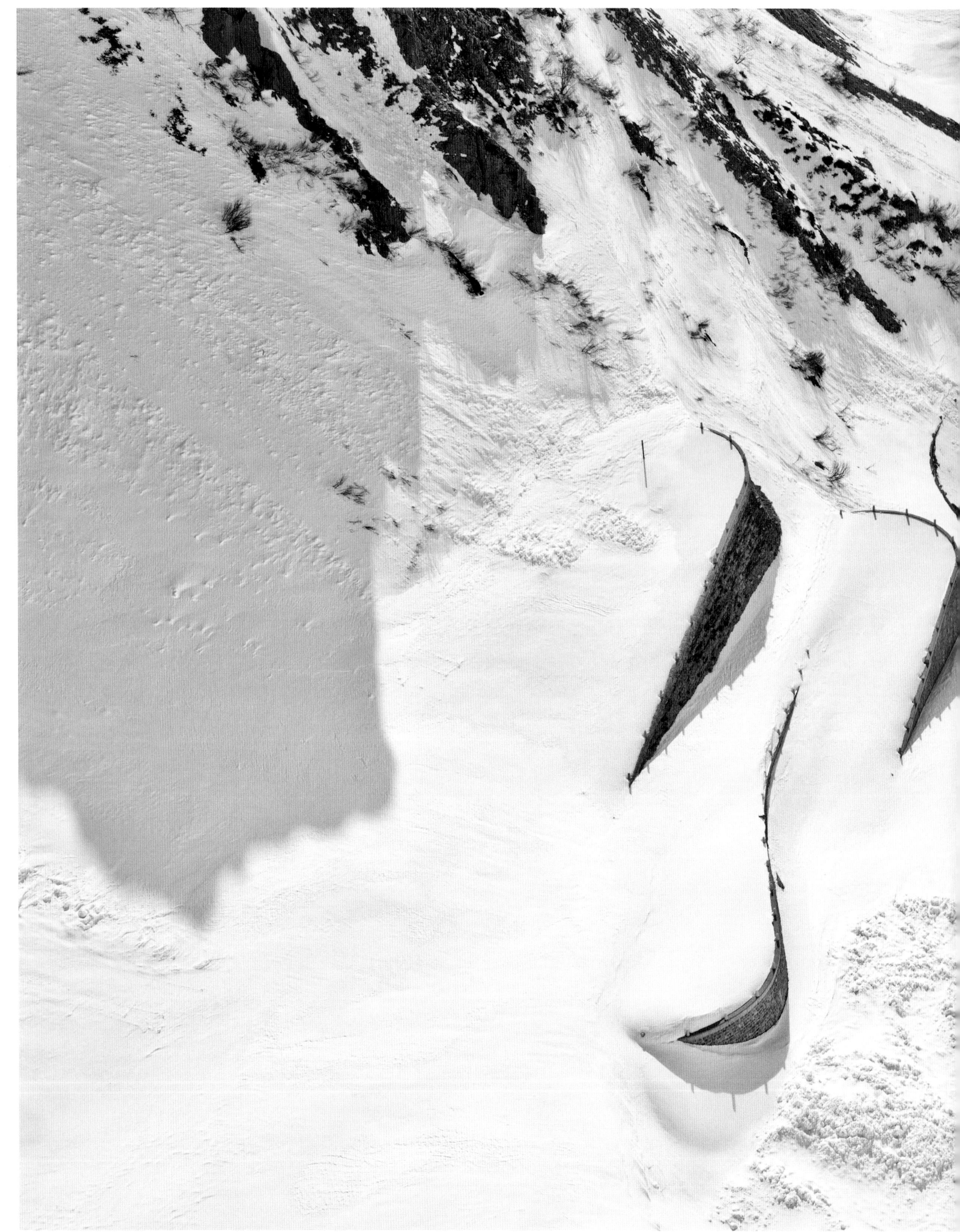

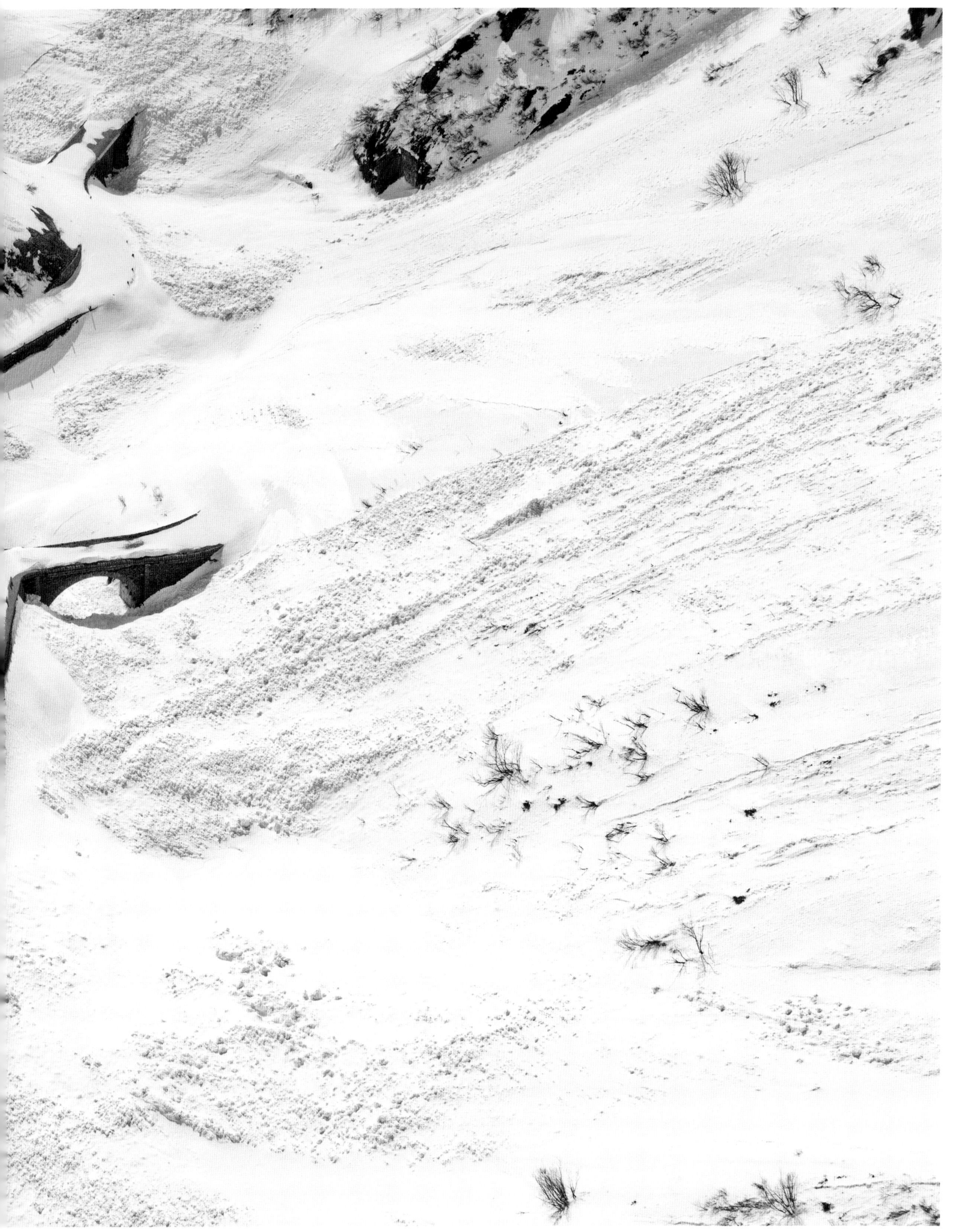

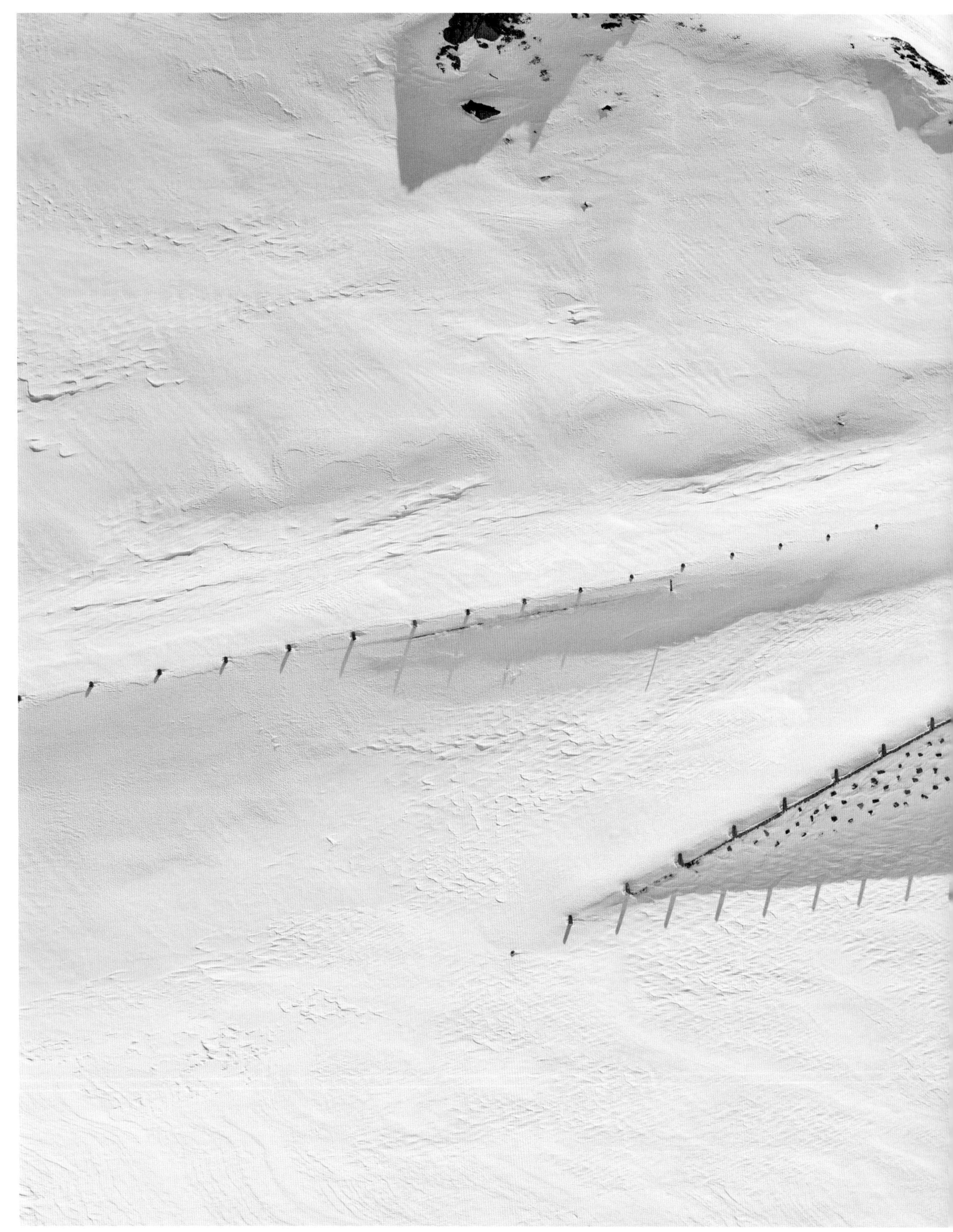

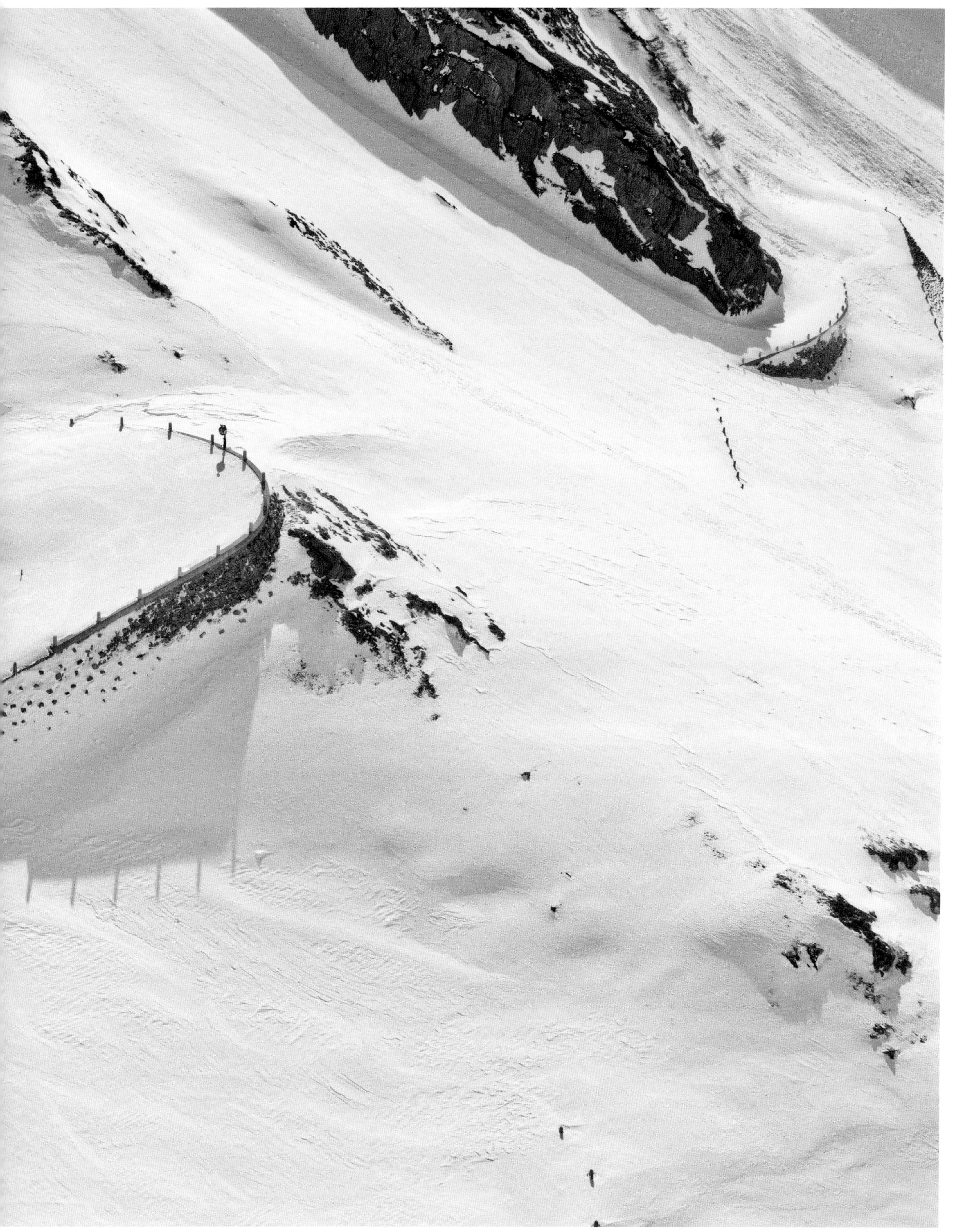

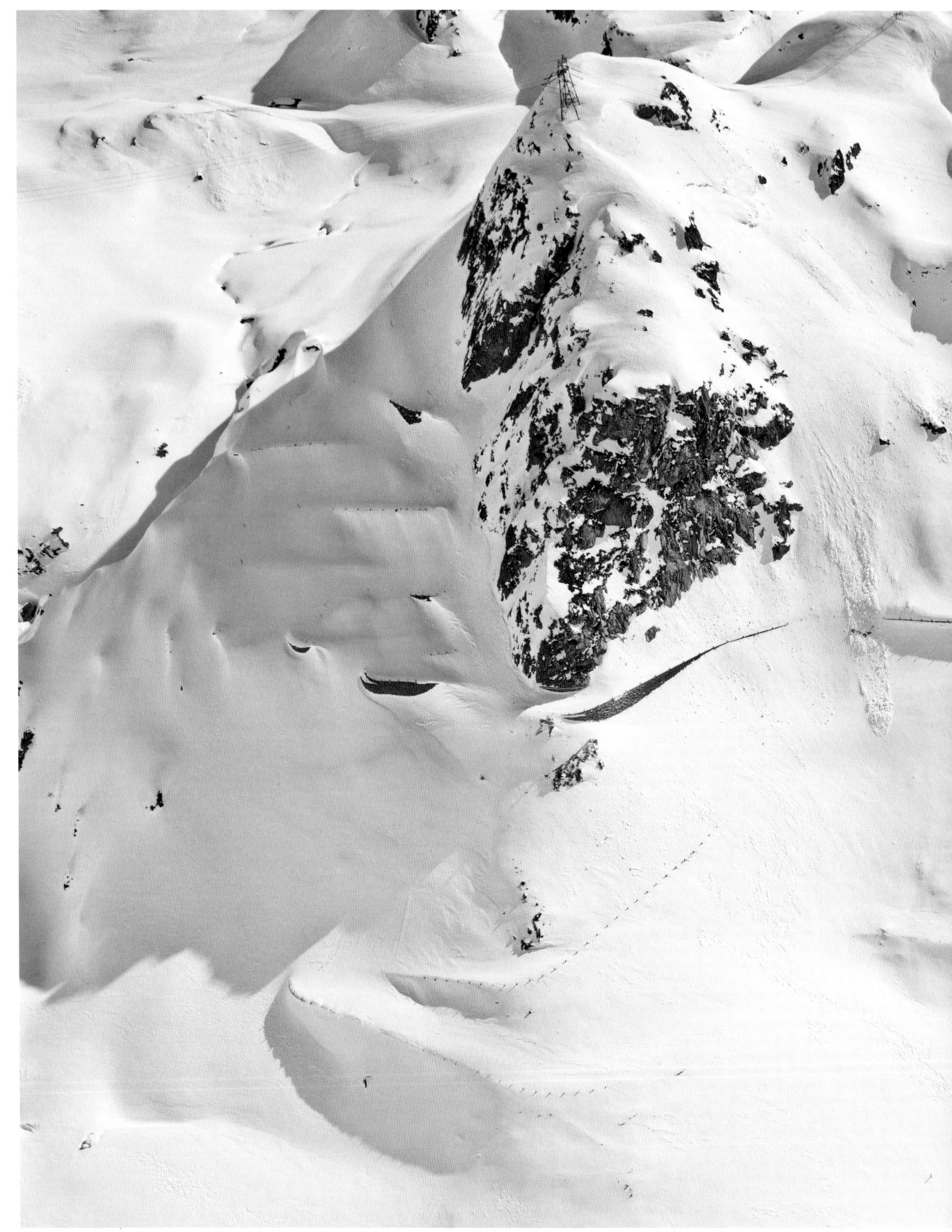

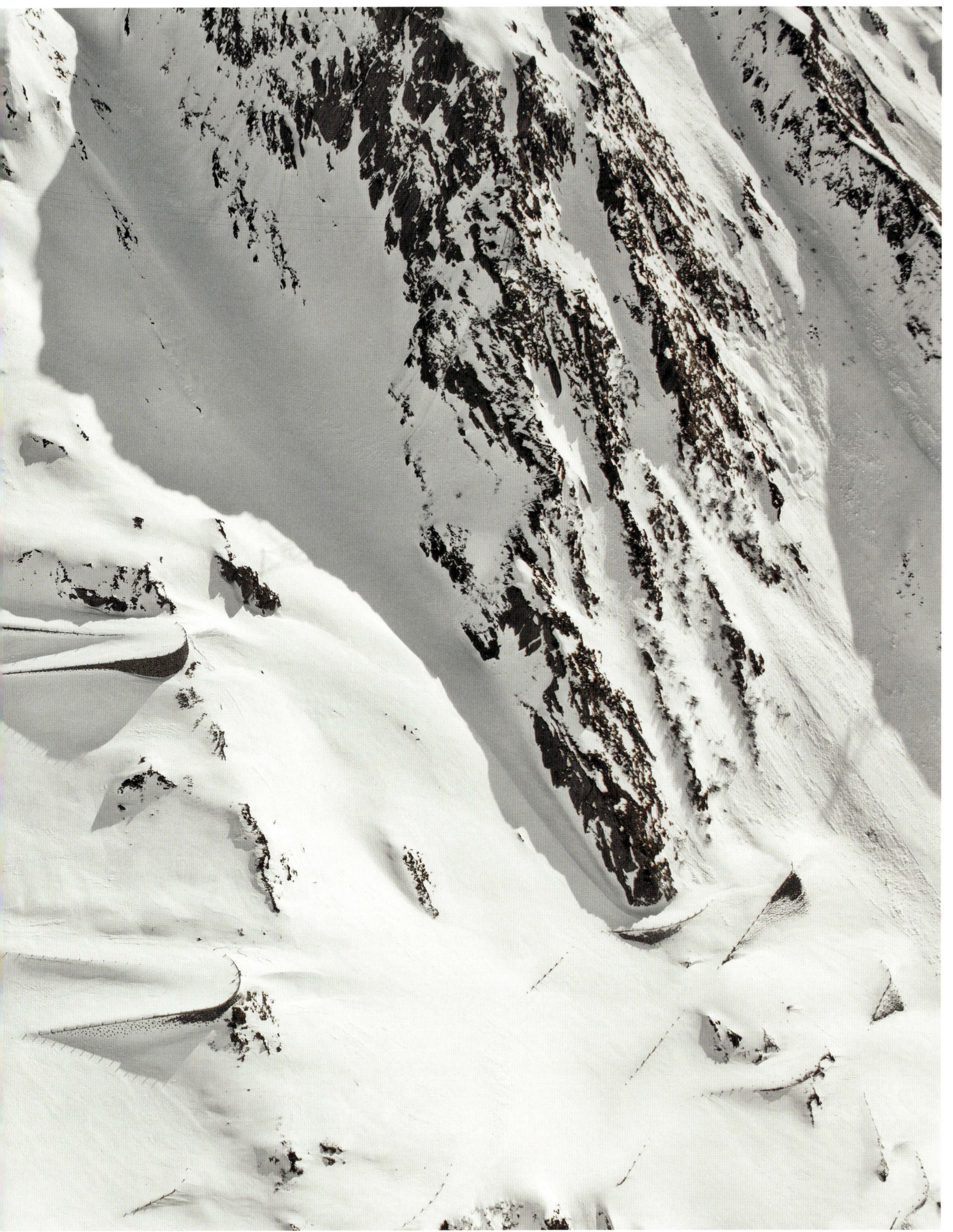

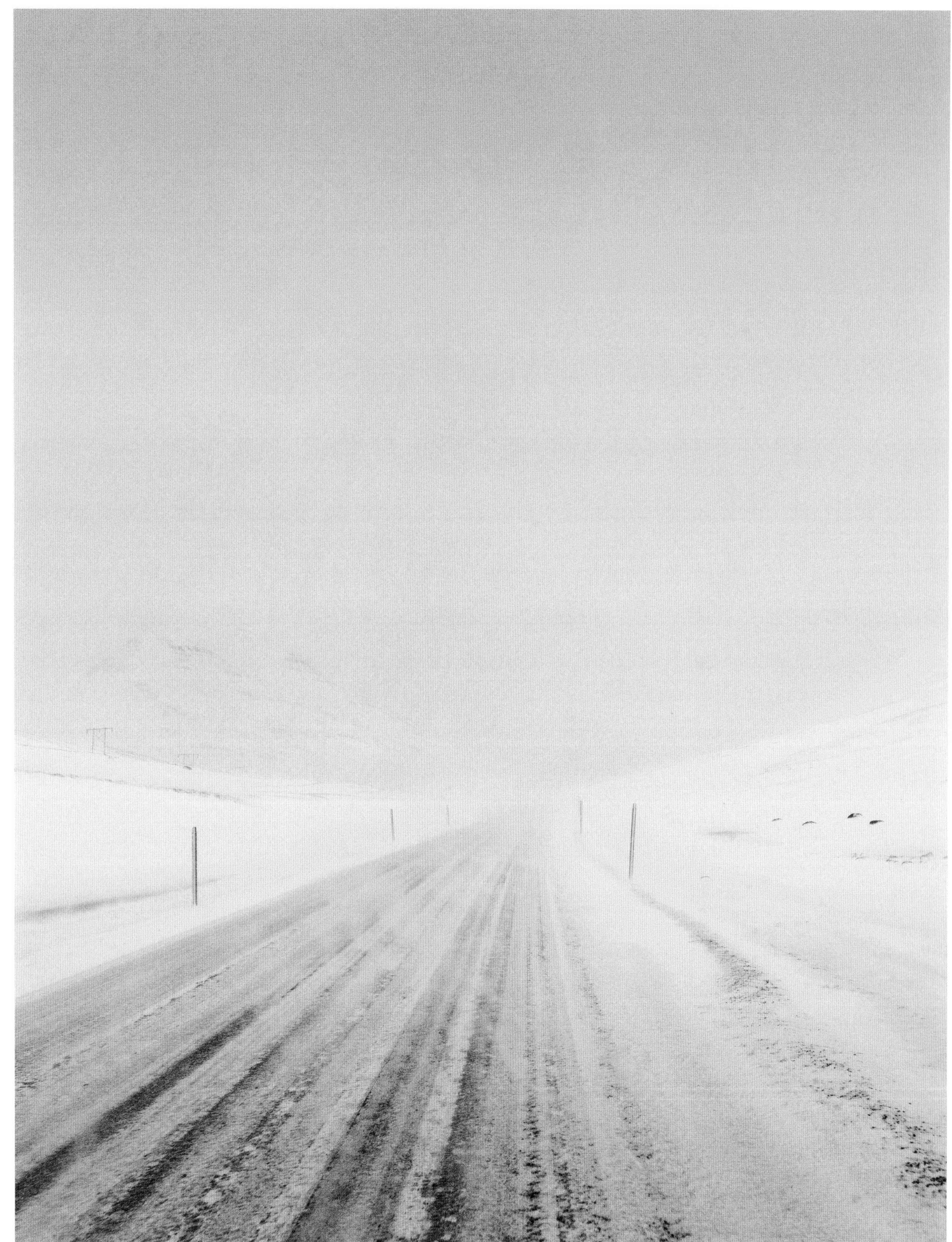

Hringvegur

— Ringstraße, Island, 1332 Kilometer
Meist ganzjährig geöffnet

Seit 1974 kann man Island auf der als Hringvegur oder Þjóðvegur 1 bekannten, 1332 Kilometer langen Ringstraße bequem umrunden. Im Winter ist die Fahrt um die Insel derweil etwas abenteuerlicher: Ohne Allradantrieb und Schneeketten geht auf vielen Teilstücken nichts, gegen die ewige Polarnacht helfen auch die stärksten Scheinwerfer nur bedingt – und wer gar abseits der Ringstraße die Vulkanlandschaft erkunden möchte, sollte bei den Mietwagen gleich die Rubrik „Monstertrucks" studieren. Kein Wunder, dass in dieser Mondlandschaft die eisigsten Szenen des Fantasy-Spektakels „Game of Thrones" gedreht wurden. Winter is coming ...

— Ring Road, Iceland , 1,332 kilometres
Usually open all year round

Since 1974, drivers have been able to easily circumnavigate Iceland on the country's 1,332-kilometre ring road, known as Hringvegur, or Route 1. In winter, the drive around the island should not be underestimated: without all-wheel drive and snow chains, many passages prove impassable, and even the strongest headlamps are not much help in the eternal polar night. Those eager to explore the volcanic landscape to the left and right of Route 1 should definitely check out the rental car category "monster trucks". It's no wonder this lunar landscape was used for the iciest scenes in the fantasy film "Game of Thrones". Winter is coming ...

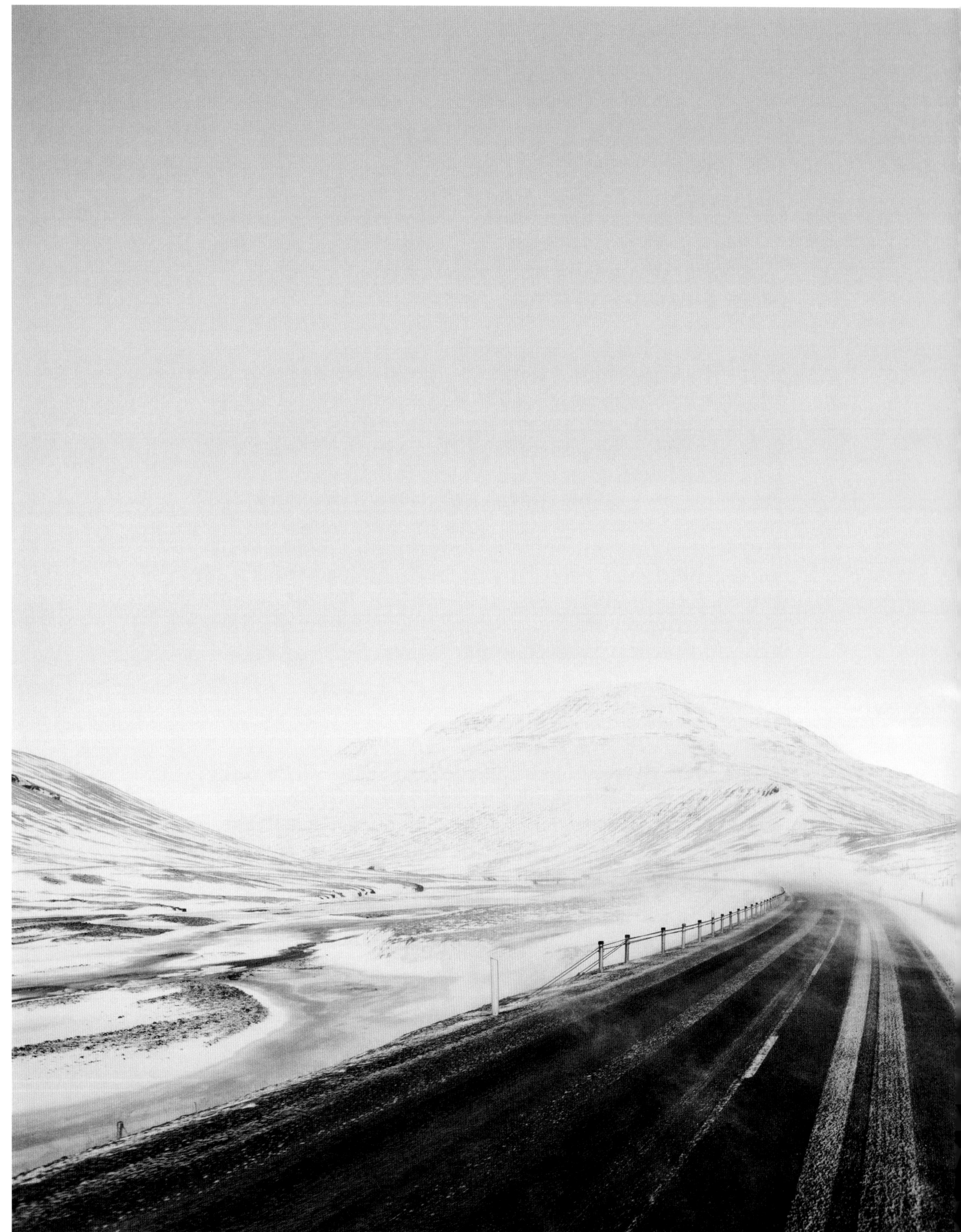

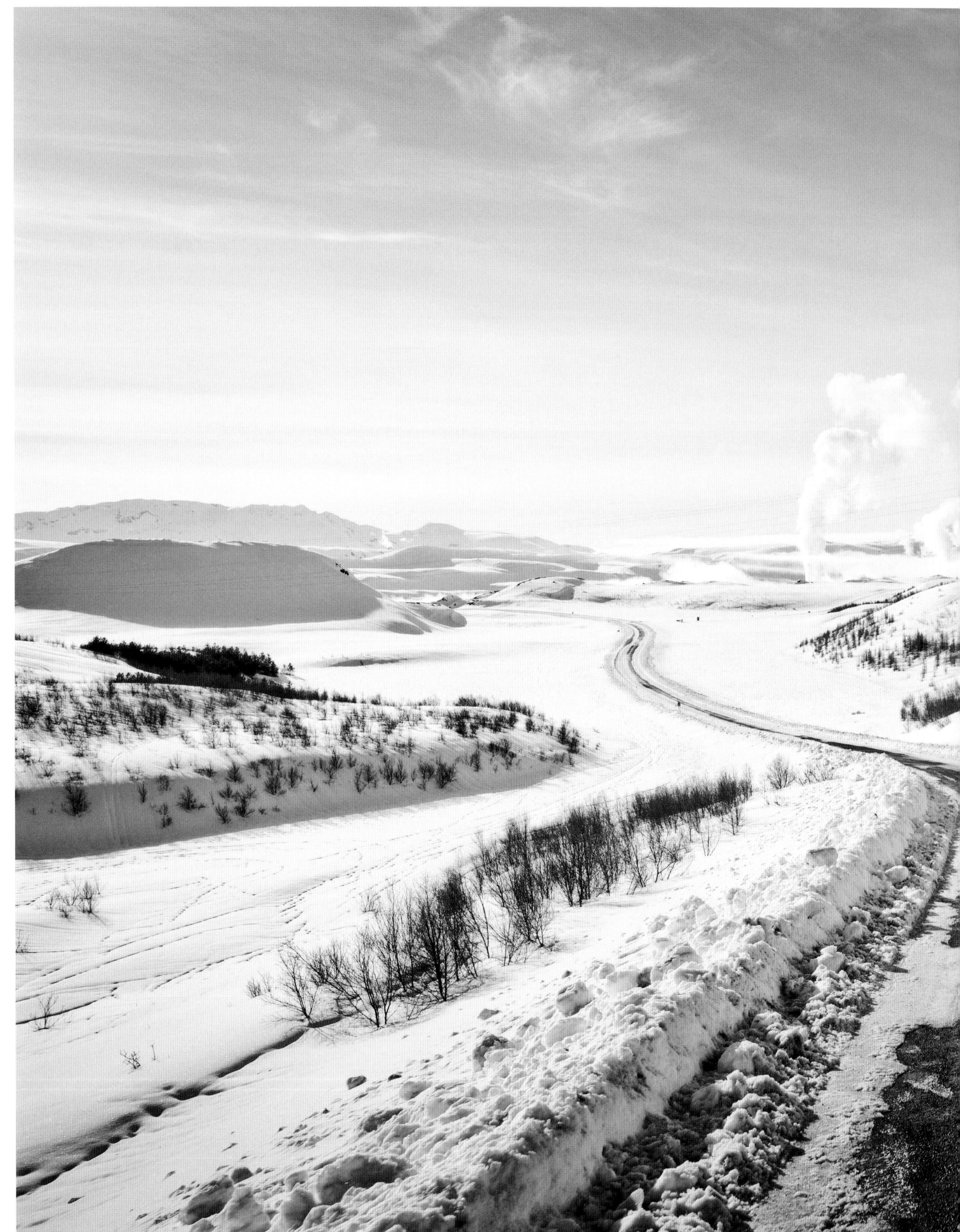

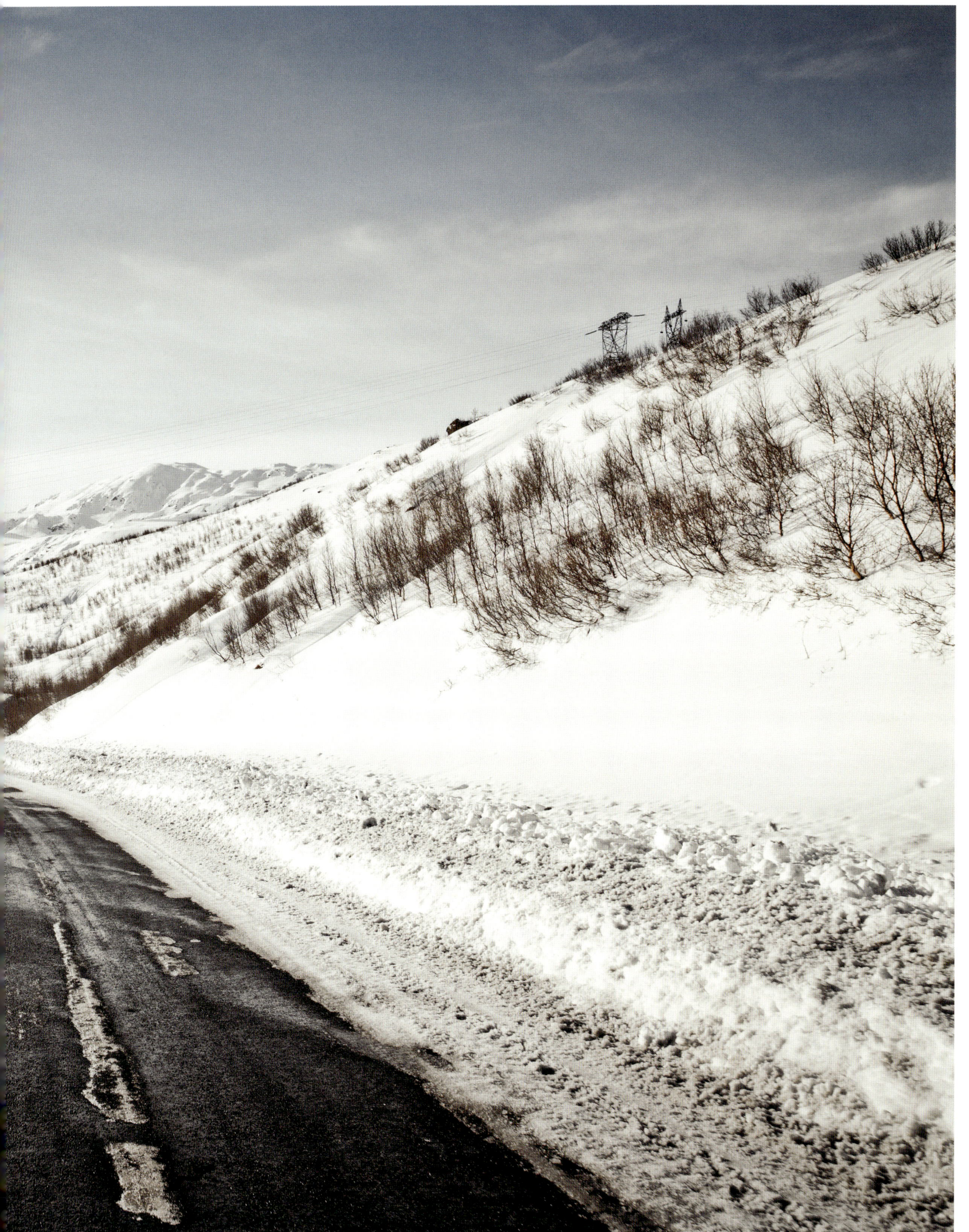

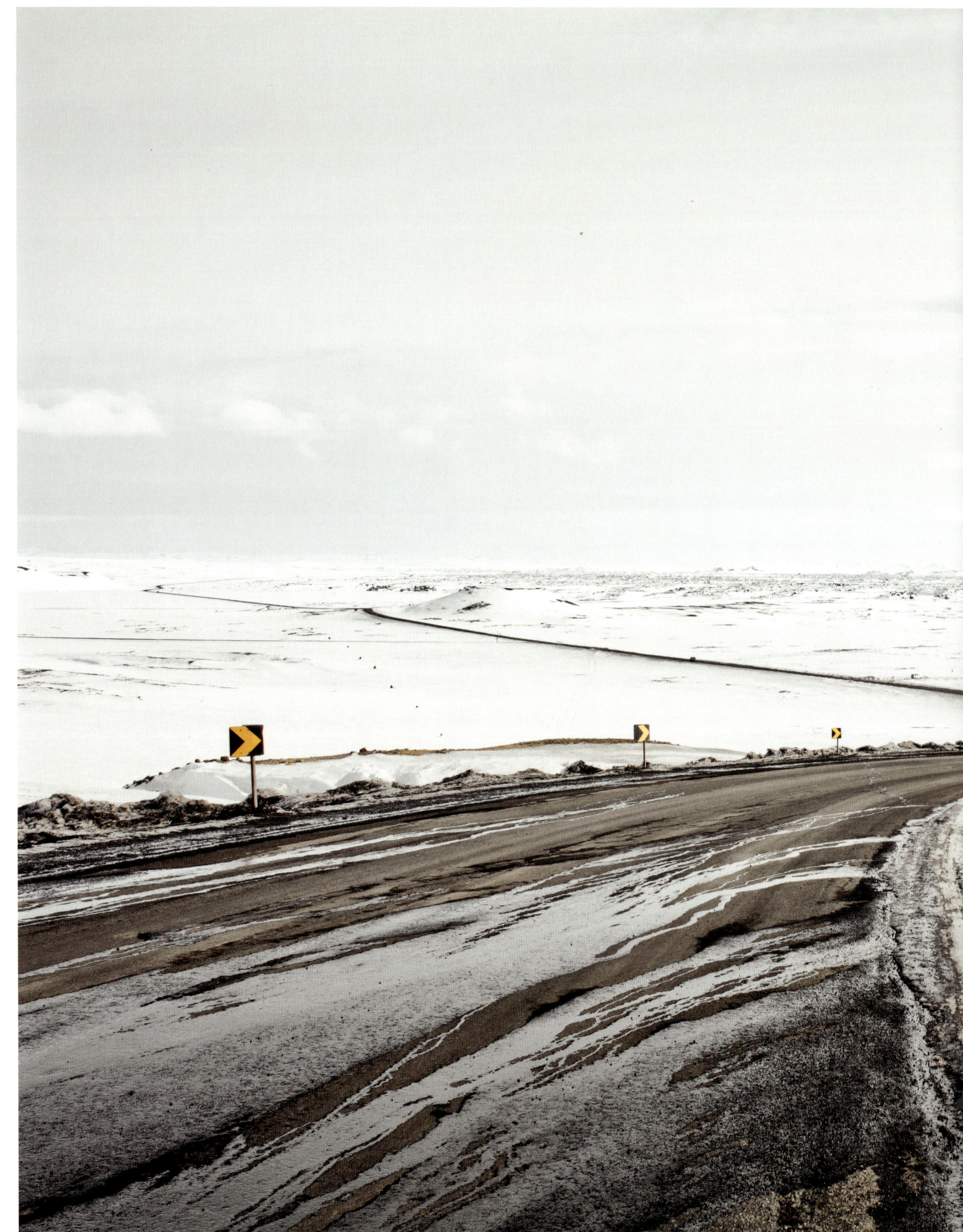

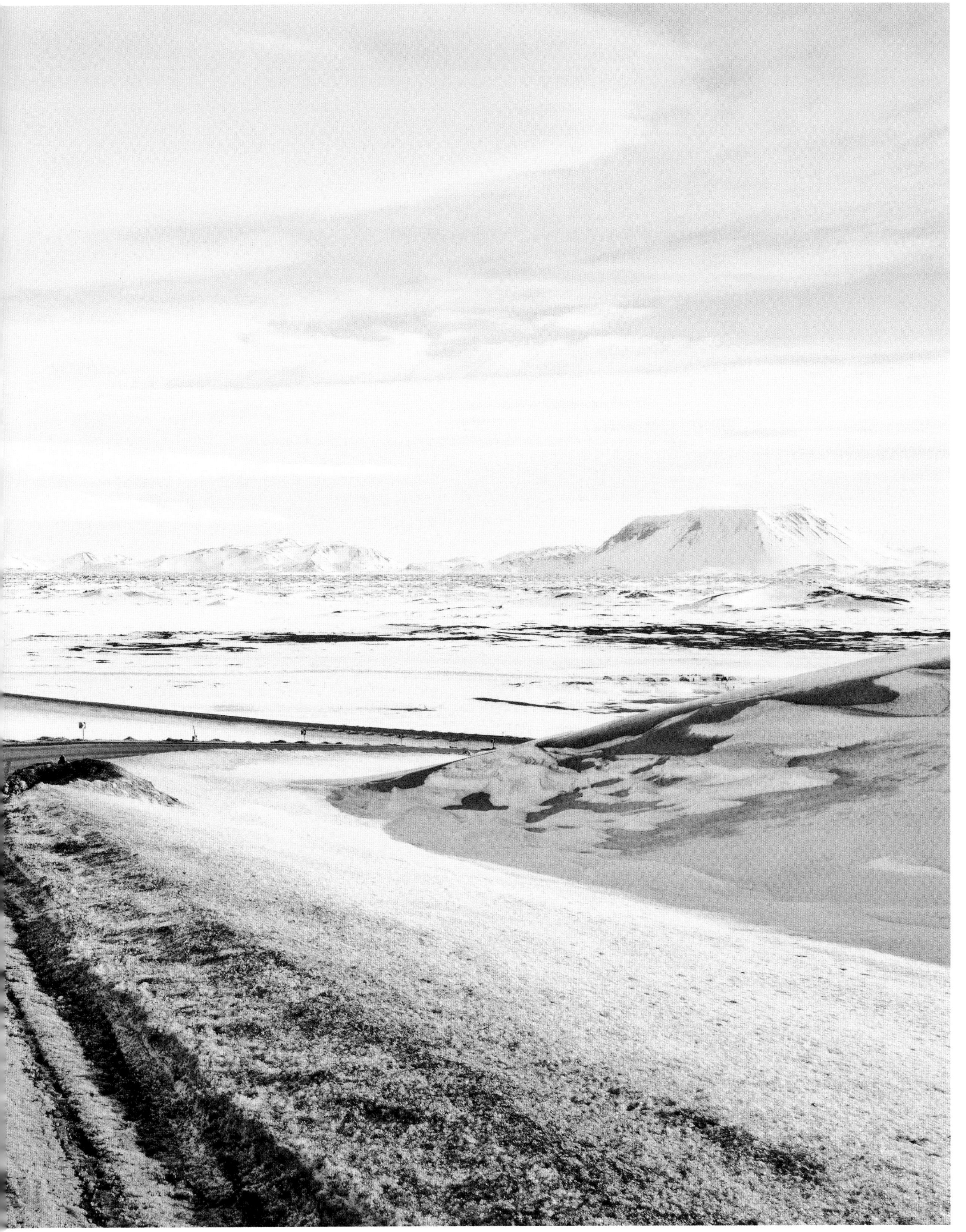

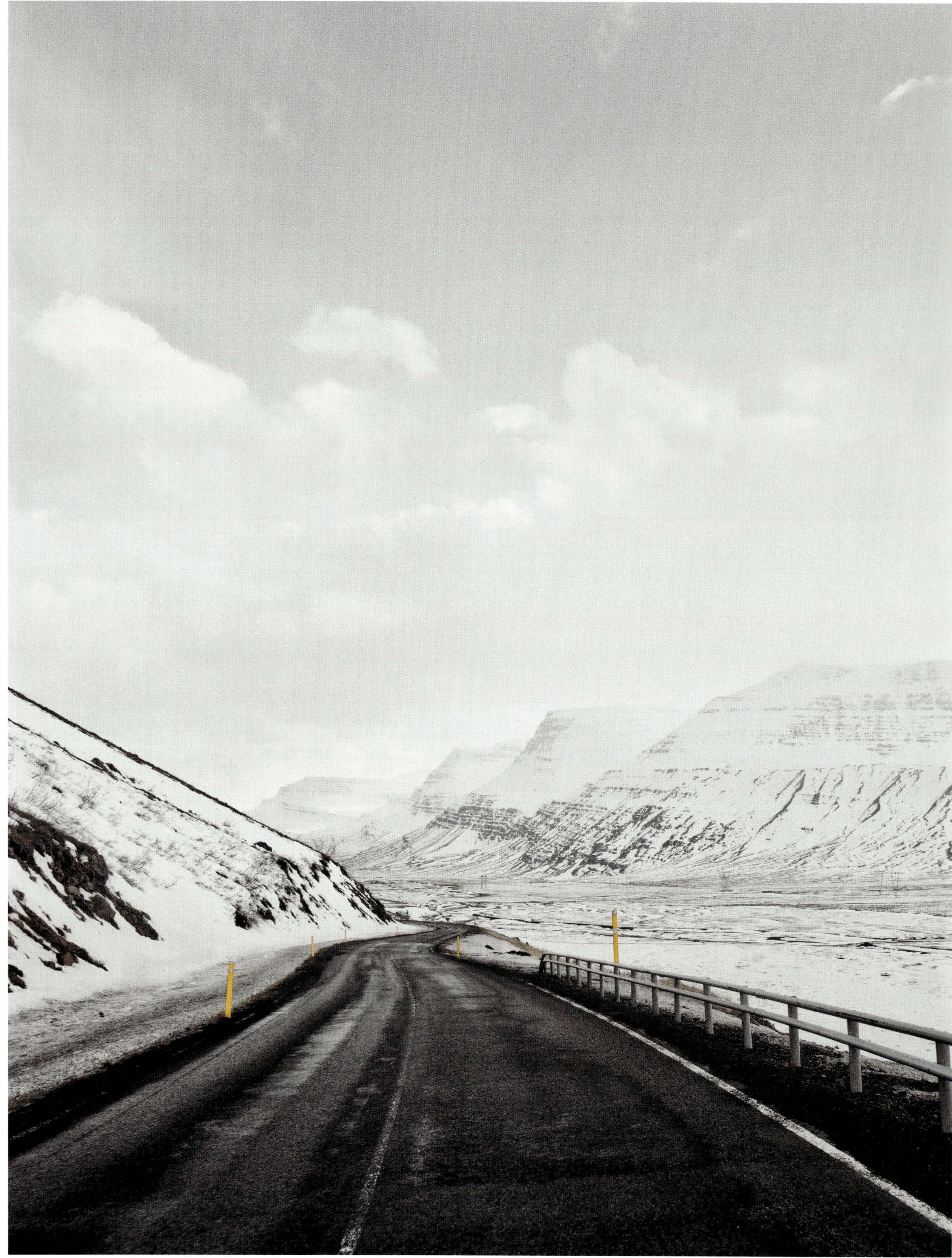

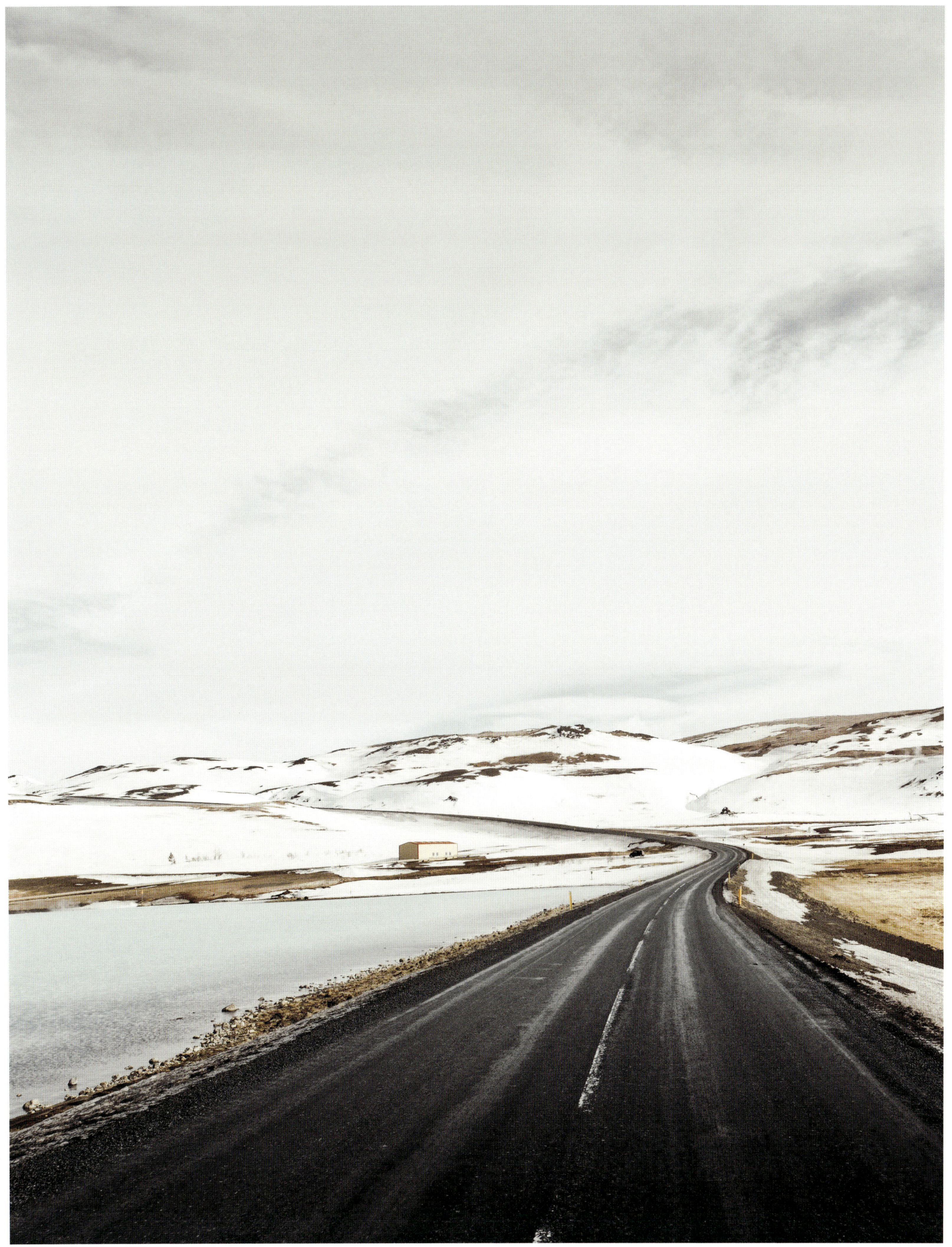

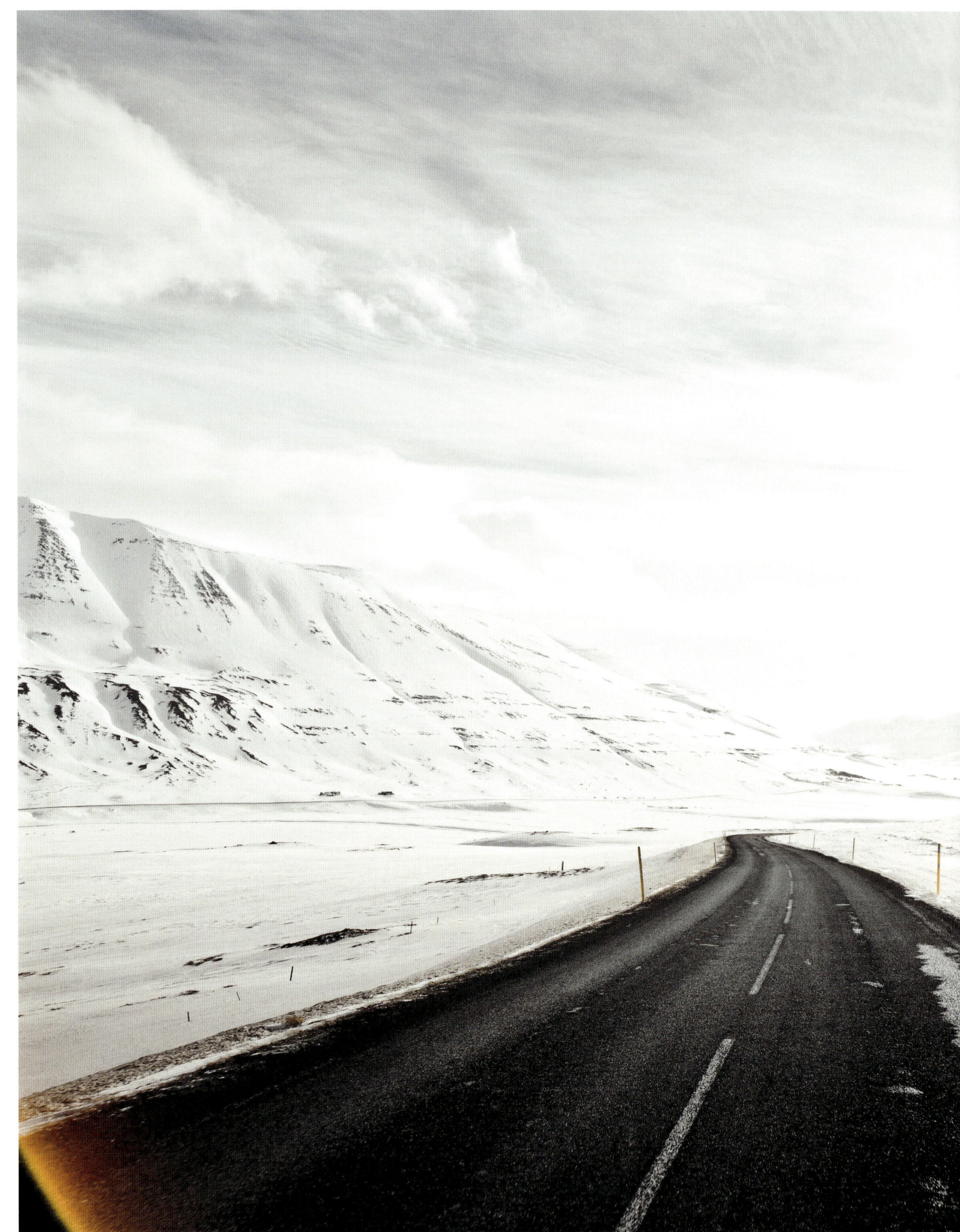

Danke / *Thanks to:*
Wucher Helicopters, Elikos, Swiss Helicopters, Heli Bernina & Icelandair
Andreas Henke, Michael Dorn, Sigfús, Pétur & Pétur, Bílabúð Benna & Porsche Deutschland.

Jan — Für Anton, in Vorfreude auf die ersten gemeinsamen Serpentinen. Und für Laura, meine liebste Reisebegleiterin, und meinen Vater Florian Baedeker, der mir Fernweh und Autobegeisterung in die Wiege gelegt hat.

To Anton - I am looking forward to our first shared switchbacks. To Laura, my dearest travelling companion. And to my father Florian Baedeker to whom I owe my itchy feet and the passion for cars.

Stefan — Ich danke allen, die mich in den letzten Jahren begleitet und unterstützt haben. Ob daheim oder on the road - ohne euch wären die Magazine und Bücher der letzten Jahre und dieses Buch nie entstanden. Meinem Freund Michi, der auf allen Reisen ein guter Weggefährte ist, und meiner Frau, die die Kurven des Alltags mit den Kids daheim bewältigt. Danke an alle Fans auf Skateboards, Bikes, Motorrädern und in Autos, die die gleiche Leidenschaft teilen.

I express my thanks to all those people who have accompanied and supported me in the past few years, either at home or on the road. Without you, this work and all the other books and magazines of the last years wouldn't exist. To my friend Michi, good companion on all journeys. And to my wife, who masters the curves of everyday life with the kids at home. Thanks to all enthusiasts on skateboards, bicycles, motorbikes and in cars who share the same passion.

BIBLIOGRAFISCHE INFORMATION DER DEUTSCHEN NATIONALBIBLIOTHEK
DIE DEUTSCHE NATIONALBIBLIOTHEK VERZEICHNET DIESE PUBLIKATION
IN DER DEUTSCHEN NATIONALBIBLIOGRAFIE; DETAILLIERTE BIBLIOGRAFISCHE
DATEN SIND IM INTERNET ÜBER HTTP://DNB.DNB.DE ABRUFBAR.

BIBLIOGRAPHIC INFORMATION PUBLISHED BY THE DEUTSCHE
NATIONALBIBLIOTHEK
THE DEUTSCHE NATIONALBIBLIOTHEK LISTS THIS PUBLICATION IN
THE DEUTSCHE NATIONALBIBLIOGRAFIE; DETAILED BIBLIOGRAPHIC
DATA ARE AVAILABLE IN THE INTERNET AT HTTP://DNB.DNB.DE.

-

1. AUFLAGE / 1ST EDITION
ISBN 978-3-667-10717-6

KONZEPT/CONCEPT: STEFAN BOGNER
TEXT: JAN KARL BAEDEKER
ÜBERSETZUNG/TRANSLATION: KAYE MUELLER
LEKTORAT/EDITOR: BIRGIT RADEBOLD
FOTOS/PHOTOS (EINSCHL. TITEL/INCL. COVER): STEFAN BOGNER

KARTEN/MAPS: ISLAND/ICELAND: ISLAND PHYSIKALISCHE KARTE VON SAMÚEL EGGERTSSON. GEZEICHNET UND GEDRUCKT VON DER TOPOGRAFISCHEN SEKTION DES DÄNISCHEN GENERALSTABS IM MASSSTAB 1:500 000. SAMBAND ÍSLENSKRA BARNAKENNARA, REYKJAVÍK / ICELAND PHYSICAL MAP BY SAMÚEL EGGERTSSON. DRAWN AND PRINTED BY THE TOPÓGRAPHICAL SECTION OF THE DANISH GENERAL STAFF AT A SCALE 1:500 000. SAMBAND ÍSLENSKRA BARNAKENNARA, REYKJAVÍK.

ALPEN/ALPS: OFFIZIELLE STRASSENKARTE DES ADAC IM MASSSTAB 1:850 000. ALPEN MIT ÖSTERREICH, SCHWEIZ UND OBERITALIEN. MAIRS GEOGRAPHISCHER VERLAG STUTTGART / OFFICIAL ADAC ROAD MAP AT A SCALE 1:850 000. ALPS WITH AUSTRIA, SWITZERLAND AN NORTHERN ITALY. MAIRS GEOGRAPHISCHER VERLAG, STUTTGART.

EINBANDGESTALTUNG UND LAYOUT/COVER DESIGN AND LAYOUT: STEFAN BOGNER
LITHOGRAFIE/LITHOGRAPHY: MICHAEL DORN
DRUCK/PRINTED BY: FIRMENGRUPPE APPL, APRINTA-DRUCK, WEMDING
PRINTED IN GERMANY 2016

DELIUS KLASING VERLAG, SIEKERWALL 21, 33602 BIELEFELD, GERMANY
TELEFON/PHONE +49 (0)521 559-0,
TELEFAX/FAX +49 (0)521 559-115
EMAIL: INFO@DELIUS-KLASING.DE
WWW.DELIUS-KLASING.DE